The Technical Communicator's Handbook

Dan Jones
University of Central Florida

D0209630

Allyn and Bacon
Boston ■ London ■ Toronto ■ Sydney ■ Tokyo ■ Singapore

In memory of my father, Frank Jefferson Jones

Vice President: Eben W. Ludlow
Series Editorial Assistant: Grace Trudo
Marketing Manager: Lisa Kimball
Sr. Editorial Production Administrator: Susan McIntyre
Editorial Production Service: Nesbitt Graphics, Inc.
Composition Buyer: Linda Cox
Manufacturing Buyer: Suzanne Lareau
Cover Administrator: Linda Knowles
Design and Electronic Composition: Nesbitt Graphics, Inc.
Indexer: Karen Lane

Copyright © 2000 by Allyn and Bacon
A Pearson Education Company
160 Gould Street
Needham Heights, MA 02494
Internet: www.abacon.com

Library of Congress Cataloging-in-Publication Data

Jones, Dan (Dan Richard), 1952–
    The technical communicator's handbook / Dan Jones.
        p.   cm.
    Includes bibliographical references (p.   ) and index.
    ISBN 0–205–27932–5
    1. English language—Rhetoric Handbooks, manuals, etc.
2. Communication of technical information Handbooks, manuals, etc.
3. Technical writing Handbooks, manuals, etc.   I. Title.
PE1475.J567   2000
808'.0666—dc21   99–23167
                                                CIP

Printed in the United States of America
10 9 8 7 6 5 4 3 2 1   04 03 02 01 00 99

CONTENTS

PART *Production*   211

PART *The Internet   245*

PART *Interpersonal Skills* 279

PART *Index 421*

PREFACE

The Technical Communicator's Handbook is intended for college students, regardless of their majors, who want to sharpen their communication skills and their knowledge of a wide variety of technical documents. Those who write on the job will also find this a valuable reference. In general, anyone—whether in school or on the job—who wants to communicate technical information more effectively can benefit from this handbook.

HOW THIS HANDBOOK IS ORGANIZED

This handbook is organized into nine major parts plus an index, according to the basic knowledge and skills that are essential for effective technical communication: planning and research, writing, design and illustration, editing, production, the Internet, interpersonal skills, correspondence, and technical documents. The first five parts—from planning to production—are commonly referred to as the technical documentation process. These are the steps you typically go through to create many kinds of technical documents. You'll find it useful to become thoroughly familiar with this process and to apply it to your assigned tasks.

The next four parts cover topics essential for any college student or technical professional. Part Six, "The Internet," provides an up-to-date overview of the Internet and the World Wide Web with the goal of enabling both novice and experienced users to achieve the maximum benefit from this increasingly important resource. Part Seven, "Interpersonal Skills," considers many aspects of dealing with other people in a professional setting, a crucial skill as students prepare for graduation and beyond and workers interact with other employees, supervisors, and clients. In Part Eight, "Correspondence and the Job Search," you will find information about writing effective letters, memos, and e-mail as well as information on many kinds of job correspondence. In addition, this part covers the essential skills of creating effective résumés and managing your portfolio. Part Nine, "Technical Documents," presents typical technical documents such as instructions, manuals, reports, and proposals, with information on how to create them, what to include, and how to present them.

Finally, Part 10, "Index," while it doesn't offer new material, is perhaps the heart of the handbook, since it organizes all the topics and subtopics comprehensively to help you locate related material in different areas of the handbook. The index is your entrance to the material in *The Technical Communicator's Handbook* when you want to find information on a specific topic quickly.

The handbook presents standards, rules, and descriptions in each of the first nine parts. Some topics also include a helpful Tips box that you will find handy as you work on your own projects. Each of the first nine parts also concludes with a Resources section, an abbreviated bibliography pointing the way to the best sources for learning about some of the topics in greater depth. Often, you will find the handbook coverage to be just what you need to know about a given subject. But for those times when a handbook isn't enough, these resource sections will point you to further readings. And many parts of the handbook include a discussion of a variety of computer tools that can help you save time and create much more effective documents.

HOW THIS HANDBOOK DIFFERS FROM OTHER HANDBOOKS

Like many other handbooks, *The Technical Communicator's Handbook* offers prescriptive or reference information on numerous topics concerning planning, research, writing, style, mechanics (spelling, punctuation, capitalization, abbreviations, numbers), grammar, design and illustration, and various kinds of technical documents from abstracts to reports. The handbook genre is after all primarily a reference source that covers basic or standard information and rules in the discipline.

However, document requirements vary depending on the circumstances. There are often many possible solutions to a particular communication challenge. By providing a variety of approaches to many communication tasks, this handbook encourages readers to experiment and select the one best suited to their needs. In that sense, *The Technical Communicator's Handbook* is also descriptive or suggestion based. So in addition to describing or defining over 300 topics and offering the basic information necessary to understand them, this handbook offers readers the opportunity to explore a range of possibilities for creating various documents.

The Technical Communicator's Handbook also overcomes the usual limitations of handbooks in this discipline by offering a more systematic approach to its topic. It avoids alphabetical arrangement and awkward groupings of topics. Because topics are organized systemically, the handbook may easily be used as a main text for some technical communication courses or as a supplementary text. The handbook format also provides a convenient resource for practicing professionals.

The Technical Communicator's Handbook offers a more comprehensive approach than other available handbooks to the many essential skills students and practicing professionals must know. Some other technical communication handbooks focus narrowly on technical documents and grammar, mechanics, and punctuation. *The Technical Communicator's Handbook* discusses these topics too, but unlike other technical communication handbooks, it also includes in-depth discussions of brainstorming, mind mapping, and freewriting; evaluating Internet sources; style and tone; designing and illustrating for online and print documents; major editing strategies; many kinds of production; using the Internet (for example, mailing lists, newsgroups, search engines); developing a wide variety of interpersonal skills (interviewing, negotiating, collaborating, listening, presenting oral reports); and creating less commonly covered technical documents (including annotated bibliographies, newsletters, and white papers).

This handbook also offers unique features such as tip lists on a variety of topics from brainstorming to writing reports, an overview of computer tools to help readers do everything from planning to creating more effective Web pages, and lists of current resources for further reading. Some of the tip lists have been contributed by highly experienced practicing professionals. The discussion of tools covers over fifty computer tools to help readers perform their daily communication tasks more effectively and efficiently. The lists of resources cite some of the most current and helpful sources available for further reading on many topics.

Finally, information in *The Technical Communicator's Handbook* is easy to find, read, and use. The book is written in a plain style appropriate for its audience; it's printed in a convenient, compact size; and it features ten external tabs to provide quick access to each of the major sections and the thorough index.

HOW TO USE THIS HANDBOOK

As a reference work, this handbook is designed so that all of the sections and subsections are self-contained. You don't have to read topics covered early in the handbook to understand topics covered later, and no topic depends on any other topic to make it usable. Cross-referencing is provided where appropriate to give you more information on related topics.

Perhaps the best way to use this handbook is to start with the table of contents or the index to find the topic you're looking for. When you find it, you'll find a description or definition or tips or a discussion on how to write a particular document. However, because writing is a process, you may find it helpful to consult other sections to benefit fully from the coverage of your topic. For example, while you can find out the basic content of a good résumé in Part Eight, "Correspondence and the Job Search," you should also review the essential principles of design in Part Three, "Design and Illustration," to help make your résumé look as professional as possible. For more complex documents, such as technical reports, you can find the essential elements in Part Nine, "Technical Documents." However, you'll also need a general knowledge of many of the major topics covered in the first five parts—from planning to production—to write an optimally effective report.

The Technical Communicator's Handbook can help you to find information on a variety of situations. If you have been given the task of preparing a paper, report, proposal, or other document, familiarize yourself with the documentation process described in Parts One through Five (the technical documentation process); check out Part Six, "The Internet," for information on using the Internet to research your subject; and then look at Part Nine, "Technical Documents," to see whether your document is addressed. If you are beginning a job search, consult Part Eight, "Correspondence and the Job Search," but also consult Part Seven, "Interpersonal Skills," for advice on handling the interview. If you are involved in a research project, Part One, "Planning and Research," and Part Six, "The Internet," will help to guide you through the process. If you want to create a Web site to advertise your company's business, look at Part Three, "Design and Illustration." For guidance in style issues, look at Part Two, "Writing." To review basic rules of grammar and the mechanics of punctuation, spelling, capitalization, numbers, and abbreviations, turn to Part Four, "Editing." And if you need to prepare a presentation, go to Part Seven, "Interpersonal Skills," for guidance.

Major Features

Major features of the handbook include:

- thorough coverage of the documentation process from planning to production
- authoritative and concise coverage of over 300 topics and skills

- illustrative examples for many topics and skills
- listings of over 100 essential and current resources for further reading
- discussions of many of the best computer tools to aid you in your work
- numerous lists of tips to show you what to do or to avoid
- useful coverage on researching on the Internet and evaluating Internet resources
- an overview of asynchronous and synchronous communication via the Internet
- many helpful tips on designing, illustrating, publishing, and promoting Web pages
- many helpful tips on creating online help files
- a discussion of over thirty common documents with many document examples
- a discussion of the major interpersonal skills you need to succeed in college and on the job
- an easily readable prose style
- tabs and spiral binding for easy use
- a detailed and highly usable index

Acknowledgments

This handbook represents, in many ways, a distillation of the essential topics I have covered during twenty years of teaching a wide variety of technical communication courses at the undergraduate and graduate levels and during ten years of teaching corporate seminars to over twenty companies. Of course, a handbook must cover so many topics that it's impossible for one person to know everything there is to know about all of them. For that reason, a good handbook cannot be an individual effort.

I am indebted to the many authors whose works I list in the Resources section at the end of each part. Their work provided me with a broad understanding of the major areas of technical communication. Practicing technical communicators Gary O'Hara, Christine Peacock, Janice Hitchcock, John Oriel, Eugene Gray, Dan Voss, and Carl Feigenbaum provided many valuable suggestions. Frank Jones, a graphic artist, offered helpful advice about design and production. I am also indebted to my colleagues who teach technical communication in our long-established undergraduate and graduate programs at the University of Central Florida, particularly Madelyn Flammia and David Gillette. And I am grateful to my students Debbie Andrisani, Cassie

Avery, Nikki Burley, Anna Gaal, Michelle Pachoski, Nichola Piber, and Michelle Randall, who tested parts, provided helpful information, or created sample illustrations. In many ways I am indebted to all of the students I have taught over the past twenty years. Their comments and questions helped me determine what is needed in a handbook.

I also want to thank the reviewers: Carol M. Barnum of Southern Polytechnic State University, David Mair of the University of Oklahoma, and Henrietta Nickels Shirk of the University of North Texas. They offered many good suggestions for improving this handbook throughout all nine parts.

I am grateful to Eben Ludlow, who two years ago invited me to research and write a handbook for technical communicators. His interest in this project and the assistance of his staff—particularly Linda D'Angelo, Grace Trudo, and Tania Sanchez—are much appreciated.

I am also grateful for the expertise and advice of Susan McIntyre, Senior Editorial Production Administrator at Allyn and Bacon. And I want to thank Tom Conville and Peg Markow at Nesbitt Graphics. The excellent copyediting provided by Nesbitt Graphics exemplifies what quality copyediting should be.

I owe a special thanks to Karen Lane. I could not have completed this handbook without her valuable assistance. She provided helpful information and sources on many topics, edited various drafts, helped to prepare the final manuscript, and created the index. Many of the strengths of this handbook are due to her talents, energy, and diligence.

Finally, I want to thank my wife, Carol, who is also a technical communicator, and my son Sam for their understanding and patience. They gave me the time and the support I needed to complete this book on schedule.

Dan Jones
Department of English
University of Central Florida
djones@pegasus.cc.ucf.edu

Illustrations

Tips

PART 1

Planning and Research

*P*lanning saves time. It is the foundation for any endeavor that is properly done. You may have heard the woodworking expression "Measure twice, cut once." The result of thorough planning is an efficiency of execution. A very small project or an everyday activity may take only as much planning as the decision to do it, but projects of any substance normally require a more elaborate process of preparation. Whether you're writing a brief letter or a long report, you'll need to devote at least some time to planning your document, and frequently you will have to research all kinds of topics in numerous ways.

Often, planning even short projects requires many of the same steps that a much longer document does. And, of course, the more complex the document, the more necessary planning becomes. For both short and long documents, you'll need to determine the focus, discover ideas, plan the organization, set or follow the standards, manage the document, and perhaps create a document plan. To save time and to help you meet deadlines, you'll need to use a variety of tools to help you achieve all of these planning tasks.

Some documentation projects are so massive that they require far more complex planning. If you are a documentation manager and find yourself responsible for a major project (for example, creating a series of software user manuals for a sophisticated software program), then you will benefit from a much more extensive discussion, such as JoAnn Hackos's *Managing Your Documentation Projects* (see "Resources" at the end of this section). Hackos's book is thorough, not only in addressing the planning phase but also in discussing the phases of content specification, implementation, production, and evaluation.

When you take a photograph, you focus on your subject so that your picture will show exactly what you intend. Similarly, you refine your document's focus to maximize its effectiveness. How you focus determines what information your audience will glean from your text. And just as in taking photographs, the decisions you make about what you are trying to show, who you expect will be viewing your work, how close you will be to your subject, and how broadly or narrowly you will frame your subject all contribute to the character and impact of the piece.

While you are planning and adjusting your focus, always keep your audience in mind. All the elements of planning, from determining your focus to creating a document plan, center on the audience for your document. Your reader should be the determining factor for what you cover in your document from the beginning, through the middle, and to the end. The information in this book will help you to plan the many kinds of documents you will typically write for a variety of audiences.

Along with planning, researching is an important early stage in many projects. Knowing how to research enables you to find the information you or someone else needs. Researching is an essential skill, not only for college students but also for people working in industry at all levels.

College students, of course, typically research a dozen or more major projects during their four years of undergraduate education. And depending on their field of study, graduate students are required to conduct even more research. Much of the college experience can seem like a series of research projects whether you're in the humanities, physical sciences, life sciences, or social sciences. During most semesters or quarters, you work on multiple research papers or projects, and often you collaborate with others on a research project.

This emphasis on research doesn't end at graduation. In your activities in the workplace, you'll be called on repeatedly to find information on corporations, trends, people, products, services, and more. A lot of your research won't be limited to the library. You'll have to interview people in person, on the phone, or via e-mail. You'll have to take notes at meetings of all kinds. You'll have to track down information in the company library (if one is available), in nearby libraries, on the Internet, or through various surveys. In brief, you'll have to continue to be resourceful to find the information that you and others need. Researching never has been and never will be an activity limited to those who work in academe.

PLANNING

Determining the Focus

Determining the focus of a document means narrowing the subject, stating the purpose, establishing the scope or depth of detail, and understanding the audience. You begin by deciding what you want your audience to know, do, or accept and then analyzing how you will achieve this goal. You will have properly determined your focus when you have done all of these effectively.

SUBJECT

The subject of a document is the main area of concern or main topic. Typically, the subject is initially a broad area that requires further focusing. In a technical report, your subject may be deforestation, acid rain, laptop computers, modems, or fractals. In technical instructions, your subject may be how to install a new disk drive in a home computer or how to create a simple Web site. In a letter, your subject may be your qualifications for a job opening or a complaint about a company's products or services. All of these are good subjects, but all of them lack a proper focus.

In academe, you are sometimes assigned a broad subject for a document, and then you are challenged to narrow your focus appropriately. Deforestation, for example, is a broad subject. What does your audience want to know about the subject? Do you have an audience other than the professor who gave you the assignment? How much do you already know? How much more do you need to know to write the document? You must answer these questions and many others before you will have a valuable focus.

In college, you are sometimes given complete freedom to choose your subject. For example, in a typical introductory technical communication class, you may be asked to write both a proposal for and a technical report on a subject in your discipline. You may also be asked to write to a layperson or a peer about your subject. Of course, these requirements don't simplify matters completely. You still have thousands of subjects to choose from within your discipline. You may be asked to consider some new technology, issue, or trend. You may be asked to consider a subject that you have recently studied in one of your other classes. The challenge lies in properly defining your focus for both yourself and your audience.

In industry, the subject is often more narrowly defined and a specific audience is more often readily apparent, but you still face the problem of achieving a focus or a statement of purpose. For example, you may be required to write a letter in response to a customer's complaint or to submit a status report to your supervisor about the project you've been working on for the past few months. Or you might be required to choose one kind of office equipment over another and to recommend a course of action to your superiors. Typically, you will also be required to respond to a great deal of pedestrian correspondence and e-mail asking you for updates, actions, favors, or information. Determining exactly what you want to write in your letter to the customer, status report to your supervisor, or routine correspondence is a matter of refining your subject in keeping with your intended focus.

PURPOSE

Determining your focus also involves stating your purpose clearly and concisely. To establish your purpose for writing, you need to answer some questions: "Why am I writing this document?" and "What do I want my readers to know, do, or accept?"

For example, the subject of computers is impossibly large and unmanageable for a technical report. However, an evaluation of the feasibility of converting from a PC-based shop to a Macintosh-based shop or deciding among several possible printers for your company is a narrow enough problem to allow for a meaningful report. Your purpose has given you a good focus for your document. If you decide that the structure of a feasibility report is the best approach for discussing advantages and disadvantages, then this kind of document further helps you to focus your subject. The structure of a feasibility report requires you to compare two or more courses of action and to recommend one of them.

In a cover letter for a job, your purpose may be stated simply at the beginning of the letter: "I am writing to apply for the position as a technical writer that you advertised in the *Orlando Sentinel*." You simply want your reader to know that you are applying for the job opening. In a progress report in brief memorandum format, you may write, "I am providing an update on the work completed so far and an overview of the work remaining to be done." In a complaint letter, you may state, "I am writing to call your attention to the poor service I received at your repair shop last Saturday." In a proposal, you may write, "I propose to write a report comparing Compaq and Toshiba laptop computers." In a technical report, you may state, "This report demonstrates how design engineers apply Hooke's Law to determine stresses on materials."

Many people advise you to state your purpose clearly in the first sentence or two in a brief memorandum or letter. Of course, when and where you state your purpose depend on your purpose, your audience, and the kind of document you are writing. There are many kinds of openings, and many different strategies for making your purpose clear.

Often you should state your purpose as soon as you can in the opening of a brief document. However, many times you should not take such a direct approach. For example, you will often begin a complaint letter by establishing a common ground with your reader. Your specific purpose is to have this person act on your request, but in a complaint letter, you may not want to state this underlying purpose in the opening sentence of the letter. See Part Eight, "Correspondence," for a discussion of direct and indirect patterns in correspondence for some other examples.

Knowing that you need to define your purpose and following through on it consistently are not simple tasks. Perhaps the two biggest problems many people face are determining their purpose for writing in the first place and controlling that purpose effectively throughout the document.

SCOPE

Defining your scope in a document is another essential step in establishing your focus. Scope is the depth of detail with which you are covering a subject. It's the specific range, limits, or boundaries of the subject you are discussing. Considering your scope helps you to narrow your subject so you can discuss it adequately.

Scope is in large part determined by audience. How much does your audience need to know about your topic to do, know, or accept something? Knowing the answer to this question helps you to determine the depth of detail. If you are writing a technical report on the disappearance of rain forests, for example, you would save yourself a great deal of time by limiting your scope to rain forests in Brazil rather than also including rain forests elsewhere in South America and on other continents. In fact, your report would be even more manageable if you limited the scope to examining deforestation in Brazil over the past ten years.

Scope is determined by other factors as well. For example, time, budget, personnel, urgency, and ethics may be factors shaping scope. If you have only a few days to research a topic and write a document, you will probably have to narrow the scope considerably to finish on time. If you have a severely limited printing budget, you may be forced to write a short manual rather than a long one. If creating the document is a team project, the scope will be limited by the number of people available and their particular talents. If the document is needed urgently, you might have time to cover only the bare essentials. Finally, if a code of ethics prevents certain topics from being explored, this may limit your scope as well.

AUDIENCE

If determining your focus is deciding what you want your audience to know, to do, or to accept, then you need to know the best means for getting your audience to understand or accept your purpose for writing. Many writers of technical prose are so preoccupied by what they know about a subject and what they want to write about it that they forget the most important reason for writing: to communicate with the audience. Perhaps you've heard the statement about the secret to a successful business: location, location, location. One of the major secrets to successful writing is audience, audience, audience. This point may seem obvious and simple, but unfortunately, most writers focus on their own view of the subject, purpose, and scope and not on their reader's view.

You need to devote a lot of time to audience considerations if your focus is to be properly defined. In particular, you need to consider the *discourse community* for which you are writing, and you need to be familiar with categorical and heuristic approaches to understanding your audience.

Discourse Communities. A discourse community, simply defined, is a group of people who share a common interest. A discourse community may be students in one class, students in one major, students on one campus, or all college students. A discourse community may be professionals in a discipline such as engineering, law, medicine, or technical communication. Or a discourse community may be hobbyists with a particular pastime such as ham radio, scuba diving, or chess.

Members of a discourse community tend to share a common language or vocabulary, a common outlook, common values, and shared assumptions. They have certain expectations or standards for determining what constitutes knowledge in their community. So one major part of your evaluation of audience will be an evaluation of the discourse community your audience is a part of.

Some typical questions you need to consider are:

- Are there special preferences in this discourse community to which you need to give particular attention? Do you know what your readers expect when they read in their field?
- How similar are your readers? Are they all members of one discourse community? How much variation is there in their backgrounds and knowledge levels?
- Are you sensitive to the values, expectations, and shared assumptions of this particular discourse community?
- Have you familiarized yourself with the documents people in this discourse community read?
- Are there particular political elements (tensions, relationships among people, histories, backgrounds) you need to pay particular attention to?
- Have you considered all the other important elements necessary for an effective document in this particular discourse community?

Categorical Approaches. A categorical approach to audience can be an effective way to gauge their knowledge level. Using this approach, observers often divide audiences into four categories: laypeople, executives, technicians, and experts. Two additional categories may also be useful: paraprofessionals and professionals.

It's important to keep in mind that these divisions are not clearcut. All of us are laypeople in many subjects, and relatively few of us are experts. Still, these categories can give you some helpful information for at least an initial approach to your audience.

Laypeople have little or no knowledge of the subject matter you are discussing. They are unfamiliar with the basic terminology of the subject. They want to know enough to understand how to use or understand a new product, for example, but typically, they don't want a very detailed discussion of the subject.

Paraprofessionals are people such as college students or apprentices who are still learning the essential discourse of their discipline or trade. They typically know much more about a subject than laypeople do, but they lack the experience of professionals, the decision-making authority of executives, the preoccupation with detail of technicians, and the extensive knowledge of experts. Paraprofessionals are still novices about many of the fine points of a discipline or trade. In industry, they are the inexperienced employees who will become much more seasoned after a few years on the job.

Professionals are members of a discipline, trade, or profession who have been practicing in their field for at least a few years. They have a college degree or have completed an extensive apprenticeship in a trade or craft. They are thoroughly familiar with the standards and demands of their professions.

Executives are a specialized group of professionals. They are the ones who have the decision-making authority in a company or organization. They have many years of experience in their field, and they typically want to know the bottom line so that they can make informed decisions. They will decide whether and how a new product should be marketed. Sometimes they have only a layperson's knowledge of a subject, and sometimes they have expert knowledge.

Technicians are another specialized group of professionals. They may or may not have a college degree, but they have extensive knowledge about a technology. In general, technicians build, repair, and modify technologies of all kinds. Engineers are one of the many kinds of technicians. Engineers, of course, have a college degree and design everything from bridges to operating systems. As a group, technicians often have a very practical outlook. They want to know how things work or why they don't work. They want to see the schematics or other details of a problem.

Experts are typically people who have advanced degrees in their field or fifteen or more years of experience in it. (Of course, many people have advanced degrees or extensive work experience and are not experts.) Experts are well acquainted with the complex concepts and terminology of their field. They typically have a keen interest in a practice or product and thorough knowledge of the underlying theory. They often work in academe or in the research and development areas of government and business. Although some experts communicate well with laypeople or nonspecialists, many experts have difficulty in this area.

Combined audiences are mixed groups of people. Some may be laypeople, and some may be paraprofessionals, professionals, executives, technicians, or experts. Combined audiences are common in both academe and industry, and communicating effectively with such diverse audiences in one document requires a variety of skills and approaches.

It's helpful to know what categories are part of your audience. If all the categories apply, you'll have more challenges than you would if you were writing just to one audience. For example, a report will typically be read by experts and technicians as well as executives. Knowing this will help you to shape different parts of the report for the different parts of the audience.

You may also discover that even an audience of one category will contain individuals with widely varied backgrounds. You may decide to ignore the least informed and appeal to the most informed, or you may include supplemental information to help the less informed part of your audience and even refer them to various beginners' books on your topic.

Knowing the categories of your audience is not enough. You also need to know whether and why your readers are motivated to read your document, whether they have certain apprehensions about the topic you are discussing, and whether other psychological factors may be relevant. You'll also need to know how your audience will use your document and what they will or will not want to read about. You may need to gather demographic information about their age, gender, political preference, location, and so on.

Heuristic Approaches. When you gather information about your audience through direct questioning and observation, you are taking a heuristic approach to analyzing it. Heuristic strategies include gathering quantifiable information about your audience and their job descriptions; asking questions of and observing them at work; examining any relevant documents pertaining to your particular audience and the document you are working on; and working alongside your audience, for example to gain the kind of hands-on experience they have with a particular software program.

You can use a variety of strategies to learn more about your audience. Obtain information about the audience's ages, spending habits, education, work experience, location, hobbies and interests, gender, language(s), physical limitations, and so on. Look at organization charts and job descriptions to see who may know a great deal about your subject or to get a better idea of how your readers will interact with your document. Interview a few members of your audience to verify what you are finding out about the audience through other means. Examine various company documents to see what your audience is already familiar with or having difficulties with and to see what kinds of design expectations you may be facing. Good documents to examine include notes from customer training, reports from the sales staff, maintenance reports discussing problems, style guides, documents discussing reader expectations, and publications that your audience typically reads. Finally, use the product or system that your audience is expected to use. Take notes on any difficulties or questions you have as a new user.

Through all of these means, you can gather valuable information that will help you to obtain a much better understanding of your audience.

Discovering Initial Ideas

Whether your writing task is a specific reaction to an urgent request or a response to an assignment allowing you to choose any topic that interests you, you still face the problem that all writers face: finding the best ideas for your purpose and audience. Discovery—or invention, as it is also called—plays just as important a role in technical communication as it does in other kinds of writing.

Knowing that you must write a particular kind of document does not solve your problem of discovery. You still need to find the topics and subtopics you will discuss. Part Nine, "Technical Documents," covers a variety of common technical documents. That section also discusses many of the elements commonly found in these documents. The structure of a formal set of instructions, for example, typically includes an introduction, a discussion of theory, a list of equipment or tools needed, and a series of steps to be performed. The challenge lies in discovering and selecting the topics and subtopics you should discuss for these instructions, discovering the best approach to your subject, narrowing your subject properly, and communicating with your audience effectively.

This phase of planning your document is not the outlining phase. The rigid structure of a topic or sentence outline is something that comes later in the planning process. In the outlining phase, you will more formally organize the ideas you acquire in the discovery phase of your planning. The emphasis in this part of the planning phase should be on creativity. Think of as many ideas, topics, approaches, or avenues as you (or a group) possibly can. You can use one or more of the common techniques of brainstorming, mind mapping, and freewriting to help generate valuable information.

BRAINSTORMING

Brainstorming is used for everything from coming up with ideas for writing to developing creative solutions to problems. If you are brainstorming about a topic for writing, then you need to think of as many ideas as you can that are associated with this topic. If you are brainstorming for a solution to a problem, then think of as many deliberately unusual solutions as possible and push the ideas as far as possible.

TIPS: Brainstorming Alone

- Use whatever materials you prefer: self-stick notes, paper, note cards, a whiteboard, a blackboard, a flip chart, or any other device that will allow you to change, erase, connect, or move your ideas easily.
- Give yourself time just to think. Do nothing for the first ten minutes.
- Jot down ideas as quickly as you can. Use small self-stick notes for example, one per idea. Use only a word, a symbol, or a phrase.
- Look for ways to piggyback ideas or to draw ideas out of other ideas.
- Don't try to evaluate or criticize your ideas.
- Stress quantity, not quality. Try to think of as many ideas as possible.
- List practical and impractical topics or solutions. Have fun.
- Limit the time of the session if you are tired; otherwise, continue for as long as you can think of new ideas.
- Set aside the work you've done and do some other activity for a while.
- Challenge yourself to be even more creative when you return to your brainstorming.

TIPS: Brainstorming in a Group

- Reach an agreement about which tools to use to record the group's ideas. You may simply choose to use a chalkboard or a flip chart. Let everyone see all of the ideas that are being developed.
- Appoint someone to lead the brainstorming session. This person should encourage everyone to participate, to be enthusiastic, and to be uncritical. The leader should also keep the group from focusing on one topic for too long.
- Set a definite period of time for the brainstorming session.
- Participants should try to think of as many ideas as possible.
- Participants should also use the ideas of others to come up with new ideas or topics.
- Allow time for people to brainstorm individually.
- Avoid letting a few members of the group dominate the session.

(continued)

Brainstorming in a Group *(continued)*

- Avoid any criticism of ideas or topics. Criticizing any ideas can stifle the creativity. During brainstorming, no idea is wrong.
- Enjoy the brainstorming experience. Brainstorming should be a fun experience. Try to think both of topics that are practical and of others that are impractical.

MIND MAPPING

The major difference between brainstorming and mind mapping is that in brainstorming, you start with a clean slate, and with mind mapping, you begin with a core topic or idea and use strategies to generate ideas that branch out from this core idea. Of course, some mind mapping occurs during the brainstorming session as well. In mind mapping, you typically place your central idea on a page, poster, or self-stick note and then branch other ideas off of this central idea. Mind mapping is also commonly referred to as clustering because you cluster related ideas together using this approach.

You can mind map in any way you choose. Use the materials and the approach that are best for you.

TIPS: Mind Mapping

- Using whatever material you prefer—a page, poster, whiteboard, blackboard, flip chart, self-stick notes, or blank note cards—write the major idea or topic in the center.
- Draw lines or branches from the major idea or topic to show topics that are related to your main idea or topic.
- To begin, ask who, what, why, where, when, and how about your topic.
- As you begin to list topics, branch off into smaller related topics.
- Don't worry for now about the organization of the topics.
- Try different colors of paper, cards, or pens to show related ideas.
- After you have finished brainstorming, organize and evaluate your ideas or topics. Put the notes into groups of related ideas. Try different combinations.
- Erase the topics that are too far afield from your topic. Throw away what you don't want.
- Draw lines to connect related ideas.
- If necessary, move ideas from one area to another until the placement of your ideas begins to make more sense.
- Don't hesitate to add ideas if you have missed something.

- Continue to work until you have mapped your entire document.
- Step back and look at what you have created.
- After evaluating your mind map, decide what your main topics are and which ideas fit under them.
- Construct a tentative table of contents using your final mind map.
- Share the tentative table of contents with others who can help you spot problems in your approach to the topic.

FREEWRITING

Like brainstorming and mind mapping, freewriting is a useful technique for generating ideas in the planning stage. It's also a helpful technique to use later in the writing stage of your project. The key to freewriting is to avoid editing or revising your work as you write. Ignore the rules and guidelines on punctuation, spelling, grammar, organization, and so on. Just start writing and keep writing without correcting your prose in any way.

Think of freewriting as a kind of brainstorming in prose. You continue to write on your topic or any topic that comes to mind. Later you'll review what you've written and start looking for key ideas. You can, for example, use freewriting to generate ideas about your topic and then use mind mapping to give more structure to your ideas and to find connections.

You can also use freewriting after you've brainstormed or used mind mapping. Use the topics you've generated with these techniques to create topic sentences and rough paragraphs.

Later, when you are writing the first draft of your document, you may want to use freewriting to overcome writer's block. Just start writing. Then when you have something written, you'll find it easier to write more because you will be looking for ways to revise.

TIPS: Freewriting

- Write without stopping.
- Write without editing.
- If you're writing at the computer, turn off your monitor while you're freewriting so that you are not tempted to edit or revise too soon.
- Don't criticize your own freewriting or allow others to criticize it.
- Let the freewriting you've done sit for a while. Come back to it later.

(continued)

Freewriting *(continued)*

■ Look over what you've written to see whether any topics stand out.
■ Look for any connections among topics and possible subtopics.
■ Remember that the purpose of freewriting is to help you discover ideas. What you've written is not a rough draft or first draft of your document. However, don't discard what you've written either. Some of the material you have written may be useful later.

Organizing

Brainstorming, mind mapping, and freewriting will help you think of many ideas to discuss, but these are chiefly informal approaches for determining what to cover in your document. Now that you've spent some time coming up with these informal ideas, you need to organize your thoughts more formally into an outline.

One of the most successful strategies you can use for communicating with your audience is to set up a logical, effective organization for your document. The keyword here is *audience*. Remember, you want to organize and present your ideas in a way that makes the most sense to your audience. Yet the most effective presentation for one audience may turn out to be only mediocre for another. Many writers who know their subject well think primarily of how *they* would like to present the information. They think of the key topics they would like to cover and the points they would like to use to support each key topic. The problem with this approach is that these writers are thinking only about what they want to cover, not about what the audience needs to know. For your document to be effective, you need to organize it using an outline that helps the readers to see the subject from their point of view.

OUTLINING

You've no doubt been told by writing instructors to outline your thoughts before you begin writing. Of course, outlining what you want to cover requires time, and many people don't want to take the time to create an outline. They would rather just begin the process of writing and shape their thoughts as they proceed. But devoting even a few minutes to deciding what you want to write (on the basis of what your audience needs to know) before you begin writing will save you time. Outlining, like most planning, is a time saver.

Experienced writers create an outline of what they want to write about. The outline may be a quick, informal list of topics and sub-

```
How to Use This Manual
Lab Access
Lab Hardware
Lab Software
Lab Manuals
Lab Procedures
Policies for Lab Assistants
Policies for Students Using the Lab
```

FIGURE 1.1 An Informal Outline for a Typical Lab Guide

topics on the back of an envelope, or it may be a more formal outline using the outline feature of a full-featured word-processing program such as Microsoft Word or Corel WordPerfect. Use whatever approach most appeals to you.

Informal Outlines. An informal outline may be simply a listing of topics you choose to cover in a memo, letter, report, or other document. The listing allows you to see your topics on screen or paper. For many kinds of writing tasks, especially brief documents, the informal outline will be the most common. (See Figure 1.1.)

Formal Outlines. The formal topic outline is one common type of outline in which you use keywords or phrases for major headings and subheadings. Each key topic should represent one main idea, and the subtopics should represent supporting topics for each key topic. In a brief document, each key topic may represent one paragraph, and each subtopic may represent a supporting idea in that paragraph. In a longer document, each key topic may represent a main idea, and each subtopic may represent a separate supporting paragraph for that topic. (See Figure 1.2 on page 16.)

The formal sentence outline is another common strategy. At its simplest, each key topic sentence may be the main topic sentence of the paragraph, and each subtopic sentence may be a supporting idea for that main topic sentence. A more complex sentence outline may show key ideas in each topic sentence on the outline, and each supporting sentence is typically a topic sentence for a supporting paragraph. (See Figure 1.3 on page 17.)

I.   Instructional Issues
    A.   Course Descriptions
    B.   University Process of Course Delivery
    C.   Copyright Issues

II. Aesthetics Issues
    A.   Web Page Layout
    B.   Web Page Length
    C.   Graphics
    D.   Types of Graphics
    E.   Color in Graphics

III. Technical Issues
    A.   Hypertext Markup Language
    B.   Ordered Lists
    C.   Unordered Lists
    D.   Definition Lists
    E.   List Attributes
    F.   Tables
    G.   Frames
    H.   Footers

IV. Accessibility Issues
    A.   Accessibility
    B.   Blindness
    C.   Impaired Vision
    D.   Hearing Impairments
    E.   Sensitivity Issues
    F.   Multicultures
    G.   Gender
    H.   Disabilities

FIGURE 1.2   A Topic Outline for a Web Style Guide for Instructors of Distance Learning Classes

TREES

Many experienced writers prefer to visualize their topics in a different fashion. The outline is primarily a word-based organizer for your document. Other, more visual methods may turn out to be more suitable for your needs. If you are primarily a visual thinker or if you are working on a project with more of a visual component (such as a

I. Topic Sentence.   **FIGURE 1.3** A Sentence Outline
   A. Sentence.
   B. Sentence.

II. Topic Sentence.
   A. Sentence.
   B. Sentence.
   C. Sentence.

III. Topic Sentence.
   A. Sentence.
   B. Sentence.

video script), these more graphical organizing techniques may work better for you.

Trees are commonly used to show relationships among topics and subtopics. Using a tree to outline your document has several advantages over more traditional word-based outlining.

One advantage of a tree is that you can see relationships immediately, since the structure is represented visually. You can see where a point fits in, see corresponding points on the same level, and see what points are logically above or below one another. You can also step back and see the document as a whole, noting proportion and balance among its various parts. With the outline, structure is invisible and has to be imagined. Relationships are represented indirectly, in numbers and letters such as I.C.2.b, so you have to infer structure by interpreting the numbers and letters.

A second advantage of a tree is that it helps you to keep the big picture in mind. The outline form encourages you to go to the deepest level of detail almost immediately, for example, I.A.1.a. It's easy to get lost and forget the larger point you are trying to make. By contrast, the top-down approach in which you fill out a tree one level at a time helps you to keep your sense of context and direction.

A third advantage is that a tree helps you to manage your thinking. Constructing a large, coherent conceptual structure is not a natural mental act. You can't hold all the details and relationships in memory. You need to get these details and relationships out in front of you so that you can see and work with them. The tree, particularly when done with self-stick notes, provides a flexible, efficient medium around which to think and in which to organize your ideas.

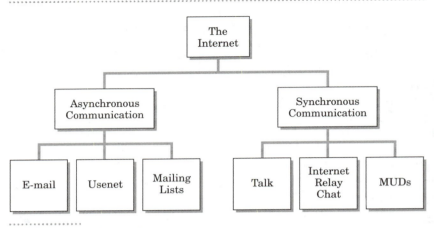

FIGURE 1.4 A Sample Tree

To help you manage your thinking during this step, use a tree diagram that looks like a family tree or an organization chart. This kind of tree differs from its real-world counterparts in being drawn upside down, with its "root" at the top and its "leaves" at the bottom. (See Figure 1.4.)

TIPS: Outlining Using a Tree

- Put the one overriding point you want to make at the top (root) of the tree.
- Use self-stick notes and write one point per note.
- Transfer some self-stick notes from the brainstorming or mind-mapping steps.
- Divide your focus into its three or four (or however many) largest, most important points.
- Place the main points under the top point, left to right in some logical order.
- Focus on each of these points in turn. Divide each into its largest component points.
- Place those subpoints under the appropriate main point, left to right.
- Repeat the process, one level of the tree at a time.
- Stop when the bottom points correspond to manageable blocks of ideas or information that you can easily write about, for example, a paragraph or so for unfamiliar material or a page or so for material that you know well.

FLOWCHARTS

Flowcharts are widely used in many disciplines and professions for all kinds of planning and problem-solving purposes. (See "Flowcharts" in Part Three, "Design and Illustration.") Like trees, flowcharts show immediate relationships, but flowcharts are specifically designed to show processes and procedures. They help you to visualize the organization of your document as well as your transition from one key concept to another. Unlike trees, flowcharts enable you to follow a sequence of steps or ideas; trees emphasize relationships and a hierarchy. (See Figure 1.5.)

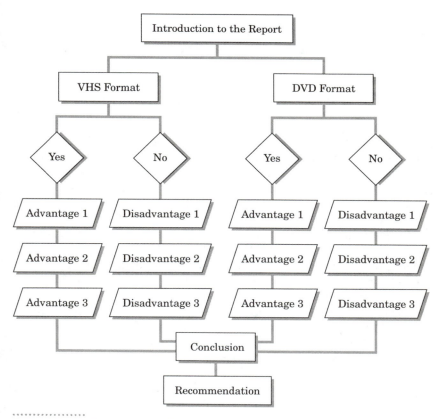

FIGURE 1.5   Flowchart for a Report

STORYBOARDING

Storyboarding is another widely used planning technique. In the film and television industries, storyboarding is used to plan practically every shot for a film or television show. Storyboarding can also be used effectively to organize the structure of a short story or novel or to plan a Web site. Information may be represented hierarchically or linearly or both. In storyboarding a Web site, for example, you would sketch and rough out each Web page. Start with your main or home page. Give each Web page a title and a rough description of the content and kinds of images you will provide. Then sketch out the links between the Web pages, and use arrows or lines to show the direction of the link. (See Figure 1.6.)

Setting and Following Standards

While you are in school, professors are constantly setting standards for the kinds of documents you write. They may require, for example, that your major research paper be typed double-spaced and have one-inch margins and numbered pages. Your professor may also require you to use a particular documentation style such as the style of the Modern Language Association (MLA) or the American Psychological Association (APA). (See "Documentation Styles," page 71.) Your professor may also require you to write in a particular academic style and to use a serious or professional tone. The professor may be particularly demanding about proper mechanics—grammar, punctuation, spelling, numbers, capitalization, and abbreviations. All of these requirements are common for documents written for college courses.

Professionals in the workplace often have very specific standards for the kinds of documents they write, too. Many companies have a style guide and standards manual specifying everything from preferred spellings for various words to the overall design and layout of memos, letters, manuals, and other documents.

Style guides are documents that tell members of an organization or employees of a company what rules or guidelines they should or must use for internal and external communications. When you create a style guide, you may be trying to ensure internal consistency for one document or for a whole suite of documents. Therefore, many style guides are only a few pages long, while others take up an entire volume. Bear in mind that style guides are subject to frequent revi-

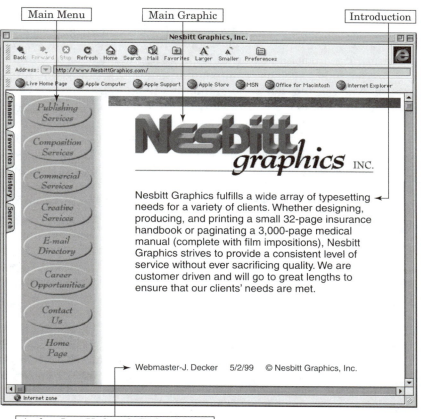

Main Menu   Main Graphic   Introduction

Author, Last Update & Copyright info

FIGURE 1.6   An Example of Storyboarding

sion. Even such respected style guides as the *MLA Handbook* began as a few pages stapled together.

Typically, style guides contain rules or guidelines pertaining to conventions in design or format, mechanics, style, and graphics. Procedures concerning editing and editing cycles, printing, and other matters may also be covered. Design topics might include the format of formal reports and manuals (the format of front matter such as the cover, title page, copyright page, table of contents, and introduction), the elements of page design for the body of the report or manual (page size, margins, headers and footers, headings, pagination, fonts), and the order and content of back matter (appendixes, glossary, index, works cited). A style manual may also specify the format for internal memos and external letters. If guidelines for mechanics are offered, topics such as grammar, punctuation, spelling, and numbers may be treated. For style, numerous topics may be covered, including tone, diction, technical terms, sentence length and variety, and active and passive voice. Guidelines for graphics might cover tips for preparing your own graphics, placement of graphics, and guidelines concerning the kinds of graphics to include. A section on editing may clarify distinctions among substantive editing, copyediting, and proofreading and may spell out how the editing cycle is done so that work is reviewed and edited properly by peers or an editor. Procedures for preparing a document for printing, methods for printing, and other print issues are typically covered as well.

TIPS: Creating Style Guides

- If your organization or company does not have a style guide, determine the need for one by polling various people within the company. If your organization or company already has a style guide, evaluate how effectively it meets current needs.
- Before beginning any writing on your style guide, go through the necessary steps of planning and researching this topic.
- Creating a style guide should be a collaborative effort. Even if you are solely responsible for creating a style guide, you should consult with those who will use it to find out what kinds of topics they think should be covered in it. With the help of others, you'll need to create a thorough document plan describing the purpose, scope, content, organization, style, and design of your proposed style guide.
- Focus on making your style guide a document that people will want to use. You may have to determine whether it will be more effective in print, online, or both.
- Cover only the essential topics. Wherever possible, refer your readers to more established or authoritative sources for some topics. Different style guides are suitable for different fields of knowledge or discourse communities. Choose a style guide from among the established ones in your field. *The Chicago Manual of Style, Words into Type, The Microsoft Manual of Style for Technical Publications, MLA Handbook, Publication Manual of*

the American Psychological Association, *The CBE Manual*, *The New York Public Library Writer's Guide to Style and Usage*, and others cover a range of topics. Find out which one is appropriate for your particular discipline or industry and use it as the foundation for your own style guide.

- Follow the principles of good writing and design in your style guide. Your style guide should be a good example of a product that adheres to the guidelines you establish.
- Remember to provide some helpful reader aids: a table of contents, tabs, sectional tables of contents, a glossary, and an index. Make your style guide as easy to use as possible. Try to anticipate how your readers will use it and provide the information where it can be found easily.
- Choose a paper quality and a binding method that are suited to the constant use your manual will have. If the guide will be revised or amended, make it easy to add new pages without having to replace the entire volume. For an online manual, follow the guidelines for creating effective online information. (See Part Three, "Design and Illustration.")
- Make sure your style guide is thoroughly reviewed and evaluated by others so that it will be more usable.
- When the style guide is finished and printed or made available online, make sure it is used. Don't let your peers and others slip back into their practice of ignoring established conventions concerning format, mechanics, style, editing, design, and so on.

DETERMINING THE STYLE

Tone and style are complex topics that are covered in more detail in Part Two, "Writing." However, it's important to keep in mind that tone—your attitude toward your reader, your subject, and yourself— and style—your manner of expressing yourself in your prose—need to be planned to some degree even before you begin writing.

Achieving the proper tone is, of course, closely related to your focus, purpose, and audience. You should give some thought before you write to the primary attitude you want to convey—whether, for example, you want to be serious and professional, humorous, or especially tactful. Tone is also a matter of distance, so you need to consider how personal or impersonal you're going to be. Are you going to address the reader as "you," for example, or are you going to use a third-person point of view?

Your writing style for a particular document can be planned to a degree. You need to decide in advance whether you are going to discuss your subject in a plain style or a more complex prose style. You need to give some thought to your ethos—character, credibility, authority—and how you can strengthen this ethos in your document. You can plan these and other elements of your style somewhat, but there's also a quality in your style that's beyond your control. In part, the kind of person you are defines your writing.

DETERMINING THE MECHANICS

The mechanics of your document involve grammar, spelling, punctuation, numbers, capitalization, and abbreviations. Part Four, "Editing" offers a detailed overview of all of these topics. Although you cannot create your own set of grammar rules to follow in your document, you can decide which preferred spellings you will use, what kinds of punctuation marks you will use to indicate various elements, and what, if any, abbreviations will be used and where. Any problems in mechanics will ultimately be dealt with during the revising stage, but you can avoid some basic problems by ensuring some consistency during the planning stage.

DETERMINING THE DESIGN

Design deals with arrangement and appearance rather than content. It concerns everything from margins to fonts and illustrations. Increasingly, writers of even the most basic memos and letters face many design choices before they begin writing. Part Three, "Design and Illustration," provides a thorough discussion of the many design considerations you must take into account for the different kinds of documents you will work on.

Unfortunately, many writers give little consideration to how even their briefest documents look. Typically, the font is too small, there are too many kinds of fonts on any one page, headings are nonexistent or the heading hierarchy is confusing, margins are crowded, there is too much text on the page, the paragraphs are too long, and so on. These faults and many others show that many writers are more concerned with what they want to write than with the visual appeal of what they write. All of these symptoms and others show that many writers are more preoccupied with themselves than they are with their readers.

Perhaps it's overstating the matter, but in many respects, the visual appeal of your document can be as important as the text. The design, illustrations, and white space are, like the text on the page, active participants in your efforts to communicate with the reader.

Most writers use a full-featured word-processing program such as Microsoft Word or Corel WordPerfect, a modest publishing program such as Microsoft Publisher, or a sophisticated desktop publishing program such as Adobe PageMaker or Adobe FrameMaker. These programs and others allow writers to make decisions about the formats of their documents before they begin writing. Some of these programs offer templates, which are essentially predesigned documents. The writer provides the text, and the program helps to create or automatically creates a new memo, for example, in a contemporary, elegant, or professional style. Many of these programs allow writers to create a style for the entire document before typing the first word. The style that is created may automatically set up the document's margins, headings, heading hierarchy, pagination, and fonts, among many other elements.

Managing the Document

Managing your document means planning for all of the details necessary to make sure it is completed, submitted on time, produced appropriately, and, if applicable, produced within your budget. The level of management a document requires varies with the kind of document and its length. For many brief documents—such as letters, memos, and e-mail—little management is involved. For longer projects, the management required is typically extensive.

If you are the sole author of a document, you still need to divide the writing process into smaller tasks before you can begin work on it and then finish it appropriately. If you are collaborating with others, you need to make many decisions on almost everything from assigning tasks to determining a complicated production budget.

ESTIMATING TIME AND SCHEDULING

Whether you are writing a brief memo or collaborating on a large multivolume user guide for a new software program, estimating and scheduling are essential tasks. For example, for the memo, you might estimate that writing and distributing it will take 15 minutes; for the multivolume user guide, you might estimate that the project will take you and three other writers six months.

In a typical junior- or senior-level technical writing course, students must estimate how long it will take them to complete all of the various assignments in such a course, such as an application letter and résumé, a letter of adjustment, a set of instructions, a proposal, a technical report, and an oral presentation. Some of these assignments can be done in much less time than others. Sometimes you may be asked to create a schedule providing proposed completion dates and tasks. This schedule may take many forms, ranging from an informal table to a Gantt chart. (See Part Three, "Design and Illustration.")

In college or in industry, you will often find yourself working with others on a writing project. See Part Seven, "Interpersonal Skills," for a discussion of the skills you will need to collaborate with others successfully. If you are collaborating with several others to complete a document, you need to manage this document early in the process rather than waiting until later.

TIPS: Estimating Time

■ Research the complexity of the project. Make sure you find out as much about the project as you possibly can. If possible, talk to others who have worked on similar projects.

■ Evaluate the complexity of the project. Will it require a great deal of planning, researching, writing, editing, revising, designing, illustrating, reviewing, and evaluating? Will production time be significant?

■ Establish specific goals or objectives for the project.

■ Create a specific document or project plan. See "Creating the Document Plan," page 27.

■ Divide the project into tasks. Write down the tasks that you face; if they are large, break them down into their component elements. If these still seem large, break them down again. Do this until everything that you have to do is listed.

■ Create detailed "to do" lists or assign a team member to write them. These lists show the tasks that must be carried out by you or your team members to achieve your project or document goals. List the tasks in order of importance and assign a priority to each one, for example, from A (very important) to F (unimportant). If too many tasks have a high priority, run through the list again and demote the less important high-priority items. Once you have revised the list, rewrite it in priority order. The result will be a precise plan that you can use to deal with problems in the order in which they need to be eliminated. This strategy allows you to separate important jobs from the many time-consuming trivial ones.

■ Establish a schedule.

■ Become familiar with various published estimates of time for tasks. Interview others about the time a particular task might require.

■ Work in "downtime" to allow for various possible unexpected delays or risks. When you have to guess time, particularly when you are likely to be held to a time estimate, ensure that you allow time for other tasks that have priority over this one, accidents and emergencies, meetings, holidays, vacation and sickness of essential staff, contact with other customers, time necessary to arrange the next job, breakdowns in equipment, letdowns from suppliers, interruptions, quality control rejections, and so on.

- Use a log to track the time for the various documents you write. A log may simply be a listing of the dates, tasks, and the amount of time you devote to each of the tasks. Keep track of every task in quarter-hour increments.
- Use progress or status reports.
- Stick to your schedule, deviating from it only for a good reason, but be prepared to make adjustments when necessary.

PLANNING THE PRODUCTION

Part of managing your document requires you to plan how you will produce it. Will you simply use a word-processing program or desktop publishing program and print it using an inkjet or laser printer? Will you require so few copies that you can print them on your home printer? Is the printing job so large that your company requires you to contract the job with a commercial printer? Will you use color? expensive paper? provide a cover? How will you distribute the document? These and many other questions must be considered during the planning phase. See Part Five, "Production," for a more detailed discussion of the many choices you must make for producing documents of all kinds.

Creating the Document Plan

Some documents require a document plan and some do not. If you are writing a brief memo or letter, you don't usually need to create a document plan. However, if you are in a college course and you are required to collaborate with others to create, for example, a software user's guide, you'll need to plan this guide extensively. You'll need to go through all the steps discussed so far: determining the focus, generating ideas for topics, organizing, setting standards, and managing your document. If you are required to submit a formal document plan, you will most likely provide a detailed discussion of your subject, purpose, scope, audience, outline, standards, schedule, responsibilities, qualifications, and so on.

Some people consider the steps you need to perform in the planning phase as creating a document plan. Others refer to the steps as creating a kind of proposal for a document. Still others refer to this part of the process as creating the document blueprint. Whatever you call the planning stage, it's essential even though it is time-consuming. By some estimates, during this planning phase, you will have completed 60 percent of the work required for your document. A thorough document plan written at the beginning of the project will save you and your team time and effort later and will help you and your team stick to the schedule.

In industry a document plan is essentially a blueprint for the entire project—from the research to the plan for distributing the document after it is completed. You might be required to complete a detailed formal document plan form, typically covering subject, purpose, scope, assumed audiences, audience use of the document, elements needed for the document (from title page to index and from letter of transmittal to warnings or notes), graphical elements, style, reviews needed (for example, legal, marketing, and research), production (including budget details), timetable for the project, outline for the project, and project distribution date.

With this kind of detailed document plan, all the team members know what their tasks and deadlines are and can plan accordingly. Creating such a detailed document plan requires a great deal of time, but it is time that is well invested. If done effectively and thoroughly, such detailed planning gives the project a clear purpose and a clear priority. (See Figure 1.7 for a typical document plan schematic.)

Using Planning Tools

Today's software programs offer a sophisticated array of features for helping any college student or professional achieve much more work in much less time. To help you with your work, you now have more computer tools to choose from than ever before. Keep in mind, however, that even the best software won't make you a better writer, planner, or designer if you don't already know a great deal about writing, planning, and designing. Planning tools don't guarantee quality, and they don't, by themselves, create quality.

The planning tools briefly summarized here concern project scheduling, personal information management, outlining, creativity, flowcharts, storyboarding, database management, presentations, and cost estimating.

Date Assigned:

Author(s):

Tentative Title:

Subject:

Purpose:

Scope:

Audience:

How Audience Will Use the Information:

Other Audience Considerations (environment in which document will be used, document format preferences, special needs):

Elements Needed for Document:

Illustrations Needed for Document:

Style for Document:

Reviews Needed:

Production (type, time, cost):

Timetable for Project:

Outline for Project:

Project Distribution Date:

FIGURE I.7   Document Plan Schematic

Project-scheduling software helps you to plan a project more carefully and efficiently. One of the most widely used project-scheduling packages is ivan. The essential feature of ivan is that it is entirely graphical.

Personal information management programs and their hardware equivalents, personal digital assistants, help you organize both your personal and business schedules effectively. Typical software programs include Sidekick, DayTimer, Outlook, and Time and Chaos. And, of course, numerous print versions of personal organizers are available too, including DayRunner and DayTimer. These programs and print organizers offer effective ways to prioritize your daily tasks, note appointments, keep track of contacts, provide reminders, and so on.

Outlining programs may be built into a full-featured word-processing program or come as a separate program. An effective outlining feature or outliner can help you easily list your major thoughts and rearrange them quickly with a few keystrokes or clicks of a mouse. Some typical programs include Symantec Grandview (a DOS-based outliner), Ecco (NetManage's product for outlining and exporting files to other programs), Lotus AmiPro (a solid and easy-to-use outliner), and WordPro (a much slower replacement for AmiPro). Other programs offer combinations of an outliner and some other kind of program. Microsoft Word, for example, offers a fairly sophisticated outlining capability in addition to its word-processing features. Inspiration Software Inspiration is a combination outliner and flowchart program. You can look at your work in either format.

Creativity programs help you to enhance your creativity in a variety of ways. Perhaps the most full-featured idea-generation software currently available is IdeaFisher. IdeaFisher, a computerized aid to brainstorming and problem solving, helps speed up the creative process. The program consists of an interactive database of questions and idea words and phrases. Questions are organized to help you define, clarify, modify, and evaluate your needs. You can even add your own questions to the thousands already supplied. Another outstanding creativity program is Inspiration, which facilitates mind mapping. Instead of listing your ideas from top to bottom, mind mapping and Inspiration work from the center out. You can do this on paper by starting with a central idea and adding your thoughts, connecting them like spokes from an axle. Playfully link ideas, expand others, and explore any associations that strike you.

Flowchart programs enable you to document a process. Some word-processing programs help you create basic flowcharts, but many programs are available that specialize in creating flowcharts of all kinds. The Visio products <http://www.visio.com/> are some of the most widely used flowchart programs in business and industry. Visio Standard can create flowcharts, work-flow diagrams, organizational charts, project timelines, geographic maps, mind-mapping diagrams, and many others. Visio Professional is designed to facilitate the creation of network diagrams of various sorts, Web site and software diagrams, and user interface designs. RFFlow <http://www.rff.com/> is a powerful program that can create a wide variety of flowcharts, organization charts, and diagrams. RFFlow provides many predetermined shapes and symbols to support a variety of applications. Text effects are particularly easy to create with this program. Among the graphic elements it can produce are three-dimensional flowcharts, circuit diagrams, data flow diagrams, database definitions, quality management charts, physical flows, structure charts, and system designs. SmartDraw <http://www.smartdraw.com/> is a program for Windows that helps you draw and edit flowcharts, diagrams, and other business graphics. Its many symbols and templates walk you through what can otherwise be a very tedious task. The diagrams and flowcharts that you create in SmartDraw can be imported into a variety of other programs for incorporation into your other documents.

Storyboard programs enable you to plan or brainstorm. A variety of tools are available to help you create mock-ups of your multimedia projects. Some are created specifically for the storyboarding process; others, such as Apple HyperCard, Adobe Persuasion, and Microsoft PowerPoint, are presentation tools that can be readily adapted to this use.

Database management enables you to make large amounts of information easier for people to understand. Databases do not necessarily require the use of a computer. You probably already maintain several databases without even knowing it. Any time you keep track of information with some type of organized system, even a list, you are managing a database. Your calendar, personal phone book, and "to do" list are all databases, even if you commonly write them in pencil. The telephone book is a database. Many computer programs help you manage your databases. The choice of software depends on the kind of database and the type of manipulation you have to perform. Read-only databases, predefined databases, flat-file databases, and relational databases are examples.

Read-only databases are those that have already been established, usually by a professional programmer or team. Some examples are various Internet search engines, online card catalogs, and airline reservation systems. Predefined databases are commercially available for people who want to manage a database but don't want to build it themselves. Often they focus on a specific topic in the business or home market, such as wedding preparation, video-library management, invoice handling, inventory, or payroll. A flat-file database is capable of handling only one table at a time. For home use, this is often just fine. You usually don't need any more than this to deal with your Christmas card list or stamp collection inventory. A relational database is perhaps the most useful database in the corporate world. It is much like a flat-file database, except it allows the use of multiple tables. For example, you can keep track of inventory, customers, and transactions in three distinct tables that share some fields. These tables are distinct but related, so the database is called a relational database. The big database management systems such as Oracle, Corel Paradox, and Microsoft Access are relational databases.

Presentation programs enable you to plan, prepare, and present effective presentations for a group. Programs such as Microsoft PowerPoint and Corel Presentations are just two of the many programs that can help you create slides, transparencies, handouts, and more for your technical presentations. More specific information about some of these programs is discussed in Part Seven, "Interpersonal Skills." These programs are mentioned briefly here because they can also be helpful when you are in the planning stage of a project. Microsoft PowerPoint can be easily adapted to for brainstorming, mind mapping, or organizing your ideas. You can easily add, import, and rearrange slides until you have a complete visual map for your document.

Cost-estimating software offers quick and convenient ways to determine your costs before you begin a project. Cost-estimating software can drastically cut your estimating time as well as save you money. CNC Concepts Profit Planner <http://www.cncci.com> and other programs can help you determine exactly what the costs will be before you submit your proposal for a particular job.

RESEARCH

Approaches

Your approach to research depends on your field of study. The physical and life sciences rely on empirical research or hypothesis testing. The social sciences rely on clinical observations or questionnaires as part of the empirical research in these disciplines, and the humanities rely on case studies, observations, and individual and group interviews. Of course, all of these disciplines rely on additional secondary sources both within and outside the discipline. The research method that you use depends on the purpose of your project, the type of information you must gather, and how you will use the results.

Most kinds of research are either qualitative or quantitative. Stated simply, qualitative research tells you why; quantitative research tells you how many. Qualitative research gives you impressions and descriptions; quantitative research produces countable results. In marketing research, for example, if you are trying to improve a product or service, identify different market segments, or develop a persuasive advertising or sales message, then qualitative research is the best approach to use. If instead your goal is to determine how many people like an idea, measure the size of a market, or prepare a volume estimate, then quantitative research is what you need.

QUALITATIVE RESEARCH

You should use qualitative research when you really want to understand in detail why an individual does something. It is often used in marketing, for example, to understand why an individual buys a particular product or service. This type of research can determine real market segments or groups of people who purchase for the same reasons. Qualitative research is particularly useful as a tool for determining what is important to people and why it is important. Qualitative research helps you identify issues and put them in perspective.

Qualitative research should not be used when you need to learn how many people will respond in a particular way or how many hold a particular opinion. Qualitative research is not designed to collect quantifiable results. After learning why one person would buy or respond in a certain way through qualitative research, it is relatively straightforward to determine through quantitative research how many other individuals there are like that person. Qualitative research is often followed by a quantitative study.

QUANTITATIVE RESEARCH

The major reason for conducting quantitative research is to learn how many people in a population have a particular characteristic or group of characteristics. It is specifically designed to produce accurate and reliable measurements that permit statistical analysis.

Quantitative research is appropriate for measuring both attitudes and behavior. In marketing, for example, if you want to know how many people use a product or service or are interested in a new product concept, then quantitative research is what you need. It is also used to size a market, estimate business potential or volume, and measure the size and importance of segments that exist in a market.

An important point to remember is that qualitative research and quantitative research are not mutually exclusive. With the necessary precautions, you can measure and code qualitative data using quantitative methods. And quantitative research can be generated from qualitative data. However, it's also important to remember their essential differences. The major difference stems from the researcher's underlying strategies. Qualitative research is considered to be exploratory and inductive. Quantitative research is viewed as confirmatory and deductive.

Planning Research

IDENTIFYING RESEARCH QUESTIONS

Whatever research methodology you use, plan your strategy before beginning your research. You need to identify the key question or questions that you are trying to answer. Sometimes these key questions may be assigned to you, but at other times, part of the process will be to determine what you are looking for. One of the major reasons for your research is to help you find a solution to the problem, but it is not unusual to grapple with questions that have no easy or readily apparent answers.

One way to begin work on either a specific assigned research topic or a problem that has no readily apparent solution is to use brainstorming, mind mapping, and freewriting, discussed earlier in this section. Discover subtopics connected to your assigned topic, or ask why the problem is a problem. Make a list or a mind map or just do freewriting about the major and minor questions you have about the topic or problem. Often while reviewing your questions you'll think of other questions. After these planning sessions, you should have a good list of subtopics or questions that you want to find the answers to. Remember that your research will continually help you to revise your list of subtopics or questions.

Eventually, your questions will lead you to the issue of significance. You'll have to deal with the "so what?" part of your research. For example, perhaps you are interested in why and how the prose style for Web pages is different from the prose style for printed pages. In your brainstorming you'll ask what the differences are, if any. You'll try to group the differences as you discover them. You'll wonder whether readers read information on the screen differently from the way they read information on a printed page. If so, you'll wonder what these differences are. You'll have many other questions about these and related issues. Then you'll need to know who else has researched this topic and what these researchers have answered or left unanswered.

Once you've focused your research adequately, you'll want to consider why it's important to answer your major research question. So what if the prose style is different for Web and printed pages? Why is this question significant? It's no easy task to focus your research topic adequately and to address the issue of significance properly. No one formula works every time, even for the same researcher. Each research problem presents its own questions, and answers to these questions have their own significance.

A discussion on audience earlier in this section covered how to consider the needs of your audience as you narrow your focus in your discussions of the purpose and scope of your document. To plan your research project, you must determine which aspect of your subject you will examine and the scope or level of detail your research will encompass. Determining your scope provides you with a sense of the complexity and depth of research awaiting you.

CHOOSING RESOURCES

Once you have focused your research, you need to consider the kinds of sources you must use or want to use. In some college classes, you may be required to consult certain sources. In a corporate environment, the company library and certain subject matter experts may provide you with all the sources you need. Make sure you know what your requirements are.

Choosing your resources also means identifying subject terms and keywords before you begin using a library's traditional card catalog or online catalog or before you begin to use a search engine or index for your Internet search. Of course, you will typically add to your list of subject terms and keywords as you refine and expand your search.

Taking Notes

Many researchers use 3 x 5 or 4 x 6 cards to store much of their information during their research. They create bibliography cards for sources they want to look at or have already looked at. They also create a variety of note cards with comments about the sources they read. Some researchers prefer to use bound notebooks for writing down source information and material collected from their sources.

In an age of laptop and notebook computers, however, many researchers are storing all of the same information electronically.

Whatever method you prefer, the important point is to use one method. Inexperienced researchers typically have no system or method, loosely keep track of the sources they consult, carelessly use material from various sources, and often waste time later trying to determine which sources they used in their work. The method described below uses 3 x 5 cards, but the approach can certainly be modified to suit your own preferences and other media.

BIBLIOGRAPHY CARDS

Many students are reluctant to use bibliography cards for research. They question the need to organize their research so formally and so tediously. Who wants to write down all that information on a lot of cards? Isn't it easier to photocopy everything? Why is it necessary to keep track of research with cards, a notebook, or a laptop computer? Of course, whether or not you use 3 x 5 or 4 x 6 cards for your sources or use a computer program to keep track of them (see "Research Tools," page 71), you need a system to record complete documentation of all the sources you are consulting. You need to keep an accurate record of all your sources so that you can fully document them later without having to track them down again for missing information. This is important so that you can give credit to others where credit is due, and so your readers can see how you have supported your views with the authority of others. (See Figure 1.8.)

NOTE CARDS

Keeping good notes is at the core of good research, whether you're gathering your information from personal experience or using a variety of libraries. For the sake of time, keep your comments—your own, a paraphrase, an indirect quotation, or a direct quotation—brief. However, be sure to write down as much as you need to, so you don't have to spend time later tracking down the original source again.

Paraphrase. A paraphrase is a careful rewording of the ideas and sentences of another author in your own words. Creating a good paraphrase takes time. You have to read carefully and have a good understanding of the original quotation. Then you must have the ability to rewrite the information in your own words. Taking the time now to paraphrase will save time later in the writing phase of

Author's name

Title of work

Publication information

Other publication items (for example, editor, edition, journal issue)

Library call number (if applicable)

Your notes about content (if any)

FIGURE 1.8   A Sample Bibliography Card

your project. When you take extra care during this phase of your research, you avoid inadvertent plagiarism later.

A typical paraphrase card consists of author, source, page number (if available), the paraphrase, and a note indicating that what you have provided is a paraphrase. (See Figure 1.9.)

Author   Topic

Page number (or other source info.)

Paraphrase of original source (rewording of original into your words and style)

FIGURE 1.9   A Sample Paraphrase Card

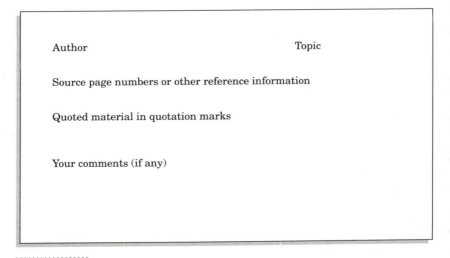

Author   Topic

Source page numbers or other reference information

Quoted material in quotation marks

Your comments (if any)

FIGURE 1.10   A Sample Direct Quotation Note Card

Quotation. The two types of quotations are direct (which repeats the exact words used) and indirect (which describes the information conveyed). For a direct quotation, be sure to copy the quoted material verbatim, even including any errors that are part of the original text. (See Figure 1.10.)

Even after just a few hours of intensive research, you may find that you've accumulated a good number of bibliography and note cards. Because this approach can be time-consuming, you'll need to be careful about the sources you use. As you work, try to determine whether the comments you are paraphrasing or quoting are valuable, insightful, or interesting. You'll need to be a discriminating reader and a highly selective note taker, lest you allow the important information to be overwhelmed by a sea of inconsequential facts.

Research Ethics

The subject of ethics is covered in more detail in Part Seven, "Interpersonal Skills," but, of course, ethical concerns must be considered throughout the process of creating technical documents.

As you research and use information, you need to be aware of ethical issues concerning your work. The following discussion provides an overview of research etiquette, plagiarism, strategies for avoiding plagiarism, and copyright.

RESEARCH ETIQUETTE

Most discussions of research ethics focus only on the responsibility of the researcher to avoid plagiarism. Another topic should be emphasized first: research etiquette. Researchers have an obligation to act in a certain way as they obtain the information they and others need.

TIPS: **Research Etiquette**

- Respect and show appreciation for the time and efforts of others as you work to uncover the information you need. Be prompt for interview appointments. Don't use more time than necessary for an interview. If you have acquired access to a special collection, use the access time appropriately. If you ask a research librarian to help you, don't rely excessively on this person's time and good nature.
- Respect the feelings and privacy of others as you phrase your questions for informational interviews, questionnaires, surveys, and so on. Of course, the kinds of questions you ask depend on the purpose of your research, but whether or not the subject of your research is of a personal nature, phrase your questions so that they are appropriate and not offensive. When you report the results of questionnaires and surveys, present the information in such a way that you do not reveal unnecessary personal details about the individuals who provided your data.
- Take care of borrowed materials. Sources that you borrow from libraries, companies, and individuals are not yours; they need to be cared for appropriately. If you have borrowed sources, remember to return them within a reasonable time.
- Obtain the proper permissions if you are using someone else's words or ideas in a commercial work. When necessary, begin the permission process during your research.
- Obtain the proper clearances for any security-sensitive information. In informational interviews, you may be given access to or knowledge of sensitive information. You may be required to sign a confidentiality agreement before gaining access to such information. Make sure you abide by the terms of the agreement.
- Protect your sources if the information you were given was provided to you on the condition of anonymity. If you gathered information under certain conditions or agreements, you have an obligation to live up to these terms.
- Place direct quotations in the proper context. You have an obligation to your sources to represent their ideas accurately.
- Make sure your information is accurate. Provide an effective research trail. Make it possible for both yourself and others to check on your

(continued)

Research Etiquette *(continued)*

notes and sources in case there are any disputes about the accuracy of information.
- Make sure your information is complete. Take the necessary steps to acquire all the information you need. Schedule follow-up interviews when necessary. Revisit some sources again if you need more information.

PLAGIARISM

Plagiarism is representing the work or ideas of someone else as your own. Unlike fraud—falsifying data and conclusions based on those data—plagiarism is solely based on representing someone else's work or ideas as your own.

In the college community, plagiarism takes many forms. When students copy someone else's answers during an exam, this action constitutes plagiarism. When students turn in papers written by someone else and claim that they wrote the paper, they are committing plagiarism. When students use material (for example, prose or illustrations) from other sources and fail to acknowledge these sources appropriately, they are plagiarizing. In sum, plagiarism is stealing the work or ideas of others.

Of course, college students aren't the only ones who plagiarize. College professors have been known to steal the work of their colleagues or graduate students, employees have been known to steal the work of their fellow employees or that of employees at other companies, novelists have been known to lift entire passages from the work of other novelists, and so on. Plagiarism and copyright infringement overlap to an extent (see "Copyright"), but they can and often do occur independently of each other.

Plagiarism is obviously an ethical issue. You have an obligation to use and represent your sources accurately and honestly.

TIPS: Avoiding Plagiarism

Because plagiarism is unethical and may have serious consequences, it's important to know how to avoid plagiarizing.

- Keep careful track of what work and ideas are yours and what work and ideas belong to someone else.

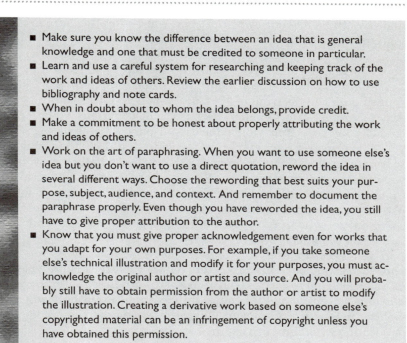

- Make sure you know the difference between an idea that is general knowledge and one that must be credited to someone in particular.
- Learn and use a careful system for researching and keeping track of the work and ideas of others. Review the earlier discussion on how to use bibliography and note cards.
- When in doubt about to whom the idea belongs, provide credit.
- Make a commitment to be honest about properly attributing the work and ideas of others.
- Work on the art of paraphrasing. When you want to use someone else's idea but you don't want to use a direct quotation, reword the idea in several different ways. Choose the rewording that best suits your purpose, subject, audience, and context. And remember to document the paraphrase properly. Even though you have reworded the idea, you still have to give proper attribution to the author.
- Know that you must give proper acknowledgement even for works that you adapt for your own purposes. For example, if you take someone else's technical illustration and modify it for your purposes, you must acknowledge the original author or artist and source. And you will probably still have to obtain permission from the author or artist to modify the illustration. Creating a derivative work based on someone else's copyrighted material can be an infringement of copyright unless you have obtained this permission.

COPYRIGHT

Perhaps no topic is as misunderstood as copyright. Copyright is something you have probably heard about and perhaps thought about as a student, and you may have an idea that it also has something to do with plagiarism, but the details may be fuzzy.

Copyright is one of those topics that have legal consequences—What you don't know can hurt you. Perhaps you think that if you are a student, copyright issues don't apply to you. You may have heard that educational uses of copyrighted materials are considered "fair use," and you don't have to concern yourself with them. Or you might have seen someone's list of rules and believed that as long as you follow them, you will be fine. Unfortunately, copyright is much more complicated than that. Let's examine some of the popular myths about copyright.

If you document where you find information, you don't have to worry about copyright. Documenting your sources is an ethical issue. You can face unpleasant consequences if you try to pass off work done by others as your own. You open yourself up to charges of plagiarism and perhaps dismissal from school or your job. But plagiarism is a separate question from copyright. Even if you document your sources, you still have to consider the copyright issue.

If you are a student, copyright issues don't apply to you. You are counting on the doctrine of "fair use," spelled out in the Copyright Act, to protect you against the charge of copyright violation. Although it is true that your educational purpose weighs heavily in your favor, some circumstances can negate that protection. You need to be informed about what fair use really is and whether it applies in your particular circumstance. Because this is often not clear, the Conference on Fair Use (CONFU) has formulated guidelines for fair use in several areas as an aid to users—primarily in the academic community—who grapple with obeying a law whose provisions can be subject to judicial interpretation.

If you're not making any money from your borrowing, you don't have to worry about copyright. Here again, you are depending on the fair use doctrine to excuse your reuse of copyrighted material. A popular misconception is that copyright violations occur only when the violator makes a profit. Profit making is an almost negligible factor in deciding whether a use is fair.

If you use only a short excerpt from a work, your use is "fair." This belief is widespread and wrong. Courts have judged even tiny borrowings as violations, depending on the circumstances. *Nation* magazine was found guilty of copyright violation for printing 300 words from Gerald Ford's autobiography when the court ruled that the excerpt was so central to the book that quoting it constituted copyright infringement.

If you change the borrowed work, you have not violated the copyright. Unfortunately, this belief is totally unfounded. Only copyright owners have the right to modify copyrighted works. They can license that right, but they don't have to. Without a license from the copyright owner, you do not have the right to create derivative works and call them your own.

Even if you violate a copyright, you won't be prosecuted, particularly if you are an individual and not a big corporation. Many people have discovered, too late, that this belief is simply not true. Even the "little guy" gets prosecuted, and fines can range into the hundreds of thousands of dollars. Recent changes to the copyright laws have even provided for criminal penalties in cases of copyright infringement.

If you didn't know the work was copyrighted, you're not responsible for violating the copyright. Again, this is not true. Copyright is a "strict compliance" law, which means that whether you know about the copyright or not, you are still responsible for respecting it.

The work couldn't have been copyrighted because there was no copyright notice on it. A work needn't necessarily have a copyright notice affixed to fall under copyright protection. You, as the borrower of material, are responsible for ascertaining whether the work is covered by copyright. Since 1989, copyrighted works do not have to have a copyright notice. If they qualify for copyright, they are automatically covered by the copyright laws, with or without a notice.

Why is copyright important? It's important because it is a legal concept that you must deal with every day. In the workplace or in school, you face copyright issues whenever you use a document or create one of your own.

Using Copyrighted Material. If you have ever read a book, viewed a film, listened to a song, watched television, or surfed the World Wide Web, you have used copyrighted materials. What do reading, viewing, listening, watching, or surfing have to do with copyright? And why should you know about copyright to perform these activities? You should know about copyright because you have certain rights as a user of copyrighted materials. These rights don't depend on your being a student or a nonprofit organization, nor do they depend on your using a small or large amount of the original work.

You are using the copyrighted work if you read, view, listen, watch, or surf the work as an individual. You may have purchased your copy, or you may have borrowed it from someone else. As long as you have acquired it legally, you are entitled to use it. Using it means enjoying it in the manner intended by the work's creator. If you want to pass it around to your friends, you may do that. If you want to make one copy for your friend to have and enjoy, you *might* still be within the area considered personal use. If you want to make multiple copies and pass the copies around to your friends, you are probably violating the copyright owner's rights.

In the case of written documents or audio recordings, after you have purchased them, you may dispose of them as you wish. This means that you may keep them or resell them, and the copyright owner has no further interest in them. You may not use them in ways that infringe on the copyright owner's exclusive rights, however. For example, you may not use them to create a derivative work without permission.

Computer software is a difficult issue for consumers, because most software is licensed, not sold. You may think you are buying the software you see in the store, but you are buying the license to use it. Under the terms of the license, you are permitted certain uses and forbidden others. The license spells out the conditions under which you can sell the software. Remember that once you've bought a book or audio recording, you can sell it as you wish. With software, you own the license—not the product—so the software itself is not yours to sell. Read the license when you buy software and be aware of your rights and responsibilities.

Incorporating Copyrighted Material into Your Work. A compiler is someone who puts together preexisting material, perhaps combined with original material, to form a new work. You are a compiler if you create a brochure that uses some of the promotional material created by your company for its other advertisements. You are a compiler if you write computer documentation and use screen shots as graphics. You are a compiler if you put together a multimedia presentation made up of music and video clips that someone else created. You are a compiler if you write a literature review for one of your classes.

When you combine previously written or recorded information into your new work, you will be concerned about the copyright issue. What material are you entitled to use? How do you find out?

If your employer asks you to write a manual or a brochure using previous versions of the manual or other promotional materials developed by your company, you can be fairly certain that you are entitled to use and reuse the materials you're given. In this case, your employer owns the copyright to the material and has given you permission to use it.

If you are writing a paper for a class and want to cite someone else's work to support your argument, you can do so, even quoting the other person's exact words, provided that you credit your source (an ethical requirement) and take only enough of the source material to make your point and no more. In this case, you do not have permission to use the copyrighted material, but you are taking advantage of the doctrine of fair use to stay within the law.

These two situations, the workplace and the classroom, both permit reuse of copyrighted materials, but for different reasons. In the workplace, you must have permission of the copyright owner before you can use copyrighted works to form a new work. Your employer can give you permission for works owned by the company. For works that are under copyright to others, you must request permission and perhaps pay a licensing fee. It can be very difficult to ascertain who owns the copyright to some works, especially multimedia works that involve many rights (text, video, and music, to name a few). In some cases, notably music, agencies that specialize in licensing copyrighted works can handle the arrangements for you, collecting the fee from you and crediting it to the copyright holder. In other cases, finding out who owns the copyright can be difficult, and it is best to consult an attorney who specializes in intellectual property rights (such as copyrights, patents, and trademarks) to make sure you have not infringed, however inadvertently, on someone else's copyright.

When you use copyrighted materials in a school or other setting and don't obtain permission from the copyright owner, you are relying on the doctrine of fair use to excuse you from what would otherwise be an infringement of copyright. Fair use is a provision of the copyright law that exempts you from having to get the copyright holder's permission under certain circumstances. If you claim that your use is fair and the copyright holder takes you to court, the court will determine the case based on how your use fits into the four provisions of the fair-use exemption. According to the Copyright Act of 1976, the court will consider:

- The purpose and character of the use, including whether such use is of a commercial nature or is for nonprofit educational purposes;
- The nature of the copyrighted work;
- The amount and substantiality of the portion used in relation to the copyrighted work as a whole; and
- The effect of the use on the potential market for the copyrighted work.

You will notice that no specific numbers of words are mentioned in these tests. Some people will tell you that you can borrow others' words if you haven't used more than 250 words, or 500 words, or 1,000 words. Many universities publish guidelines that you can follow. These rules of thumb have no firm basis in law. They are suggestions for keeping you within reasonable limits for the third test, that of "amount and substantiality." Remember that the "amount and substantiality" clause is just one of four tests. When in doubt, seek the advice of an attorney who specializes in copyright and other intellectual property.

Copyrighting Your Work. If you are a writer, student, webmaster, illustrator, multimedia specialist, computer programmer, or other content developer, you are creating copyrightable material. Who then owns that copyright depends on your particular situation. If you are a student and have written a paper for a class, the copyright is yours. No one else can reprint your paper without your permission. If you write computer documentation for a company that employs you to do that writing, your employer owns the copyright for your work. You hold no rights to the finished work, even though you wrote it. If you are an independent contractor who is hired to create a particular project, Web site, book, brochure, computer-based training module, or multimedia compilation, you may own the copyright or it may be owned by the person or firm that contracted with you to create the work. Your contract should spell out who is to own the copyright for the finished work.

The requirements for registering a copyright have changed over the years, but contrary to popular belief, a work does not have to show a copyright sign or notice, nor does it have to be registered, to have copyright protection. The U.S. Copyright Office still registers copyrights, however, and your copyrights can be registered by sending in the appropriate form for the type of work, a small filing fee, and two copies of the work. Your ability to recover damages from infringers depends on prior registration of your work. You can request the forms you need from the Copyright Office by calling the Copyright Hotline at (202) 707–3000. These forms are also available on the World Wide Web at <http://www.loc.gov/copyright/forms/>.

Research Methodologies

Once you have narrowed the focus of your research and have your key question(s) and keywords, you are ready to begin your research. Many methods are available for acquiring the information you need. This section covers some of the most common approaches to research. Depending on your assignment or discipline, you will more than likely use a variety of methods, including personal experience or knowledge, the library, the Internet, questionnaires and surveys, samplings, interviews, letters of inquiry, protocols, and experimentation.

PERSONAL EXPERIENCE AND KNOWLEDGE

Sometimes the best place to start your research is with yourself. Perhaps through previous or current personal experience you acquired firsthand knowledge about your topic. You learned or are learning about your topic by doing. Of course, you also have knowledge that is not based on direct personal experience. You learn about many topics without having experience of those topics. Knowledge is what you have learned from observation and research.

Depending on your topic, you might find a wealth of information from simply brainstorming or mind mapping your own experience or knowledge. Many people don't realize how much they know about a topic. They are often surprised at how much information they come up with through extensive self-analysis and use of a listing or visualization technique.

USING THE LIBRARY—PRINT SOURCES

Today's libraries offer a wide variety of print and nonprint sources. College and university libraries are not the only kind of libraries where you can conduct your research. Most communities have good branch libraries, and many companies have extensive corporate libraries. There are also numerous government libraries and other kinds of special libraries. Whatever kind of library or libraries you decide to use, you need to have a basic familiarity with how to find what you need.

Knowing your library means much more than knowing what print and electronic sources are available. Libraries offer many services that can save valuable time in your research efforts. For example, many inexperienced researchers make the mistake of thinking that if a book, journal article, or other source is not available in their library, then they are out of luck or will have to travel to another library. But most libraries have interlibrary loan offices, which can usually locate almost any book or article in relatively little time for little or no expense.

Your local college or community library is typically the largest depository of printed material of all kinds—books, journals, magazines, newsletters, pamphlets, government documents, and so on. By becoming familiar with a few basic resources, you can learn how to navigate through all this material to find the information you need.

Card Catalogs. Many libraries have converted their traditional card catalogs into online catalogs. Some libraries maintain the card catalog while making this conversion. Many other libraries have only card catalogs, so it's necessary to be familiar with a card catalog system in case you find yourself doing research in such a library.

Traditional card catalogs are large filing cabinets of 3 x 5 cards organized by subject, title, or author. Some libraries combine organization of these categories in various ways.

Reference Works. Most libraries have one central area for their reference holdings. In this area, you'll find everything from abstracts, encyclopedias, and dictionaries to biographical publications, print databases, guides to other reference materials, and general reference materials. In a big library you'll find thousands of reference works of all kinds. Typical widely used references works include *Books in Print, Information Sources in Science and Technology, Guide to Reference Books, Ulrich's International Periodicals Directory*, and *Library of Congress Subject Headings*. Looking carefully through the sources pertaining to your topic will give you a good sense of what may be available. Unlike working your way through electronic sources, checking through printed matter can often yield unexpected benefits in the form of nearby materials that are useful to your research.

Print Abstracts and Indexes. You should start with print abstracts and indexes if you want to identify recently published research articles on your topic. Online services are increasingly providing this same information. Some typical abstracts are *Biological Abstracts, Chemical Abstracts, Dissertation Abstracts*, and *Applied Science and Technology Abstracts*. Useful indexes for medicine and science include *Index Medicus* and *Science Citation Index*.

Periodicals. Periodicals are the journals, magazines, newspapers, and newsletters in a library. Most disciplines have at least one major journal, and many have numerous journals in which research findings are reported. Magazines include trade and special interest publications. Trade magazines are typically for professionals in a field; special interest publications may be for hobbyists or professionals in fields ranging from computers to photography. Newsletters are brief publications (four to twelve pages) focusing on the current events in a company or discipline. Finally, most libraries carry a variety of local, state, national, and international newspapers.

Books and Monographs. Books may be on a variety of special interest topics or may be textbooks, monographs, or trade books. A library that uses the Library of Congress Classification System will have books in twenty-one classes. (See Figure 1.11.)

A library that uses the Dewey Decimal System organizes its titles in a different way. (See Figure 1.12.) Some libraries use a combination of these two major methods.

A   General Works
B   Philosophy. Psychology. Religion
C   Auxiliary Sciences of History
D   History (includes Travel)
E   America
F   United States. Canada. Latin America
G   Geography
H   Social Sciences
J   Political Science
K   Law
L   Education
M   Music
N   Fine Arts
P   Language and Literature
Q   Science
R   Medicine
S   Agriculture
T   Technology
U   Military Science
V   Naval Science
Z   Books in General

FIGURE 1.11   Library of Congress Classification System

Government Publications. Many libraries are depositories for federal publications and have a dedicated area as well as one or more staff members who specialize in archiving material and assisting people with these holdings. Typical government publications range from books, periodicals, and special reports to all kinds of technical reports. Increasingly, much of this material is available in electronic format.

000   Generalities
100   Philosophy and psychology
200   Religion
300   Social sciences
400   Language
500   Natural sciences and mathematics
600   Technology (applied sciences)
700   The arts
800   Literature and rhetoric
900   Geography and history

FIGURE 1.12   Dewey Decimal Classification System

Special Collections, Maps, Archives. Special collections consist of the papers and personal documents of people who are important in a particular field. Map collections may consist of every kind of world, country, state, and local map. A university library may archive many of the documents of its own institution, such as its annual report, planning documents, various technical reports, and so on. In addition, in the archives you may find scholarly and rare collections, a local history collection, and university archives: faculty and staff publications, master's theses, doctoral dissertations, graduate research reports, official university documents, university press publications, memorabilia including campus history and drawings of the campus, catalogs, directories, yearbooks, and the student newspaper.

USING THE LIBRARY—ELECTRONIC SOURCES

Today's college, community, corporate, government, and special libraries are nothing like libraries of even just a few years ago. Today's libraries offer a wide variety of electronic sources and services, both within the library and through access to other sites.

Online Catalogs. Most university libraries and many major city libraries now make their holdings available through online catalogs. These catalogs offer far more than just a listing of holdings of books, periodicals, indexes, and so on. Many provide search capabilities and databases as well as access to sophisticated research services. (See Figure 1.13.)

Electronic Reference Sources. Many libraries have extensive electronic reference sources available for their patrons. Many of these may be searched only through a library's CD-ROM local area network (LAN). Typically, the LAN is a series of interconnected workstations in the reference area of a library where students, faculty members, and other researchers can access a multitude of reference databases. Typical databases that may be available through a CD-ROM LAN are *America: History and Life, Business Abstracts, Government Printing Office Catalog of Government Publications (GPO), Corporate and Industry Research Reports (CIRR), Criminal Justice Abstracts, MLA International Bibliography, General Science Abstracts, Historical Abstracts, Humanities Abstracts, National Trade Data Bank (NTDB), National Technical Information Service (NTIS),* and *Social Sciences Abstracts.*

Other useful databases that you can find in electronic format are the *Oxford English Dictionary,* the electronic counterpart to the

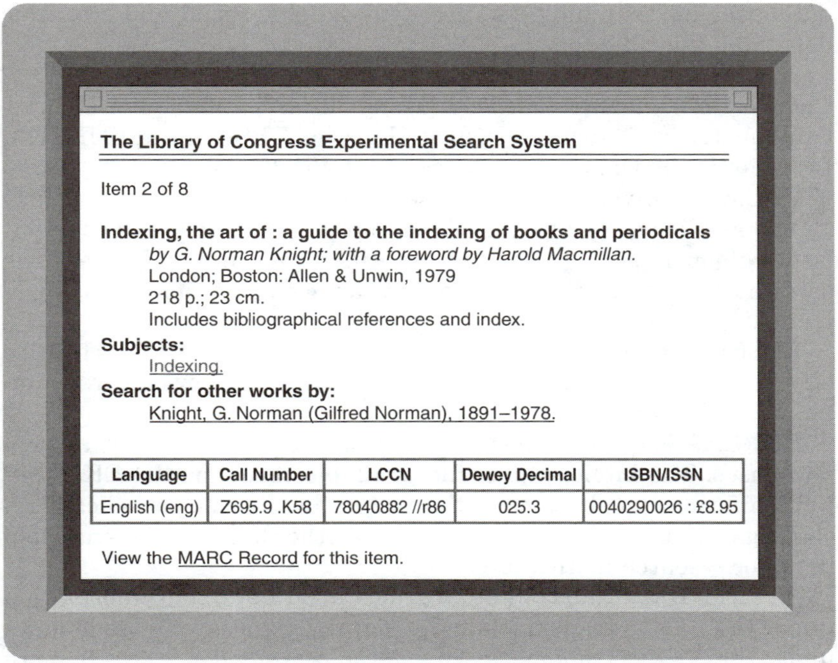

The Library of Congress Experimental Search System

Item 2 of 8

Indexing, the art of : a guide to the indexing of books and periodicals
   by G. Norman Knight; with a foreword by Harold Macmillan.
   London; Boston: Allen & Unwin, 1979
   218 p.; 23 cm.
   Includes bibliographical references and index.
Subjects:
   Indexing.
Search for other works by:
   Knight, G. Norman (Gilfred Norman), 1891–1978.

Language	Call Number	LCCN	Dewey Decimal	ISBN/ISSN
English (eng)	Z695.9 .K58	78040882 //r86	025.3	0040290026 : £8.95

View the MARC Record for this item.

FIGURE 1.13   A Screen from an Online Catalog

twenty-volume second edition of the print version, and *Moody's Company Data,* which contains financial and general background information, current and historical, on more than 10,000 public corporations.

Many other databases can be searched through a library's online catalog or via gateways available through its online catalog. Typical databases available on a library's online catalog are *Biography Index, Book Review Digest, Books in Print, Business Index, Cumulative Book Index, Dissertation Abstracts Ondisc, Education Index, Engineering Index, Essay and General Literature Index,* and *General Academic.* Typical databases available through a gateway are *UnCover* (indexes over 17,000 multidisciplinary journals, which are available for document delivery directly to patrons) and *World Law Index* (an index to national laws in a variety of Spanish-speaking countries in Latin America).

Some electronic resources are available only in the government documents area of a library, such as U.S. Census Databases (various databases for searching detailed demographic social and economic data from the 1990 census).

Commercial Database Providers. Many libraries offer access to invaluable research services such as UnCover, EBSCOhost, EUREKA, LEXIS-NEXIS, Dialog, FirstSearch, and WESTLAW.

UnCover delivers more than seven million articles from nearly 17,000 English language periodicals online. It includes even current issues of journals. Once you have specified your area of interest, the service can e-mail notification of new articles that fit your requirements.

EBSCO is a full-text database that enables users to search titles of over 15,000 journals. It also provides access to newspapers, pamphlets, and brochures.

EUREKA is a software system that is used to search a variety of databases. The bibliographic file available through EUREKA contains information about more than 22 million books, periodicals, recordings, scores, archival collections, and other kinds of material held in major research institutions.

LEXIS-NEXIS collects and provides legal news and information under the LEXIS trademark and general, business, and trade news under the NEXIS trademark. This database provides access to more than 7,000 newspapers, newsletters, magazines, trade journals, wire services, and broadcast transcripts. The legal part of the database, LEXIS, consists of federal and state case law, statutes, secondary sources such as law reviews, and state legal materials. Many of the documents provide the full text of the original.

Dialog (originally developed by Lockheed and now owned by Knight-Ridder Information, Inc.) is one of the major vendors of online databases. It provides access to over 500 databases. Databases may contain citations, full text, or tabular data. The Knight-Ridder–owned newspapers and many news services articles are available in full-text format.

OCLC's FirstSearch is a database that is available on the World Wide Web. It offers many kinds of other databases, including *World Book Encyclopedia*, a full-text academic and professional journal collection, a health reference center, and a religion database. FirstSearch also offers online full text for *General Science Abstracts*, *Humanities Abstracts*, and *Social Sciences Abstracts*. Databases available through FirstSearch vary from library to library.

WESTLAW, a service of West Publishing Company, is a vast online source of legal and business information accessing over 10,000 databases that cover a variety of jurisdictions and disciplines, including statutes, codes, federal regulations, public records, insurance materials, securities law information, rules, judicial and administrative decisions, and legal articles and periodicals. News, business, and financial information are also covered.

Using these services effectively is more than a matter of just accessing them. You should be familiar with a few strategies for successful searching before you go online:

TIPS: Successful Online Searching

- Become familiar with the different kinds of databases. Bibliographic databases are indexes to the author, title, and source of the full text which must be located elsewhere. Numeric databases are tables of statistical data, which may or may not have accompanying text. Directory or dictionary databases contain factual information about products, institutions, and subjects. Full-text databases provide the complete text of magazine articles, newswire articles, encyclopedia entries, and so on.

- Familiarize yourself with other key differences in databases. Databases vary in subject content (business, medical, technical), amount of data (brief citation, abstract, full text), manner of data presentation (text, image, table), technical level (general interest, research reports, scientific data), currency or recency (can range from newswires that are updated hourly or daily to encyclopedias that are updated annually).

- Select the appropriate database(s) on the basis of your research needs. How many sources are collected? How are they selected? Do they cover the relevant journals for your topic? How many results do you need? How technical should they be? How current?

- Be prepared to conduct sequential searches. You may think of new ways to ask your question based on the results of your initial searches.

- Refine the key concepts in your topic. Select words and phrases that express each concept and are appropriate for the database(s) chosen.

- Vary your strategy to match the database type. For example, in partial databases (citations, abstracts) think of all possible search terms. In full-text databases, use proximity connectors to link terms closely.

- Use database and software features such as truncation (abbreviated search term endings); logical (*and, or, not*) and proximity (*near, same, within*) connectors; selection of terms from a thesaurus; and so on to create the most effective search statement.

Electronic Services. Many libraries offer a variety of electronic services. For example, if your library provides this service, you can order interlibrary loan materials by e-mail request or via a form you can fill out and submit over the Web. When the material you ordered arrives, the library notifies you by e-mail. In some libraries you can borrow books via e-mail or Web forms and have the books mailed to your home or business. Many libraries offer online search services at a nominal fee. At some libraries you can e-mail research questions or fill out a Web form with your question and receive a prompt answer from a reference librarian.

Many libraries offer an electronic quick reference service. A typical quick reference service at many university libraries, for example, provides answers to questions about the availability and location of books, journals, and other materials in the library; inquiries about library services; statistical data; verification of bibliographic citations; addresses for corporations, institutions, and publishers; biographical information; assistance with subject-related research; patent and trademark information; and assistance with the use of the online catalog. You can usually receive a reply within one business day.

Microfilm and Microfiche. Many libraries continue to use and provide much information on microfilm and microfiche. Many major newspapers, journals, and magazines are stored on microfilm or microfiche, and special areas in the library may be devoted to these media. Special machines read the microfiche cards or handle the rolls of microfilm. Usually you can photocopy items directly from the microfilm and microfiche readers. Microfilm and microfiche are useful when you want a printed facsimile copy of an article; unlike many other kinds of online full-text article suppliers, microfilm and microfiche maintain the exact pagination and graphics of the original.

Video and Other Media. Many libraries also have extensive video, music, and CD-ROM collections. You'll often find sources in these collections that you cannot easily find anywhere else. For example, you can find music compact discs by almost any major classical music composer and many prominent jazz musicians. In research libraries, videos tend to be educational rather than for entertainment. Increasingly, libraries also have available for loan a variety of CD-ROMs for use on your home or office computer.

The Internet. One simple description of the Internet is basically "everyone's computer hooked up to everyone else's computer." (See

Part Six, "The Internet.") This means that if you have an Internet connection for your computer at home, at a university, or at work, you have access to any other computer hooked up to the Internet and any other library in the world that has an Internet connection. Of course, some of these libraries are more accessible than others, and for some you have to provide appropriate identification as an authorized user to access them.

Because so many online catalogs are also available on the Internet, the two are increasingly providing many of the same sources. Many of the electronic sources available in your local university or community library are the same sources that you can now access via the Internet. Many libraries provide a Web page of links to reference sources available on the Internet. The reference sources include biographies, citation guides, city directories, dictionaries, handbooks, maps, news distributors, quotations, science data, statistics and demographics, zip-code guides, and general information.

Many library catalogs are available by telnet, a remote terminal application that enables you to access the library catalog from your own computer as if your computer were a terminal in the library. There are several main places to start when you want to access a particular library and you do not know its telnet address or URL. One is the master index of libraries maintained at Yale and other sites. Another source is Hytelnet. Hytelnet adds information on many other resources that are accessible through telnet, too. You can even install Hytelnet on your own computer if you want rapid and reliable access to the software.

Use these URLs to access library catalogs and telnet resources:

Yale catalog of libraries at <gopher://libgopher.yale.edu/>
Hytelnet at <http://library.usask.ca/hytelnet/>
Hywebcat at <http://library.usask.ca/hywebcat/>
Hytelnet menu at Washington & Lee U at <gopher://liberty.uc.
    wlu.edu/11/internet/hytelnet/>
Library of Congress at
    <http://lcweb.loc.gov/homepage/lchp.html>

An incredibly vast range of resources is at your fingertips, and knowing the best ways to find and access all of the availiable information databases is a formidable task. Part Six, "The Internet," covers the essentials you will need to know about Web browsers, search engines and indexes, various Internet utilities, and so on. You'll find a great deal of information in that section on how to use the Internet to find information and to do research. Of course, you also need to know how to evaluate the sources you find on the Internet. As with print sources and other media, you need to gauge the authority, accuracy, honesty, and completeness of the Internet sources you find. See "Evaluating Research Information," on page 63 for a discussion of strategies.

TIPS: Researching a Topic in the Library

- Allow enough time. Good research takes time. It is all too easy to procrastinate in doing the necessary work for a good research project. All phases of the process take a lot of time, from planning to writing and revising the paper. If you wait too long before beginning, you risk running out of time before you have found all the information you need.

- Document your research steps. Document not only the source of your direct quotations and paraphrases, but also the sources you have consulted but not used. Write a complete citation for each source you find; you may need it again later. Also write down where you found the source.

- Define the topic as precisely as possible early in the research process. Use whatever strategies you prefer for helping you to focus your topic. For example, state your topic as a question and then identify the main concepts or keywords in your question.

- Translate your topic into the subject language of the indexes and catalogs you use.

- Work from the general to the specific. Get general information on the subject and related areas. Look up your keywords in the indexes to subject encyclopedias. Read articles in these encyclopedias to set the context for your research. Note any relevant items in the bibliographies at the end of the encyclopedia articles. As you read the articles, note significant names, dates, events, organizations, and issues. You may also wish to consult other reference sources, such as annuals for current research on a topic, handbooks which outline all aspects of a specific subject, or a statistical source.

- Remember that additional background information may be found in your lecture notes, textbooks, and reserve readings.

- Ask questions. When you don't know how to find information you're looking for, don't hesitate to ask a librarian for help.

- Find a good variety of appropriate sources. Most professors who assign research papers expect you to use a variety of sources: books, journal articles, and so on. Don't rely exclusively on one medium, such as the World Wide Web, for your sources.

- For books, find your subject in the Library of Congress Subject Headings Search on the online catalog. If you know of a good book on your topic, look up that book in the online catalog and consider the subject headings listed in the record. Check bibliographies in books, encyclopedias, and other sources.
- Periodical articles will provide you with current research and information on a topic. Your instructor or supervisor may want you to use scholarly journal articles rather than magazine and other types of periodical articles. Find out the difference between a scholarly journal and other types of periodicals. Use print periodical indexes and abstracts. Use electronic indexes. Ask a librarian about online database searches. Check bibliographies in articles, books, encyclopedias, and so on.
- Current research and information are widely available on the Internet. See Part Six, "The Internet" for a discussion of using search engines, indexes, and other tools for finding the kind of information you need.
- Use other library services if necessary. You can request a book or journal article from another library through interlibrary loan. Start your research early. Interlibrary loans take time. Ask a librarian how to connect to another library's catalog on the Internet. If you are short on time, you may need to go to the library and photocopy the article yourself. Many university libraries will allow you to use the library, but they will not allow you to check out books if you are not a student, faculty member, or staff employee there.
- Make sure you don't have too many or too few sources. If you are doing research for a class project or paper, ask your instructor for guidelines on the appropriate number of sources. If you are doing research for your job, solicit this information from your colleagues or supervisor.
- Evaluate what you have found. You'll need to screen the information for its authority, accuracy, honesty, completeness, and so on. If you have found too many or too few sources, you may need to narrow or broaden your topic. Check with a reference librarian, your instructor, or your supervisor.
- Use a standard format for your bibliography. Ask your instructor or supervisor which citation style is required.

QUESTIONNAIRES AND SURVEYS

Questionnaires and surveys are a useful way to gather material and research information. Along with interviews they allow for a more personal response and are often used when personal opinions and attitudes are required. They can be used for both qualitative and quantitative research. Questionnaires can be administered personally, over the telephone, by mail, or on the Internet.

Questionnaires are valuable, but only if they ask appropriate questions that people can understand and only if they are written in a format that people can answer easily. Too often questionnaires and surveys are thrown together with little thought; these efforts often result in questionable data, leading to even more questionable interpretations. When you are designing a questionnaire, you must give it a lot of thought, prepare it carefully, and test it thoroughly. Only through methodical, thorough analysis up front will questionnaires and surveys provide the desired information. As with all writing, you should think through the process carefully. What do you want to find out, how will you go about finding it, what kind of questions will be in the questionnaire, and how will you analyze the data?

Creating a questionnaire is a process. You have to perform a series of important steps to make sure the questionnaire will bring useful results. Start by determining whether a questionnaire is the best method for gathering the data you think you need. Clearly define your focus and purpose. Prepare a draft set of questions and a draft format. Test these questions and then modify the questionnaire. When the questionnaire is as good as you can make it, distribute it, collect it, and then proceed to analyze the data.

TIPS: Creating a Questionnaire

- Use specific, concrete words instead of general, vague ones. Consider the following question, for example: "How much value do you place on serene surroundings in the computer lab?" The question doesn't define "serene," and it relies on you to determine what you think "value" is. Value means different things to different people. Consider rephrasing the question as follows: "Would you use the computer lab more often if the lab would always be a quiet place for you to work?"

- Give your readers enough choices. For example, when a multiple-choice question doesn't cover all of the possible answers, provide the possible answer of "other" and allow readers to explain their unlisted choice in more detail.

- Avoid including two negatives in any question. For example, many people would be confused by the following question: "Are you opposed to not buying more graphics software for the lab?" Rephrase the question as follows: "Would you support buying more graphics software for the lab?"

- Avoid asking two questions in one. Consider the following question, for example: "Do you feel the computer lab needs more computers and which brand of computer should be purchased?" Readers may agree with the first question and have many differences of opinion concerning the second question, or they might disagree with the first question but feel pressured to change their mind by the second question. You need to separate the questions to collect more meaningful and useful responses.

- Avoid overuse of styles for emphasis. It can be helpful to use italics, bold, or underlining to call attention to various keywords, but if your questionnaire has too many bold words, for example, readers can lose track of what needs emphasis and what does not.

- Help your readers to make effective comparisons. Avoid asking questions such as "Do you think your computer lab is a better equipped lab?" Instead, rephrase the question to "Do you think your computer lab is better equipped than the computer lab in the library?"

- Avoid mistaken assumptions. A mistaken assumption can completely undermine your question. Consider the following question, for example: "Would a workshop on Microsoft PowerPoint be more valuable to you than a workshop on Microsoft FrontPage?" This question assumes that the two programs are comparable and asks readers to guess what you mean by "valuable." Rephrase the question to: "Which software program training do you need most—learning to create presentations with Microsoft PowerPoint or learning to create Web pages with Microsoft FrontPage?"

- Phrase your questions so your readers will provide useful information in their responses. Consider the following question: "Do you think the department will need to hire more faculty over the next two years?" A "yes" or "no" answer to this question simply requires you to ask several follow-up questions. By rephrasing, you can collect more useful information: "List the areas in which you think the department should hire faculty over the next two years."

- Make the length of the questionnaire appropriate to the purpose. Determine carefully what kind of information you need, and ask questions only concerning this information. Many readers will not respond to questionnaires that take too long to complete. And if you send out a questionnaire that fails to ask key questions, you might have to send out a revised questionnaire to obtain the information you should have gotten the first time.

INFORMATIONAL INTERVIEWS

Informational interviews are one-on-one meetings with subject matter experts to help you gather information about a topic. Such interviews can provide valuable information in almost any discipline. Often you will find yourself in a situation in which you must talk to several subject matter experts to acquire the information you need. See Part Seven, "Interpersonal Skills" for a discussion of job informational interviews.

The informational interview has four phases: researching for the interview, preparing for the interview, conducting the interview, and following up on the interview.

Researching for the Interview. In this phase you find out who the subject matter experts are if you don't know to whom you should be talking. Also, you need to research the credentials of these experts. If they are in academe, find out what they have published. If they are in industry, find out what their achievements have been. A professor who is well published in a field is generally more knowledgeable than someone who is not as well published. A person who has achieved a great deal in industry is often more credible than someone who has not.

In addition to doing research about the subject matter expert, you'll have to do some research about the subject if you know little about it. You need to find out what the important issues are, who the important players are, and what changes or trends are taking place.

Sometimes you also have to research the company that employs the subject matter expert. You can find out about a company through various print and electronic sources in the library, from home pages on the World Wide Web, from your campus or local placement office, from an on-site visit or tour if the company is local, and from professional organizations in the field or industry.

Preparing for the Interview. In addition to finding out more about the subject matter expert, the subject in question, and the company and its products, you need to prepare for the actual interview. Decide in advance how you will keep track of the information you obtain: by taking notes, by audio recording, by video recording, or by a combination of these.

If you are interviewing a person at that person's company, you also need to check on any security clearances that may be necessary, and you'll need to allow enough time for paperwork, obtaining a badge, and other security matters. You may be asked to bring certain specific identification documents with you. These requirements should be made known to you when you make the appointment or when the appointment is confirmed. Make sure you have the required documents or identification cards with you; if you will not be able to bring them, find out whether you can substitute other documents. If the person is coming to your site, you'll need to make all of the necessary arrangements to make this person comfortable. For either situation, you'll also need to consider what you're going to wear: business-appropriate clothing for some situations or possibly safety gear if you are interviewing someone at a construction site, for example.

After making these preparations, plan how long the initial interview will be. Decide whether one long interview will suffice or if a series of shorter interviews would be better. Then you need to make the actual physical arrangements for seeing this person, such as scheduling the meeting, reserving the meeting room, providing for any travel, brewing of coffee or chilling of other beverages, and so on.

Prepare for the interview by creating a list of questions. Don't waste your interviewee's time by scheduling an interview and then having only a vague idea of what questions you want answered. Of course, the questions you ask depend on the subject you are researching. Generally, you can begin making your list by thinking of those questions your audience would likely ask. Focus carefully on this list of questions. Depending on the subject, you should limit the number to ten to thirty questions. You'll have to study your final list carefully, so that as the interview progresses you can ask the right questions at the right time.

Finally, in your preparation for an important interview, it's a good idea to practice the interview. You may want to invite a friend to play the role of the interviewee and then go through your list of questions. You may also want to practice your questions and work on your body language in front of a mirror. Your interviewee will be basing the answers to your questions not only on the actual questions but also on other perceptions, including your body language. You want to use body language that shows that you are confident and poised, that you are eager to learn from this person, and that you understand what you are told. Practice by sitting in a chair in front of a mirror. Find those poses that show your confidence. Rehearse nodding your head different ways to show you are listening carefully and understanding. Ask follow-up questions on any points that you don't understand fully.

Conducting the Interview. Now that you have done your research and otherwise prepared for the interview, you should be ready to conduct the interview and be confident that it will go well. You need to remember that there is an art to interviewing. You will want to make the person you are interviewing relax and feel comfortable with you.

Be aware of first impressions, both the ones you receive and the ones you provide. From the handshake to small talk to any offer to provide refreshments, you show the other person that you are concerned with creating and maintaining a friendly, yet businesslike, atmosphere.

When the interview part of the encounter begins, remember to focus on why you set up the interview in the first place. Ask your questions clearly and listen to the answers carefully to make sure you're given the answers that will be useful to you. Listen to the intonations of the interviewee's voice. Are you hearing enthusiasm, anger, frustration, humor, confidence? What can you infer from the choice of words, the tone of voice, the volume of sound? What does the interviewee's body language suggest? If you listen well, you will be able to ask relevant follow-up questions at the appropriate times.

In addition to listening effectively, focus on taking effective notes, whether you do so with a pencil and pad, a laptop computer, an audio recorder, or a video recorder. Don't focus too much on note taking, or you will adversely affect your ability to listen well to what is being said. If you're doing any recording, you'll need to obtain all of the necessary permissions in advance. Keep in mind that many people are uncomfortable about having their voice recorded and even more uncomfortable about video recording. Also, in many situations, you will have to obtain permission from others in the company to record an interview.

As the interview comes to a close, be sure to look over your notes to make sure they are as complete as you can make them. You may not have another opportunity to interview this person again or soon.

Even though the interview is over, you still have a lot of work to do. If you're going to use or publish any direct quotations, make sure you obtain the necessary permissions. You may even need to get this permission in writing. Be sure to thank the interviewee before you leave for spending time with you and providing useful information.

Following Up on the Interview. When you return to your office or company, take a few minutes to send a written thank-you letter to the person you interviewed. And don't wait long before carefully reviewing your notes; do it while your memories of the interview are fresh. You may be able to add helpful comments throughout your notes. Consider working your notes into an outline to organize them. At this time you may also think of questions you should have asked and still need answers to. If you have to prepare your notes for someone else to see, be sure to follow all of the necessary style requirements and meet the deadlines for providing the notes. The steps that you take after the interview is over are just as important as the other parts of the interview.

EXPERIMENTATION

The physical, life, and social sciences rely on experimentation as an important method of gathering information on a subject. Successful experimentation depends on your ability to follow the steps of the scientific method.

The major steps of the scientific method can be described in various ways, but in general they follow this path:

- Define the problem as specifically as existing knowledge will permit.
- Summarize the work already done. Make sure that the problem has not already been solved.
- Compare similar phenomena. The solution to one problem may be useful in finding the solution to another problem of the same kind.
- Formulate a hypothesis. Propose a tentative answer to the problem.
- Test the hypothesis. By observation and experiment, try to establish the falsity or the truth of the hypothesis.
- Modify the hypothesis. You may verify your hypothesis, disprove it, extend its original scope, or modify or restrict it.
- Publish the results. After you have thoroughly tested your hypothesis and have verified it through further observation and experiment, publish your results.
- Submit the final hypothesis for verification. When your results are published or privately submitted, you should expect others to test your hypothesis by repeating your experiments and devising new experiments.
- Establish a theory. After a published hypothesis has been thoroughly rechecked and retested and has been verified to the satisfaction of many researchers in the area, then the hypothesis becomes a theory.

Evaluating Research Information

While you are researching your topic, you must give careful attention to the quality of the information you find. Whether the source is in print, on the Internet, or available in some other medium, you need to determine how authoritative, reliable, accurate, and complete the information is. There is no easy solution to determining the quality of the information you find, and this is especially true of information you find on the Internet. However, you can ask some basic questions that will at least help you form an opinion about the quality of the information you find.

TIPS: Evaluating Print Resources

- What are the author's educational background, previous writings, and experience in this area? Is this book or topic in the author's area of expertise? Have you seen the author's name cited in other sources or bibliographies? Respected authors are cited frequently by other scholars.
- When was the source published? Is it current or out-of-date for your topic? Topic areas of continuing and rapid development, such as those in the sciences, demand more current information. On the other hand, topics in the humanities often require material that was written many years ago.
- Is this a first edition or not? The existence of further editions indicates that a source has been revised and updated to reflect changes in knowledge, includes omissions, and harmonizes with its intended reader's needs. Also, a work that has had many printings or editions is likely to be a standard source in the area and is probably reliable.
- Who is the publisher? A source that is published by a university press, for example, is likely to be scholarly. The fact that a publisher is reputable does not necessarily guarantee quality, but it does show that the publisher may have high regard for the source being published.
- Is this a scholarly or a popular journal? This distinction is important because it indicates different levels of complexity in conveying ideas and different standards of peer review. A peer-reviewed journal article has been evaluated and accepted by acknowledged authorities in the field.
- What type of audience is the author addressing? Is the publication aimed at a specialized or a general audience? Is this source too elementary, too technical, too advanced, or just right for your needs? Has the topic been oversimplified to suit a lay audience, with the possible loss of detail or precision that a journal for professional readers requires?
- Is the information fact, opinion, or propaganda? It is not always easy to separate fact from opinion. Facts can usually be verified; opinions, though they may be based on factual information, evolve from the interpretation of facts. Skilled writers can make you think their interpretations are facts.
- Does the information appear to be valid and well researched, or is it questionable and unsupported by evidence? Assumptions should be reasonable. Note errors or omissions.

- Is the author's point of view objective and impartial? Is the language free of emotion-rousing words and bias?
- Does the work update other sources, substantiate other materials you have read, or add new information? Does it extensively or marginally cover your topic? You should explore enough sources to obtain a variety of viewpoints.
- Is the material primary or secondary? Primary sources are the raw material of the research process. Secondary sources are interpretive or extrapolative and are based on primary or other secondary sources.
- Is the publication organized logically? Are the main points clearly presented? Do you find the text easy to read, or is it stilted or choppy? Is the author repetitive?
- Locate critical reviews of books in a reviewing source, such as *Book Review Index* or *Book Review Digest*. Is the review positive? Is the book under review considered a valuable contribution to the field? Does the reviewer mention other books that might be better? Do the various reviewers agree on the value or attributes of the book, or has it aroused controversy among the critics?

In the case of Internet files and documents, some of the questions listed above cannot be easily answered, but the guidelines should help you approach information with some measure of critical analysis. It is important that you apply critical standards to the information you find on the Internet. Keep in mind that the Internet is a network of networks that have widely different goals and origins. No central authority governs the information you find on the Internet. You need to consider what you are looking for in your Internet research. Are you looking for facts, opinions (anyone's or an expert's opinion), strong arguments, statistics, eyewitness reports? Are you trying to find new ideas, more support for a position? When you have established what you are looking for, then you can screen your Internet sources by testing them against your research goal.

TIPS: Evaluating Internet Resources

- Bear in mind that many sources published on the Internet have not been reviewed by a refereeing process, by peers or other authority, by an editor, or by libraries.
- Evaluate the author's reliability and credibility. Who is the author? Is the author qualified to write this article? What are the author's occupation,

(continued)

Evaluating Internet Resources *(continued)*

position, education, and experience? What are the author's credentials? How easy is it to identify the author's authority? How well has the author documented the sources of the information presented?

- Try to ascertain the material's accuracy. Are the facts correct? How does this information compare with that in other sources in the field?
- Identify the author's perspective. Does the author have a bias? Does the author express a particular point of view? Is the author affiliated with particular organizations, institutions, associations? Does the forum in which the information appears have a bias? Is it directed toward a specific audience? Where is the information "published"? When was it written? Is the author's perspective culturally diverse, or narrowly focused?
- Evaluate the author's purpose. To what audience is the author writing? Is this reflected in the writing style, vocabulary, or tone? Does the material inform? explain? persuade? Is there sufficient evidence? What conclusions are drawn?
- Consider the material's usefulness for your particular need. If you can't identify its usefulness immediately, it should be considered a low priority to save, print, or read online.
- Look for evidence of quality control. Is the information presented on an organizational Web site or a home page? If the site is an online journal, is it peer reviewed? Is the information taken from authoritative books and journal articles?
- Critically view the quality of the content. What kind of information is it? Is it factual or opinion? Is there any documentation? a bibliography? Are there footnotes? credits? quotations? Does the information support or refute your position? Are any major findings presented? How does the information compare to other related sources?
- Evaluate the quality of the format. Can you clearly identify what type of information it is? Is it a Web home page? Is it a newsgroup posting? Is it a file or downloadable software? Is it a government report? Is it an advertisement?
- Check the material for completeness. Does it have the features you need: graphs, charts, illustrations, glossaries, maps? How complete is the information?
- Analyze the content for balance, objectivity, bias, and accuracy. What is the intended purpose of the information? Why is the information being presented or made available? What is the perspective of the publication? Is the information presented accurately and objectively? How can you tell? What clues are present to help you judge?
- Check for timeliness. When was the information produced? Is it too old or too new for the needs of your research?
- Ascertain originality. Is it primary information or secondary information? Is the originality of the information important for your research?

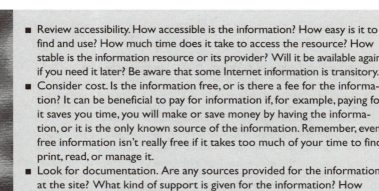

- Review accessibility. How accessible is the information? How easy is it to find and use? How much time does it take to access the resource? How stable is the information resource or its provider? Will it be available again if you need it later? Be aware that some Internet information is transitory.
- Consider cost. Is the information free, or is there a fee for the information? It can be beneficial to pay for information if, for example, paying for it saves you time, you will make or save money by having the information, or it is the only known source of the information. Remember, even free information isn't really free if it takes too much of your time to find, print, read, or manage it.
- Look for documentation. Are any sources provided for the information at the site? What kind of support is given for the information? How does the author know this information?

With so much Internet information available, it is easy to suffer from information anxiety or information glut. Use the evaluation criteria just listed to help you decide what information you really want to look for, save, and use for your research.

Annotated Bibliographies

Annotated bibliographies are organized lists of sources, with each source followed by a brief comment or annotation. (See Figure 1.14 on page 68.) Annotated bibliographies serve many valuable purposes in research. They describe the content and focus of sources; suggest the usefulness of the sources to your research or someone else's; evaluate the methods, conclusions, or reliability of sources; or simply record your reactions to sources.

Many annotated bibliographies are published as separate pamphlets or monographs. Some are published as journal articles, bibliographic essays, or literature reviews. Ideally, an annotated bibliography provides an up-to-date overview of the important literature on a well-focused topic.

You may be asked to write an annotated bibliography to show the quality of your own research, to provide additional information or background material to your reader, to explore a topic for further reading or research, or to give your research historical perspective or relevance.

There are several major kinds of annotations. Summary annotations discuss the main argument of the book or article or other source. Descriptive annotations discuss what is included in the source. Evaluative annotations discuss what the writer thinks of the source. A

Annotated Bibliography

Spiekermann, Erik, and E.M. Ginger. *Stop Stealing Sheep and Find Out How Type Works.* Mountainview: Adobe, 1993.

Type has personality in addition to size, weight, shape. Knowledge of historical and modern principles for using type enables document designers to create appropriate and elegant documents.

Stone, Sumner. *On Stone: The Art of Use of Typography on the Personal Computer.* San Francisco: Bedford Arts, 1991.

The designer of the Stone family of typefaces uses computer technology to develop and design type that benefits from his clear typographic knowledge and experience. Type designers today are empowered by computer technology and benefit greatly from the computer's powers to work more quickly and track changes automatically The goal of information design is trouble-free navigation through the text. Many design decisions (headings, titles, justification, alignment, indentation) affect the readability of text. A history of type design, a case history in typographic design, and many specimens inform and educate the novice typographer.

_____. "The Type Craftsman in the Computer Era." *Print* Mar./Apr. 1989: 84+.

Computer typography applies the old craft of making letterforms to the new context of computer science. The tools and materials may be new, but the letters are old. Today's typographer must become familiar with the original sources of the letterforms and make "faithful and lively versions" of them with the modern technology. The historical precedent of mixing alphabet styles is valid today in the mixing of different typefaces from the same family, for example, the Stone typeface family.

Stopford, Charles. "Desk Top Type: Tradition and Technology." *Technical Communication* (1992): 74–79.

Despite technological changes over the centuries, reading has remained the same, so the requirements for readable text have not changed. Only typographic literacy will ensure that those presenting written material for others to read will provide aesthetically pleasing and usable documents.

Swann, Cal. *Language and Typography.* New York: Van Nostrand Reinhold, 1991.

Visible language consists of arbitrary symbols, including orthographic and paragraphological. The conventions of visible language are the visual expression of intonation. Letterforms have connotative qualities beyond their denotative ones, and their importance becomes clear when we examine posters, advertisements, and signs.

Wheildon, Colin. *Type & Layout: How Typography and Design Can Get Your Message Across—Or Get in the Way.* Berkeley: Strathmoor, 1995.

Research undertaken between 1982 and 1990 exposes those factors that make for effective, readable, legible text.

FIGURE 1.14   An Annotated Bibliography

combination annotation provides a blend of the above, offering, for example, a summary, a description, and an evaluation. The combination annotation is the most common type.

TIPS: **Writing an Annotated Bibliography**

- Once you have focused your research and finished finding all of the sources you need, compile all of your sources. You may, of course, have taken good notes on some sources and no longer have access to the original. If you use this approach, you'll need to make sure your notes are summary, descriptive, and evaluative. A better approach, however, is to photocopy the articles or print copies of the Web sources or borrow the books so that as many originals as possible are available to you when you begin the work on your annotations.

- Determine the documentation style you must use for the sources: APA, MLA, Chicago/Turabian, CBE, or house style if your company has one.

- Determine whether you should write your annotations as phrases or complete sentences. Both are common. A phrase annotation simply offers the necessary few phrases to get the job done. A complete sentence annotation allows you to be a little more detailed. For a sentence annotation, vary the sentence length and avoid long and complex sentences. Sometimes you may be asked to provide a paragraph annotation. Each paragraph should be complete, unified, coherent, and ordered.

- Determine the kind of annotation you are required to do: a summary, a description, an evaluation, or a combination.

- To write a summary annotation, begin by stating the purpose or argument of the source. Then develop the main argument with supporting points and examples and state the conclusion. State the arguments, supporting points, and examples directly, as if you were the author. Do not include phrases such as "the author contends," "the article states," or "according to this book." Your annotation is assumed to be a representation of the contents of the work you are summarizing.

- To write a descriptive annotation, define the scope of the source, list the significant topics, and tell what the source is about. In this type of annotation, you shouldn't try to give actual supporting data or examples. Generally, you just include topics or chapter titles.

- To write an evaluative annotation, you need to determine the strengths and weaknesses of the source. Discuss why the source is interesting or helpful for you or why it is not interesting or helpful. Discuss the type, quantity, and usefulness of the information.

- To write a combined annotation, provide several sentences summarizing or describing the content and several sentences evaluating the source.

- Be brief. Highlight only the essential details or points.

Literature Reviews

Literature reviews discuss and assess some of the important published literature on a specific topic. They play a key role in most disciplines by summarizing, classifying, comparing, and evaluating the key existing work on a topic, area, or issue. Many literature reviews stand alone. Others may be part of a proposal, an article, a report, or a chapter in a thesis or dissertation. In fact, a literature review is an essential part of a good grant or research proposal. You have to persuade your audience that you are making further contributions to a field of study, and to do this successfully, you must summarize the existing significant research on your topic.

TIPS: **Writing Literature Reviews**

- Define or identify the general topic, issue, or concern to provide some context for reviewing the literature.
- Discuss why you think it's necessary to review the literature on this topic.
- Explain how you will judge the literature you discuss and how you will organize the review.
- The purpose of the review may simply be to inform your readers about the status of research on the topic, or your literature review may be a specific argument in support of a point of view. If you are simply informing your readers about the status of the research on a topic, you should comment on any specific trends identified by your research. You may also want to comment on any disagreements, controversies, or debates in the literature. You may go further and point to a specific dominant issue and discuss the merits of this issue.
- If it's appropriate to do so, you may also want to comment on why some literature on your topic has not been included in your review.
- For the body of your literature review, decide on various strategies for grouping or classifying the literature you are discussing. You may decide, for example, to cover earlier literature first and then more recent literature. You may decide to classify the discussion according to the kinds of literature (reviews, journal articles, reports, case studies, Web documents). Or you may decide to group the literature according to the conclusions of the authors or the purposes of each document.
- Summarize each work according to its importance and comment on the significance of each source you discuss. Some works contribute new thought to the topic; others add nothing new to the discussion. If the

work is particularly important, devote more discussion to it in your literature review. Devote less discussion to less important works.

- Help your reader as much as possible in the literature review. Wherever possible, provide clear topic sentences at the beginnings of paragraphs. Provide brief summaries at various points in your discussion informing your readers of the importance of what you have covered so far.
- Reemphasize in the closing of the literature review the main points of the most important sources discussed in your review.
- Provide an overall evaluation of the state of knowledge concerning your topic. Comment on any gaps in the research, any inconsistencies, and any areas requiring further study.
- Wrap up by discussing how the topic of your literature review relates to the larger discipline or profession.

Research Tools

Many of the electronic sources discussed earlier are research tools of one kind or another. Using these tools will save you a great deal of time. Other tools will help you keep track of your sources, generate your notes and bibliography automatically, and save time by keeping track of your research in other ways.

One newer tool is Q-Corp Q-Notes. This program enables you to select electronic information and source information from the Internet, CD-ROM, e-mail, and other sources for later reference. You can create note cards automatically with the information you enter, add your own comments to them, and organize them into file folders. You can keep track of sources and build references and footnotes using the bibliography prompt. You can copy and paste quotations directly from note cards into documents, and you can use the program with major word-processing programs and other software.

Research Information Systems ProCite is a program that stores bibliographic information, notes, keywords, and full abstracts in a searchable form. You can easily sort through your notes and print citations in a variety of styles. It is easy to edit and update references, and you can work within your word processor to create a bibliography from references you've cited within the text.

Documentation Styles

Every discipline has a preferred documentation style. As with all of the other decisions you make about your document, you must choose a documentation style for the notes and bibliography that is suitable for the discourse community for which you are writing. Following are some of the major documentation style guides and examples of common citations.

HUMANITIES

MLA Style. The most widely used documentation style guides in the humanities are the *MLA Style Manual* and the *MLA Handbook for Writers of Research Papers*. These style guides offer an overview of writing and publication in the humanities, a guide to the mechanics of writing, tips on preparing the scholarly manuscript, rules for preparing the List of Works Cited, guidelines for documenting sources, a discussion of abbreviations, lists of reference words and proofreading symbols, and guidance on preparing theses and dissertations.

Chicago or Turabian Style. The Chicago or Turabian style, sometimes called documentary note or humanities style, places bibliographic citations at the bottom of a page or at the end of a paper. Although *The Chicago Manual of Style* and Kate L. Turabian's *A Manual for Writers of Term Papers, Theses, and Dissertations* offer guidelines for parenthetical documentation and reference lists as well, these styles are most commonly thought of as note systems.

SOCIAL SCIENCES

The *Publication Manual of the American Psychological Association* sets the documentation style not only for psychology, but also for the other behavioral and social sciences, as well as nursing, criminology, and personnel management.

If you are following the *Style Manual for Political Science* published by the American Political Science Association (APSA), use parenthetical citations within your text to indicate the source of borrowed ideas and quotations. At the end of your paper, provide a list of all of the references cited in the paper.

PHYSICAL SCIENCES

A number system of documentation is used in many scientific fields. Because details of number documentation differ according to discipline (and sometimes according to journals within the same discipline), ask your instructor or supervisor what specific style to use. Documentation styles for chemistry, computer science, mathematics, physics, and medical sciences are summarized in James D. Lester's *Writing Research Papers,* an excellent source for a variety of documentation styles.

LIFE SCIENCES

The CBE (Council of Biology Editors) manual, *Scientific Style and Format,* describes two systems of source documentation: the citation–sequence system and the name–year system. Both systems, often with variations, are used in scientific publications.

ELECTRONIC SOURCES

The organizations that provide guidelines for bibliographic formats in their fields are developing recommendations for citing electronic sources—newsgroup postings, mailing-list (listserver) messages, Web-page articles, e-mail messages, and the like. Because these types of sources are relatively new, many citation formats have been suggested, but most organizations have not yet codified them. For guidance on citing electronic sources, consult journals in your field to see how they are handling this question. Numerous individuals have suggested ways to treat online citations, and many of them have posted their suggestions on a variety of Web pages. Some of these have been written by graduate students or librarians at various universities, so when you look for these Web pages, they may have moved to another location. Try some of these, and if you need more information, use a Web search engine for further references:

<http://bailiwick.lib.uiowa.edu/journalism/cite.html> A Guide to Citation Style Guides

<http://www.apa.org/journals/webref.html> American Psychological Association

<http://www.mla.org/main_stl.htm> Modern Language Association

WITHIN-THE-TEXT EXAMPLES: MLA, TURABIAN, APA

When you cite information in your work that you have derived from the works of others, you will have to inform your readers where the information can be found. You will do this on each occasion that you use the material you have either copied or paraphrased. Of course, you will also include a section called List of Works Cited, References, Bibliography, or a similar title at the end of your paper or report.

Page References in Paraphrased Material

MLA

> To Spiekermann and Ginger, type families share many elements with musical ones (101).

> Spiekermann and Ginger see letterforms as taking on lives of their own. "As soon as there are a bunch of letters gathered together, they fight for space, for the right to be recognized, to be read" (120).

Turabian

> To Spiekermann and Ginger, type families share many elements with musical ones (1992, 101).

> Spiekermann and Ginger see letterforms as taking on lives of their own. "As soon as there are a bunch of letters gathered together, they fight for space, for the right to be recognized, to be read" (1992, 120).

APA

> To Spiekermann and Ginger, type families share many elements with musical ones (1992, p. 101).

> Spiekermann and Ginger see letterforms as taking on lives of their own. "As soon as there are a bunch of letters gathered together, they fight for space, for the right to be recognized, to be read" (1992, p. 120).

Author and Page References in Paraphrased and Quoted Material

MLA

> Type families share many elements with musical ones (Spiekermann and Ginger 101).

Turabian

> Type families share many elements with musical ones (Spiekermann and Ginger 1992, 101).

APA

> Type families share many elements with musical ones (Spiekermann & Ginger, 1992, p. 101).

WORKS CITED EXAMPLES: MLA, TURABIAN, APA

Books: One Author

MLA

Vaughan, Tay. *Multimedia: Making It Work.* 3rd ed. Berkeley: Osborne-McGraw, 1996.

Turabian

Vaughan, Tay. *Multimedia: Making It Work,* 3rd ed. Berkeley: Osborne-McGraw-Hill, 1996.

APA

Vaughan, T. (1996). *Multimedia: Making it work* (3rd ed.). Berkeley, CA: Osborne-McGraw-Hill.

Books: Two Authors

MLA

Holman, C. Hugh, and William Harmon. *A Handbook to Literature.* 6th ed. New York: Macmillan, 1992.

Turabian

Holman, C. Hugh, and William Harmon. *A Handbook to Literature,* 6th ed. New York: Macmillan Publishing Company, 1992.

APA

Holman, C. H., & Harmon, W. (1992). *A handbook to literature* (6th ed.). New York: Macmillan.

Books: Many Authors

MLA

Lay, Mary, Billie Wahlstrom, Stephen Doheny-Farina, Ann Hill Duin, Sherry Burgus Little, Carolyn Rude, Cynthia Selfe, and Jack Selzer. *Technical Communication.* Chicago: Irwin, 1995.

Turabian

Lay, Mary, Billie Wahlstrom, Stephen Doheny-Farina, Ann Hill Duin, Sherry Burgus Little, Carolyn Rude, Cynthia Selfe, and Jack Selzer. *Technical Communication.* Chicago: Richard Irwin, 1995.

APA

Lay, M., Wahlstrom, B., Doheny-Farina, S., Duin, A. H., Little, S. B., Rude, C., Selfe, C., & Selzer, J. (1995). *Technical Communication*. Chicago: Richard Irwin.

Corporate Author
MLA

New York Public Library. *Writer's Guide to Style and Usage*. New York: HarperCollins, 1994.

Turabian

New York Public Library. *Writer's Guide to Style and Usage*. New York: HarperCollins Publishers, 1994.

APA

New York Public Library. (1994). *Writer's guide to style and usage*. New York: HarperCollins.

Reprint
MLA

Bernstein, Theodore M. *Miss Thistlebottom's Hobgoblins: The Careful Writer's Guide to the Taboos, Bugbears and Outmoded Rules of English Usage*. 1971. New York: Farrar, 1991.

Turabian

Bernstein, Theodore M. *Miss Thistlebottom's Hobgoblins: The Careful Writer's Guide to the Taboos, Bugbears and Outmoded Rules of English Usage*. New York: Farrar, Straus and Giroux, 1971; New York: Farrar, Straus and Giroux, 1991.

APA

Bernstein, T. M. (1971/1991). *Miss Thistlebottom's hobgoblins: The careful writer's guide to the taboos, bugbears and outmoded rules of English usage*. New York: Farrar, Straus and Giroux.

Chapter in a Book
MLA

Picard, Rosalind W. "Does HAL Cry Digital Tears? Emotions and Computers." *HAL's Legacy: 2001's Computer as Dream and Reality*. Ed. David G. Stork. Cambridge: MIT P, 1997. 279–303.

Turabian

Picard, Rosalind W. "Does HAL Cry Digital Tears? Emotions and Computers." In *HAL's Legacy: 2001's Computer as Dream and Reality*, ed. David G. Stork, 279–303. Cambridge: MIT Press, 1997.

APA

Picard, R. W. (1997). Does HAL cry digital tears? Emotions and computers. In D. G. Stork (Ed.), *HAL's legacy: 2001's computer as dream and reality* (pp. 279–303). Cambridge: MIT Press.

Article in a Journal

MLA

Connatser, Bradford R. "A Phonological Reading Model for Technical Communicators." *Journal of Technical Writing and Communication* 27.1 (1997): 3–32.

Turabian

Connatser, Bradford R. "A Phonological Reading Model for Technical Communicators." *Journal of Technical Writing and Communication* 27, no. 1 (1997): 3–32.

APA

Connatser, B. R. (1997) A phonological reading model for technical communicators. *Journal of Technical Writing and Communication, 27*, 3–32.

Article in a Magazine or Newspaper

MLA

Kieran, Michael. "Image Management: Covering All the Bases." *Publish* Feb. 1997: 64–70.

Turabian

Kieran, Michael. "Image Management: Covering All the Bases." *Publish* Feb. 1997, 64–70.

APA

Kieran, M. (1997, February). Image management: Covering all the bases. *Publish, 12,* 64–70.

RESOURCES

American Political Science Association. *Style Manual for Political Science*. Rev. ed. Washington: American Political Science Association, 1993.

American Psychological Association. *Publication Manual of the American Psychological Association*. 4th ed. Washington: American Psychological Association, 1996.

Booth, Wayne C., Gregory G. Colomb, and Joseph M. Williams. *The Craft of Research*. Chicago: U of Chicago P, 1995.

Buzan, Tony. *The Mind Map Book: How to Use Radiant Thinking to Maximize Your Brain's Untapped Potential*. New York: Dutton, 1994.

Chicago Manual of Style. 14th ed. Chicago: U of Chicago P, 1993.

Council of Biology Editors. *Scientific Style and Format: The CBE Manual for Authors, Editors, and Publishers*. 6th ed. Cambridge: Cambridge UP, 1994.

de Bono, Edward. *Serious Creativity: Using the Power of Lateral Thinking to Create New Ideas*. New York: HarperBusiness, 1992.

Gibaldi, Joseph. *MLA Handbook for Writers of Research Papers*. 5th ed. New York: Modern Language Association, 1999.

———. *MLA Style Manual and Guide for Scholarly Publishing*. 2nd ed. New York: Modern Language Association, 1998.

Hackos, JoAnn T. *Managing Your Documentation Projects*. New York: Wiley, 1994.

Kirsch, Jonathan. *Kirsch's Handbook of Publishing Law*. Los Angeles: Acrobat, 1995.

Lester, James D. *Writing Research Papers: A Complete Guide*. 9th ed. New York: Longman, 1999.

Microsoft Corporation. *The Microsoft Manual of Style for Technical Publications*. 2nd ed. Redmond: Microsoft, 1998.

New York Public Library. *Writer's Guide to Style and Usage*. New York: HarperCollins, 1994.

Porter, Lynnette, and William Coggin. *Research Strategies in Technical Communication*. New York: Wiley, 1995.

Pyrczak, Fred. *Writing Empirical Research Reports: A Basic Guide for Students of the Social and Behavioral Sciences*. Los Angeles: Pyrczak, 1992.

Skillin, Marjorie E., and Robert M. Gay. *Words Into Type*. 3rd ed. Englewood Cliffs: Prentice Hall, 1974.

Smedinghoff, Thomas J. *The Software Publishers Association Legal Guide to Multimedia*. Reading: Addison-Wesley, 1994.

Turabian, Kate L. *A Manual for Writers of Term Papers, Theses, and Dissertations*. 6th ed. Chicago: U of Chicago P, 1996.

Zimmerman, Donald E., and Michel Lynn Muraski. *The Elements of Information Gathering: A Guide for Technical Communicators, Scientists, and Engineers*. Phoenix: Oryx, 1995.

Zimmerman, Donald E., and Dawn Rodrigues. *Research and Writing in the Disciplines*. Fort Worth: Harcourt, 1992.

Illustrations

Tips

Writing

Many people think that writing is an easy skill to master. Unfortunately, you can't learn everything you need to know about effective writing from reading a handbook, by taking a few college writing courses, or by attending a few corporate seminars. And writing technical prose presents even more demanding challenges than writing other kinds of prose. You *can* learn to write better if you understand that writing is a craft. Like any other craft, it has its specialized vocabulary, its assumptions, its standards for critical success, and its particular tools to help you achieve these standards.

To improve your writing, you need to understand the finer elements of style (your manner of expression in your prose), tone (your attitude toward your subject, your reader, and yourself), writing, and revising.

You need to know what prose style is, that it varies within and between discourse communities, that it varies depending on the kind of technical document you are creating, and that there are many ranges of styles to choose from depending on your purpose.

Concerning tone, you need to understand how every element in your document contributes directly or indirectly to the tone. And you need to understand how many otherwise strong documents fail because they convey the wrong tone for the intended purpose and audience and how many documents fail, for example, when the writer neglects to take into account the complexities of writing to an audience from a different culture.

Finally, to improve your writing, you need to know how to present your prose in the manner that is appropriate for your topic, your audience, your focus, and your needs. This crafting process involves many choices on your part, selecting the most effective words, sentences, paragraphs, and larger segments. And these choices are part of an iterative process; you need to revise your prose as completely and as often as circumstances permit, over and over again until it becomes just the right tool to achieve the effect you are seeking.

STYLE

Style means different things to different people. For some, it is voice or tone. For others, it is a level of elegance or informality. For still others, it is a narrative or conversational quality. These attributes all contribute to style, but choosing which ones you will use when you sit down to write depends on who you are and what you perceive to be your subject, purpose, scope, and audience. See Part One, "Planning and Research," for a discussion of each of these factors.

Style, simply defined, is your manner of writing in your prose. But this definition, while simple, is also too general. Style encompasses so many possibilities that defining it in a sentence or two is impossible.

Style develops from many areas. Some of them are your use of words, phrases, clauses, and sentences and how you connect these sentences. Your tone, your preconceptions and assumptions, your connection with your audience, and your audience's requirements of you all shape your style. Style is who you are and how you reflect who you are, intentionally or unintentionally, in what you write. As you understand how all of these elements constitute style, you will have better control over all these elements and thus improve your own style.

Style and Discourse Communities

Some of your choices about how you write depend on the purpose of your writing. Others are just part of your personal style regardless of the situation. But many of your choices depend on whom you're writing for. Thus, style is directly tied to audience.

Every audience is a discourse community (see "Audience," page 6), and often each audience consists of many discourse communities. Typical discourse communities are students in the same major, members of a profession, and practitioners of a hobby. Technical communication students, lawyers, and gardeners each have their specialized language, experiences, basic knowledge, expectations, values, methodologies, idiosyncrasies, and goals.

As much as possible, your decisions about what to write and how to write about it must be directly tied to your readers and their interests. Part One, "Planning and Research," discusses ways to help you narrow the focus of your subject according to what your readers need to know. Also see the discussion of audience analysis in Part One, "Planning and Research," for ideas on how to consider the interests of the discourse community for which you are writing. If you have planned your document well, you will have a focus—a definite purpose and scope for your intended audience. You will have appropriately narrowed the discussion of your subject, and you will be telling your audience what it needs to know. However, your document may fail if you don't take into account many considerations concerning the discourse community for which you are writing. The point here is more complex than to suggest merely that you write differently to lay people, paraprofessionals, professionals, executives, technicians, or experts. Rather the point here is that *all* of these audience levels are present in almost any discourse community.

Style and Genre

Just as the discourse community directly shapes the style, so does the genre or the kind of document you are writing. Typical genres in technical communication are letters, memos, instructions, reports, and proposals. Some genres require a more formal style, and others allow a wide range of styles. A job application letter requires a serious tone and, in general, a more formal diction than a personal letter to a friend. An abstract for an article in a chemistry journal will usually be more formal than an executive summary for a technical report. A software user guide for a novice audience will generally be more informal and conversational than a technical reference guide for the same product. And an official government report on the space shuttle *Challenger* disaster requires a more serious tone than an instruction manual on how to build toy rockets.

Sometimes it is difficult to determine the best genre to use in a particular situation. For example, for some kinds of business correspondence, it may make a better impression to correspond by letter than by an informal e-mail message, since e-mail is generally a more informal medium than print correspondence. It is still more widely accepted to apply for a job by mailing a cover letter and résumé to the prospective employer. Yet many companies are encouraging applicants to send initial inquiries and résumés by e-mail or fax. See Part Nine, "Technical Documents," for a more detailed discussion of the various challenges in determining which genre to use for which occasion.

Kinds of Styles

Many writers don't give a lot of thought to the kinds of styles that are available to them before they write. If they have a habit of writing in a plain style (clear, straightforward prose), then they will typically write just about everything in a plain style. However, skillful writers will use the style that best suits the subject, purpose, scope, audience (including discourse community), genre, and context. For example, if you are writing a simple training manual to help soldiers learn the basics of a new weapon, you may choose to use a simple primer style for every sentence you write. If you are writing a scientific article for the *New England Journal of Medicine,* you will cultivate a complex style (a more intricate or involved prose style) to impress on your expert readers the sophistication of your argument and your command of the subject. If you are writing a software manual for novices and trying to teach them how to use a new software program, you will use a plain style throughout to make sure your readers can perform the necessary tasks to use the program.

You have many choices when you select a dominant or major style for your document. Often these styles are combined. For example, many people write in both a verb style and a noun style (using many verbs and nouns throughout). Often, writers whose chief style is complex make their style more difficult by combining it with an affected style (a style that is pretentious). And, of course, there are a wide variety of levels of formality, ranging from the formal to the informal, for each of these styles. By becoming familiar with these styles, you will give yourself the freedom to express your ideas more effectively in your technical documents.

You may use many styles in your writing, but by far the two most usual styles you will use and read are the plain style and the complex style.

PLAIN STYLE

A plain style, a style that is straightforward and easy to understand, is the dominant style in technical prose. Also referred to as a reader-friendly style, this style is common in technical prose because readers want prose that enables them to read quickly and to apply the information correctly. The plain style seeks simplicity over complexity. In the plain style, language is generally simple and straightforward. Short, commonly used words combined into simple, short sentences, often written in the second-person "you" style, invite the reader into the text. The tone is friendly rather than impersonal. The strategies for a plain style are almost endless.

COMPLEX STYLE

A complex style is characterized by a technical vocabulary and complex or compound-complex sentence structure, with perhaps long paragraphs exposing complex concepts. Complex styles can seem more distant and less personal than the plain style through use of a less personal tone.

Part of the reason for the difficulty in reading this style is that style and content are so closely related. If a subject is only slightly technical, it will usually be easier for a reader to understand than a subject that is moderately technical or highly technical. Many subjects are more technical, abstract, theoretical, or difficult than others. Still, a complex style, handled competently, can clarify a technical subject even to a novice audience.

A plain style is not always the right or best choice. Depending on the genre, audience, subject, purpose, and scope of your writing, a complex style may be just the approach you need to convey your information effectively.

TONE

Tone is usually defined as your attitude toward your subject and your audience. However, tone is more accurately defined as your attitude toward your subject, your audience, and yourself. You must carefully consider all three to convey the desired attitude in your documents. Your writing may have a clear purpose, be well organized, be free of errors in grammar and spelling, be free of errors in punctuation, and so on, but unless you also convey an appropriate tone, your document may fail to communicate your message to your intended audience.

Planning is just as appropriate for tone as it is for content and organization. It is up to you to plan the attitude you want to convey. Will it be angry, conciliatory, deferential, authoritative, humorous, sarcastic, pedantic, apologetic, exasperated, or something else? Conveying attitude effectively takes a great deal of skill. You must aim for consistency and appropriateness. Your readers follow your attitude toward the subject, themselves, and yourself through your handling of the vocabulary and sentence structure appropriate to the attitude.

You establish a tone in your writing by controlling your perceived distance from your readers, your attitude and emotion, and your character.

Distance is achieved through formality and impersonality; closeness occurs through informality and personability. You personalize your prose through the use of personal pronouns that bring you and your readers closer: *you, we, us*. Use an informal manner of address: *Let's look at the next message box and see what we can do here* instead of *Open the message box and read the menu items*. Depending on the context, you may need a more personal approach or a more impersonal approach. The key is to decide which will work more effectively in the context and use that tone consistently throughout the document.

Tone is, in part, a matter of attitude and emotion, the attitudes and emotions you want to convey about your subject and to your audience. Examples of attitudes are trust, distrust, sincerity, insincerity, seriousness, lightheartedness, hope, despair, belief, incredulity, acceptance, denial, desire, repulsion, and attraction or rejection. Examples of emotion are anger, love, hate, desire, envy, lust, joy, excitement, disgust, contempt, revulsion, obsession, and indifference. As you can see, attitudes and emotions overlap. What's important is to realize that all of the attitudes and emotions listed above (and many more) can be the basis of an appropriate tone in a piece of writing. But remember, the emotion and attitude that you convey in your writing should never be the accidental result of your feelings of the moment.

Before you write, you need to decide how you want to treat both your subject and your audience. For example, do you want to treat the subject seriously and convey this seriousness to your audience? Or do you want to treat your subject with anger and convey this anger to your audience? To convey a serious tone, you have to show your readers that you respect them and take them seriously, that you take the subject matter seriously, and that you also take yourself seriously. To convey this serious tone, you have to choose your words carefully, phrase your sentences and paragraphs with skill, and organize the document so that it will achieve the desired effect.

To convey anger about a subject, all of your style choices must help you suggest this anger appropriately.

Finally, you establish a tone by showing your readers your character. After all, the believability of your prose depends on your establishing a relationship with your readers, and to do that, you have to show them that they can rely on your competence. So tone is a matter of ethos or character and how you establish this ethos with your audience. Ethos comes from the Greek word *ethos,* meaning custom or character. Is your ethos really you? Yes, but sometimes instead of your true character, you will assume a mask, a *persona,* in your prose. Your reader may not know the difference, but your tone may incorporate a combination of your ethos and the persona you assume to achieve your purpose.

Developing an Appropriate Tone

Decide what kind of relationship you want to have with your audience. Do you want your audience to regard you as a friend? Do you want your audience to regard you as highly informed and knowledgeable? as trustworthy?

After you decide what tone you want to convey and after you have reviewed the possible effect of the situation on your communication, you need to choose the appropriate words for your message. If you want to establish a friendly tone, for example, avoid accusatory words and phrases. Also remember that readers can recognize and appreciate your efforts to be friendly, courteous, concerned, or insistent without being too angry.

State the ideas and thoughts that show clearly what kind of relationship you want with your audience. For example, if you have reservations, state that you have reservations. If you have confidence in the situation, state that you have confidence. Remember before you begin to write that you can add all kinds of phrases and sentences to influence the way your readers will perceive you.

Sometimes, to make your tone convincing, it is necessary to write richly detailed paragraphs. There are many occasions for being brief, but there are also times when careful attention to detail is necessary. One effective way to write an angry letter is to provide lots of details to make your points convincingly and, of course, to avoid any personal attacks on the reader's abilities or integrity. Often the reader will see that your anger is well justified.

Don't forget that the page layout and design of your letter, memo, or report also help to establish tone. Hard-to-read fonts, too many fonts, fonts that are too small, a poor choice of paper color, poor margins and spacing, the wrong binder, and many other physical properties of your document all make a negative impression on the reader. You can make all kinds of careful decisions about your tone, diction, opening, middle, and ending, but unless you also give careful attention to the appearance of your document, you are not doing all you can to communicate well. See Part Three, "Design and Illustration," for more information about enhancing the effectiveness of a document through managing its appearance.

TIPS: Using Humor

- Consider your audience. What does your audience expect from this document? Will humor enhance the message or hinder it?

- Consider the genre of your document. Some genres—e-mail, letters, memos, and others—readily lend themselves to a humorous approach, while others—such as abstracts, summaries, reference manuals, and feasibility reports—do not.

- Consider the purpose of the document. If your purpose is to explain something serious or of a delicate nature, resist the urge to use humor. But if your purpose is to put your audience at ease when introducing new and perhaps intimidating material, then you might well consider using humor to achieve your goals.

- Consider the subject matter. Some subjects just do not lend themselves to humor: injury, illness, and death should be treated respectfully and seriously unless some unusual circumstance warrants the exceptional humorous treatment.

- Be cautious when using humor to make sure that you have not relied on a culturally dependent point of view to generate humorous effect. Keep in mind that humor depends on a shared point of view. What seems funny to you may not amuse your readers if their background or outlook is different from yours.

- Use humor for a reason. Gratuitous humor will seem out of place if it is dropped into a document for no apparent reason.

- Balance humor with the rest of your prose. Even when humor is appropriate for the context, too much humor can defeat your purpose. After all, unless you are writing a joke book, your document has some other purpose besides entertainment.

- Be alert for possible offensive interpretations to your humor. Humor that is derived from negative sexual, social, cultural, national, religious, or racial stereotypes has no place in documents produced on the job or in school. Even sarcasm may be misinterpreted by your audience, so avoid it unless you have a very good reason for using it.

■ If you use humor, be consistent. If humor is part of your approach in a document, make it an integral element of the document, well absorbed into the text. The sudden appearance of humor in a serious document can be very jarring.

■ When your document can benefit from a lighter touch, don't shy away from humor. In the right setting and with the right treatment, an overly drab subject can be made more palatable with a sprinkling of humor to spice it up.

Bias

Bias is an inclination or personal judgment, a prejudice, a preconceived judgment or opinion. Essentially, to have a bias is to favor someone, something, or some idea. If you have a bias, you are not usually being fair, impartial, or objective. Of course, it's difficult to be free of bias. All of us have feelings, opinions, beliefs, and attitudes about all kinds of issues, people, and things. However, you don't have to be an automaton to be mostly free of bias. Ultimately, bias is about degrees or levels. Bias may range from a slight inclination or disposition to a blatant prejudice. When you write, you need to be aware of potential biases that can shape your writing.

Writers may reflect all types of bias in their writing. Kinds of bias include gender, corporate, philosophical or religious, political, racial, cultural, age, size, and bias against people who are physically and mentally challenged.

Gender bias is bias in attitudes, language, and actions by members of one sex against the other. Many people, of course, have biased attitudes toward one sex but don't necessarily express these attitudes in language or directly in action. Others use sexist language, language that expresses stereotyped views or assumes that one sex is superior to another. Still others express their bias toward one sex through physical acts ranging from sneers to violence.

Gender bias is a particular problem in writing, since over the centuries the English language has developed into patterns that follow the assumptions of those who have committed their ideas to paper. These assumptions have often made men the model or placeholder both in cases in which men were the usual subjects of the action and in cases in which a "generic" person was intended. Today, these language patterns are considered exclusionary and should be avoided.

Unless your purpose for writing is to reveal one bias or another or to write propaganda, you must always be aware of how your reader will perceive the words you use.

TIPS: Avoiding Gender Bias

- Unless a specific male person is referred to, do not use the so-called generic *he*.
- Do not use modifiers or suffixes that can be regarded as belittling or less than their unmodified counterparts. Use *actor, waiter,* and *doctor,* regardless of the sex of the person; use *flight attendant* rather than *stewardess*.
- Avoid gender-stereotyped or demeaning characterizations.
- Whenever possible, use examples drawn from both sexes.
- Vary the order by which you refer to people so that you're not always mentioning the men first or the women first.
- Be sensitive to possible sexism in assigning roles in examples.

International Technical Communication

International technical communication presents many special challenges concerning style and tone. Of course, mastering the intricacies of international technical communication requires a great deal of knowledge, but a few key areas can be covered here. Most mistakes in international communication involve the following areas. First, many writers are not specific enough in their documents. They fail to use simple, straightforward language to state exactly what they mean. Second, many writers use metaphors, clichés, jargon, idioms, slang and other language variations that foreign readers have difficulty understanding. Third, many writers use too many big words and long sentences. They should use simpler, short words and shorter sentences. Fourth, much international communication requires documents to be punctuated carefully, even heavily, and many writers fail to adhere to this requirement. Finally, many writers fail to respect the customs of their audience's country or culture. They fail, for example, to consider how they may be violating various social customs concerning proper address for other people, using inappropriately informal tones, or ignoring certain religious taboos. An effective tone is especially difficult to master in international communication. For example, you must often be direct and straightforward in cultures such as those of the United States, Switzerland, and Germany, whereas in cultures such as those of Japan, Arabia, and Latin America, you often need to be or should be less direct.

WRITING

You are ready to write a draft after you have done adequate planning and research (see Part One, "Planning and Research") and after you have determined your style and tone. Effective planning, of course, requires determining your subject, audience, purpose, and scope as well as outlining. After all of this preparation, writing the first rough draft is still difficult, but writing the first draft would be a lot more difficult without this necessary preparation.

Use the method of writing the first rough draft that works best for you. Many writers prefer to write the first draft quickly, concentrating on choosing the most appropriate or accurate words that will do for now and writing sentences and paragraphs that concentrate on and develop key ideas. These writers focus chiefly on unity, keeping the focus on the main purpose and supporting ideas. The goal is to create a sketch of a document that will make it easy to return later to make many further improvements. Other writers prefer to write less quickly and to labor more over every word and on making each sentence and paragraph more polished before moving on to the next paragraph.

Whatever method you use, you should keep in mind that writing and revising are different activities in the writing process. Whether you are writing quickly or laboring over every word, effective writing still requires more refining and polishing. In revising, you reread what you have written and make every effort to improve the document in every way you possibly can.

Levels of Diction

Diction is your choice of words. You will tend to make certain word choices, even if unconsciously, according to your writing habits, the needs of your audience, your purpose, your scope, and the genre. Perhaps you normally favor concrete terms, simple words, and a nontechnical vocabulary. Or perhaps you normally choose more abstract words, more literary images, and more emotionally expressive prose. Along with your personal preferences, you will certainly modify your diction according to the particular needs of the document you are working on. And in technical prose, you will have to deal with subject matter and audience considerations that may require a vocabulary that is quite different from everyday language.

Just as there are many levels of formality in style—from highly formal to highly informal—there are many levels of diction. The four generally accepted levels of diction are *formal, informal, colloquial,* and *slang.* Any of these four levels may be correct in a particular context but incorrect in another.

FORMAL

Formal diction is the language that is appropriate for a formal context. You would expect to hear formal speech at a commencement address or a State of the Union speech, and you would expect to read formal prose in a contract, government document, document abstract, or academic journal article. The choice to use formal diction is determined by the context, and the context is determined by your purpose, the genre of the technical document, the audience, and other factors. Sometimes you will be able to choose your diction, and sometimes the genre will dictate the one you must use. Some genres tolerate a range of diction. For example, you may find that rather than keeping to a highly formal diction in a documentation set, you can use an informal approach quite appropriately. The "dummies" series of computer books is an example of colloquial diction used successfully to document computer software and operating systems.

INFORMAL

Informal refers to ordinary, casual, or familiar use. Formerly, technical documents were all formal and in fact rather stuffy. To some people, "technical document" is just another way of saying "unreadable." But technical documents have evolved from the stilted works of years ago, and many of the most effective examples today are written in an informal prose style. And an informal prose style can now be appropriate for a good many technical genres, such as letters and e-mail.

COLLOQUIAL

The term *colloquial* refers to conversation or diction that is used to achieve conversational prose. It may also mean everyday speech and slang. In many situations, colloquial prose will not be appropriate as

a workplace diction choice. Occasionally, as in certain e-mail corre-
spondences, it may be acceptable.

SLANG

Slang and colloquial language overlap. Slang is sometimes thought
of as a synonym for colloquial language, but in fact it has its own dis-
tinctive features. Slang is colloquial language with an attitude. It
consists of a vocabulary shared by a particular group, often with the
purpose of excluding others. You wouldn't use slang in a document
unless you knew that it was expected by the discourse community
you were writing for.

Usage

Usage, simply defined, concerns appropriate and accurate word
choice. See the discussion of jargon below for the problems a specialized vocabulary presents to writers of technical prose. It's important
to realize that writers of technical prose have as many problems
with words in the general vocabulary as other writers do. Writers of
technical prose, as do other writers, often confuse *affect* and *effect,*
bring and *take, can* and *may, shall* and *will, comprise* and *constitute,*
and so on. To improve your technical prose, you must have a good
understanding of writing at the word level. To take control of your
text, you need to understand denotation and connotation, commonly
confused words, standard and nonstandard words, and specialized
vocabularies.

DENOTATION AND CONNOTATION

The denotative meanings of a word are the dictionary meanings of
the word. The connotative meanings of a word are the additional
meanings that we associate with the word. Connotations are the hidden agendas our words carry along with them. The words you choose
may be absolutely correct denotatively, but their connotations can either sabotage or strengthen the effectiveness of your prose. You are
responsible for the effects your words create, so be sure you take into
account the sometimes subtle messages they carry along.

COMMONLY CONFUSED WORDS

Your readers are distracted when you misuse words that are similar.
If you say *infer* and you mean *imply,* your message is distorted. If
you want *continuously* but substitute *continually,* your readers will
wonder whether your other points are equally incorrect. There are

many more words that people often confuse. (See Figure 2.1) Make sure that you know the exact meanings of the words you use.

COMMON MISTAKES IN USAGE

Common mistakes in usage include archaic expressions (words that are old-fashioned and no longer used naturally) such as *amongst;* barbarisms (word distortions) such as *disremember* and *irregardless;* improprieties (good words used inappropriately) such as *set* for

---

Affect/effect. *Affect* is a transitive verb meaning to influence or to pretend; *effect* as a verb means to bring about; effect as a noun means result.

Adverse/averse. *Adverse* means harmful or inclement; *averse* means disinclined or unwilling.

Anxious/eager. *Anxious* expresses apprehension or fear; *eager* expresses enthusiasm or impatience.

Appraise/apprise. *Appraise* means estimate the value of; *apprise* means inform.

Assure/insure/ensure. All three can mean to make sure, but *assure* means to give positive guarantee, *insure* means to take out insurance, and *ensure* means to make certain.

Average/mean/median. *Average* and *mean* are the same. Each is the sum of the items divided by the number of items. The *median* is the number that falls in the middle of the items: Half the numbers are higher than the median and half are lower than the median.

Between/among. *Between* is used for relationships concerning two entities; *among* is used for relationships concerning three or more entities. However, if one item is being considered in relationship to each of the others individually in turn or if a succession is meant, then *between* is used even though there are more than two items. *(The traffic light cycled between green, amber, and red.)*

Bring/take. *Bring* is to cause to move toward the speaker or the subject; *take* is to cause to move away from the speaker or the subject.

Can/may. *Can* expresses ability; *may* gives permission.

Compliment/complement. A *compliment* expresses praise; a *complement* completes something.

Comprise/constitute. The whole *comprises* (includes) the parts; the parts *constitute* (make up) the whole.

Continuously/continually. *Continuously* is all the time; *continually* is over and over again.

Imply/Infer. A speaker or writer may *imply* more than he or she says; the hearer *infers* what the speaker intends.

FIGURE 2.1
Commonly Confused Words

Its/it's. *Its* is the possessive of the neuter pronoun *it*. *It's* is the contraction meaning *it is.*

Lay/lie. *Lay* is usually transitive and takes a receiver of its action; its principal parts are lay, laid, and laid. *Lie* is intransitive in its usual meanings and does not take an object; its principal parts are lie, lay, lain.

Less/fewer. *Less* refers to quantity; *fewer* refers to number.

Shall/will. *Shall* is the verbal auxiliary indicating the simple future for first person (I, we) and determination or a command for second and third person (you, he, she, it, they). *Will* is the reverse, indicating the simple future for second and third person and determination for first person.

That/which. *That* is restrictive and tells "which one." *Which* is non-restrictive and describes what something is like. *We installed the software that was developed by Microsoft.* (There are other software packages, but we chose this particular one.) *We installed the software, which was developed by Microsoft.* (This is the software we installed, and it was developed by Microsoft.)

Who/whom. *Who* functions as the subject of a verb; *whom* functions as the object of a verb. *Who is going there? Whom do you see?*

sit, most for *almost, accept* for *except;* and clichés (words or phrases that have been overused until they have lost their effectiveness) such as *cold as ice* or *green as grass.* Writers may also misuse idiomatic expressions (expressions that are particular to every language and often not governed by the rules of grammar) such as *get into hot water* or *with a grain of salt.*

STANDARD AND NONSTANDARD WORDS

The issue of standard and nonstandard words is controversial to many people who study language. Is the English that is spoken, for example, by a news anchor on a major network news program more standard than the English that is used by a disk jockey in the inner city? The language that the news anchor uses would most likely be more formal. And if *standard* means "widely used by mainstream society," then the language used by the news anchor would be more standard. However, standard does not imply superior. The news anchor uses a more standard language because of the mass audience; the disk jockey uses a less standard language because of the more narrowly defined audience.

Additionally, in any language some words are standard, some are nonstandard, some are archaic, some are vogue words, and some are slang. Many nonstandard words become standard after a while, and many other words fall out of use and become archaic.

When you are writing for an audience, your decision to use nonstandard words should be made very carefully. For most mainstream writing, nonstandard words would defeat the communication purpose of your text, distracting the reader away from your message. Technical language is not nonstandard, although it may be unfamiliar. As a writer, you are expected to render the unfamiliar familiar through your prose, graphics, and design. By using nonstandard language, you impose an additional obstacle between your reader and your meaning.

JARGON, SHOP TALK, COMPUTERESE, AND TECHNOBABBLE

Technical prose style consists, in part, of a specialized vocabulary or, more broadly, a specialized language. Every discipline, profession, trade, or hobby has its special vocabulary, and using this special vocabulary appropriately requires careful attention. However, this necessary specialized language is often misused, abused, or used pretentiously, and you must understand what these misuses are to know how to avoid them.

Jargon. Like *technical terms,* the term *jargon* applies to the specialized vocabulary of a trade, profession, hobby, or science. When used within the trade or profession, it serves as a verbal shorthand to convey exact information in an abbreviated form. But when it is used outside its own discourse community, it can confuse and confound. Since this is sometimes the intent, jargon is often correctly perceived as pretentious or hostile in those contexts. Jargon is known by other names as well: *computerese, engineerese, bureaucratese, governmentese, academese, officialese, tech speak, technospeak, newspeak, doublespeak, doubletalk, gobbledygook,* and *pentagonese.*

Unless you are writing within a discourse community that is familiar with the jargon you are using, you will better serve your audience by avoiding its use. Whether the jargon is simply incomprehensible vocabulary or gobbledygook designed to obscure your meaning or absence of meaning, you are not communicating by using it. If this smokescreen is your intent, then perhaps you need to reexamine your purpose in writing. If your use of jargon unintentionally confuses, then have others review your work for clarity and revise your work where necessary.

Shop Talk. *Shop talk* is jargon that has become so specialized that it applies only to one company or subgroup of an occupation. For example, shop talk at the Kennedy Space Center in Florida includes words or phrases such as *stand by* (please wait), *Roger* (okay), *under-volt* (low electrical gauge reading), *full-scale high* (gauge needle off the scale to the high side), and *showing zip* (no reading).

Computerese and Technobabble. *Computerese* is the use of computer jargon in your writing. *Technobabble* is an offshoot of computer-ese. Computerese or computer jargon devolves into babble when it is used as filler or decoration, when it is employed intentionally for ob-fuscatory purposes, when it is employed gratuitously, when it is used obsessively, and when it is used by people who are unfamiliar with its meanings in an attempt to sound as if they know what they are talking about. Unfortunately, writers who have become accustomed to hearing and using computerese in appropriate contexts often slide into its use even when it's not intended. In-house terms, com-mon at your company, can appear in your writing even when you are writing for an audience that would not be familiar with these ex-pressions. Guard against letting your usual work vocabulary spill into documents that are intended for others.

Of course, clear technical language consists of more than just using technical terms your audience understands. An ill-defined purpose, a confusing organization, poorly written sentences, overly long paragraphs, and many other faults contribute to an ineffective document. Yet there are many factors you can consider to make your use of technical terms more appropriate.

TIPS: Achieving Clarity with Technical Terms

- Understand the meanings of jargon—necessary technical terms you must use for many occasions within a discourse community, inappropri-ate technical terms used for the wrong audience, and the pretentious use of technical terms.

- Use jargon and other types of specialized language appropriately. Define jargon when necessary; avoid it if it's not necessary. If you must use jar-gon for an audience that is unfamiliar with it, provide a glossary or de-fine terms when you introduce them for the first time in the text.

- Provide an analogy or some other kind of comparison, anecdote, con-crete example, illustration, or nontechnical explanation to help explain your point.

- Omit unnecessary modifiers and qualifiers and excessive wordiness.

- Use terminology consistently.

Sentences

The ability to construct clear, concise, and accurate sentences in technical prose is an invaluable skill. In early drafts of the writing process, many of your sentences may be ambiguous, longer than they need to be, and somewhat inaccurate in content. Effective sentences require a lot of editing and revising. And, of course, it helps to have a thorough understanding of the components of a good sentence, from structure to sentence length to variety.

STRUCTURE

A sentence typically consists of a subject and a predicate, either expressed or implied. A subject is the topic of the sentence. A predicate is what is said about the subject. Consider the following sentence, for example: *My computer has a 17-inch monitor*. The subject is *my computer*. The predicate is *has a 17-inch monitor*.

The major structural components of sentences are phrases and clauses. A phrase is a group of related words that are used as part of a sentence but do not have a subject or predicate: The television *on the table* is a Zenith. A clause is a group of related words that contains a subject and a predicate: *David delivers pizzas* after he completes his homework. A clause is either an independent clause (also called a principal clause or main clause) or a dependent clause (also called a subordinate clause). An independent clause makes complete sense when standing alone: *David delivers pizzas* after he completes his homework. A dependent clause is used as a part of speech in a sentence and usually does not make sense when standing alone: *after he completes his homework*.

FORMS

A sentence may be classified as one of four forms or constructions: *simple, compound, complex,* and *compound-complex* (or *complex-compound*). The form of the sentence depends on the type and number of clauses.

A simple sentence has only one independent clause and no dependent clauses:

> The major computer manufacturers market their products much more effectively than they did just a few years ago.

A compound sentence is equivalent to two or more simple sentences connected by coordinating conjunctions or by punctuation:

The major computer manufacturers market their products much more effectively than they did just a few years ago, and this more effective marketing has led to a major increase in sales for most of these manufacturers.

A complex sentence contains one independent clause and one or more dependent clauses:

After marketing their products much more effectively, the major computer manufacturers have experienced increased sales.

If either independent clause of a compound sentence has a subordinate clause, the sentence is called compound-complex or complex-compound:

After marketing their products much more effectively, the major computer manufacturers have experienced increased sales, and the consumer has benefited from the wider variety of quality choices.

FUNCTIONS

Sentences are divided into four classes according to function: declarative, imperative, interrogative, and exclamatory.

The most common sentence in technical prose, the declarative sentence makes a statement or asserts a fact:

As the price of computers continues to decline, the quality continues to improve.

The imperative sentence expresses a command, request, or entreaty. The subject of an imperative sentence is *you,* but this subject is usually omitted:

Save any files you download from the Internet to your Temp directory or some other temporary directory before installing the file on your computer.

The interrogative sentence asks a question:

Do you defrag the hard drive on your computer as part of your routine computer maintenance?

The exclamatory sentence expresses surprise or a strong emotion. Exclamatory sentences can be useful in technical prose, for example, to motivate novices who are becoming familiar with a software program for the first time. Many tutorial programs have online messages such as *Congratulations! You are now ready for the next lesson!*

STYLES

Sentences are also classified by the arrangement of their material as periodic, balanced, or loose. Periodic and balanced sentences are far more carefully arranged than loose sentences and are therefore particular favorites of prose stylists, but these sentences also occur in technical prose.

Periodic sentences are complex sentences in which the main clause is delayed until the end of the sentence. These sentences are tightly structured to wait until the last word or phrase to make their point. Such sentences are called periodic sentences because they do not make much sense until the reader reaches the period at the end:

> When you consider the capabilities and features of computers of just 15 years ago—processor speed, memory, screen size, resolution, and so on—and compare these with the computers of today, you may be amazed when you imagine what computers will be like 15 years from now.

Periodic sentences can be difficult for people to read because readers have to hold all of the earlier information in memory. When they reach the end of the sentence, they can fit the subordinate elements to the main point. Periodic sentences may be used in a technical memo or letter, for example, to help make a point more emphatically or persuasively. In general, however, you should choose a less sophisticated sentence style. Tightly structured periodic sentences help authors to achieve elegance in their prose, but elegance is not the primary goal of writers of technical prose.

Because balanced sentences also help writers to achieve elegance in their prose, they are also less common in technical prose than in other kinds of prose. Balanced sentences use parallel structure, antithesis, or symmetry to achieve their effect. Basically, a pattern that is used at the beginning of the sentence is repeated elsewhere in the sentence. Like the periodic sentence, the balanced sentence has a tight structure, but the balanced sentence is not as structured as the periodic sentence. However, a balanced sentence may also be a periodic sentence. When writers aim for technical prose balance, it is often achieved through the parallel structure of lists in the form of nouns, verbs, phrases, clauses, or sentences: Use Notepad, WordPad, or Word to open your files that are average, large, and huge, respectively.

In contrast to periodic sentences, loose sentences are complex sentences in which the main clause comes first, usually followed by several subordinate clauses. Such sentences are generally easier for readers to understand because the main clause is at the beginning. Because of this easier comprehension, loose sentences are very common in technical prose.

Periodic and balanced sentences may be more tightly structured, but loose sentences have a definite structure as well. They are also sometimes called cumulative sentences because they cumulate subordinate clauses after the main clause. Note the structure of the following loose sentence:

> You can now order a computer in a number of ways, for example, by calling an 800 number for many computer catalog companies and ordering by phone, by visiting your local computer retailer, or by accessing a computer manufacturer's Web site and ordering via the Web.

Some writers prefer to play with the style of loose sentences, making them appear less structured by using add-on phrases and clauses, almost as though additional thoughts just occurred to them. Consider the following loose sentence, for example:

> Most consumers today feel so overwhelmed by the many choices of computer manufacturers, their products, the product features, and warranties that they greatly appreciate any guidance anyone can provide them before they make a major computer purchase, especially during the hectic holiday season.

Knowing how to use each of these types of sentences enables you to control your presentation most effectively.

EMPHASIS

Sentence emphasis is the art of ending a sentence well. Strategies for ending sentences effectively include trimming the end, shifting less important ideas to the front of the sentence, and shifting more important ideas toward the end of the sentence.

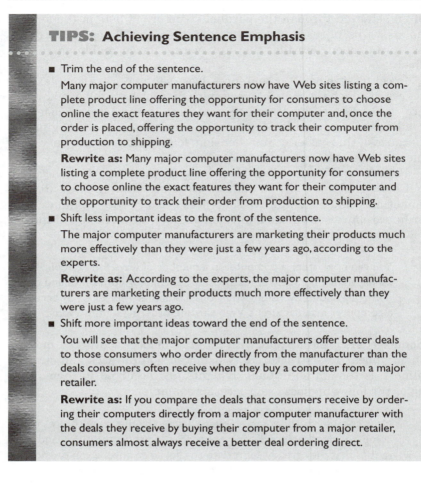

TIPS: Achieving Sentence Emphasis

■ Trim the end of the sentence.

Many major computer manufacturers now have Web sites listing a complete product line offering the opportunity for consumers to choose online the exact features they want for their computer and, once the order is placed, offering the opportunity to track their computer from production to shipping.

Rewrite as: Many major computer manufacturers now have Web sites listing a complete product line offering the opportunity for consumers to choose online the exact features they want for their computer and the opportunity to track their order from production to shipping.

■ Shift less important ideas to the front of the sentence.

The major computer manufacturers are marketing their products much more effectively than they were just a few years ago, according to the experts.

Rewrite as: According to the experts, the major computer manufacturers are marketing their products much more effectively than they were just a few years ago.

■ Shift more important ideas toward the end of the sentence.

You will see that the major computer manufacturers offer better deals to those consumers who order directly from the manufacturer than the deals consumers often receive when they buy a computer from a major retailer.

Rewrite as: If you compare the deals that consumers receive by ordering their computers directly from a major computer manufacturer with the deals they receive by buying their computer from a major retailer, consumers almost always receive a better deal ordering direct.

RHYTHM

Sentence elegance (achieved, for example, by effectively using balance, symmetry, rhythm, and metaphor) is not the intention of most writers who must write technical prose, so many technical sentences are not elegant. Nonetheless technical sentences can still have a balance, flow, or rhythm. In fact, all sentences, whether written or spoken, do have a certain rhythm. Rhythm is a recurring pattern of strong and weak elements or stresses. The strong elements are the peaks of the sentence; the weak elements are the valleys. Good sentences have a balance, a flow, and a rising and falling rhythm: *Open your Word document, delete the first paragraph, and move the last paragraph to the beginning of the document.* As simple as this sentence is, it still has a balance, flow, and rhythm.

VARIETY

Sentence variety is essential for achieving an effective style. Using too many simple sentences together may create a primer style; using too many compound-complex sentences may create a turgid style. Sentence variety is a matter not only of sentence form, but also of sentence function, style, combining techniques, and length. Varying sentence length is especially important. Place too many short sentences together and you create a choppy style; place too many long sentences together and you create a wordy style.

It's impossible to set a limit on the length of a sentence. The average length of a sentence is eighteen words, but many longer sentences are often necessary. The complexity of the information is also a factor in determining sentence length. A long sentence may have too much complex information and would be better rewritten into several shorter sentences.

There are some helpful factors to consider in determining the length of each sentence. You should consider, for example, your readers' familiarity with the material, the importance of the material, whether the sentence is offering new information or repetitive information, and the method you will use to achieve a logical flow of ideas.

Paragraphs

Good writers give as much careful attention to paragraphs as they do to words, clauses, phrases, and sentences. Technical prose differs from nontechnical prose in its use of section headings or summaries to alert readers to the topic of the succeeding or preceding paragraphs. Thus, paragraphs in technical documents may not have a typical topic sentence. Headings, headers, footers, and other aids guide readers and add to the message that is conveyed in the body text. Yet the rules for writing effective technical paragraphs are not so different from those for writing any other kind of paragraph: Limit yourself to one major topic per paragraph, develop your topic using an appropriate paragraph pattern, control the flow, and provide adequate details.

WRITING ON ONE TOPIC

Each paragraph should focus on one key idea or complete thought. One of the major reasons for ambiguity in writing is that too many writers combine several different topics into one paragraph. A carefully constructed outline will help you to determine what your key ideas are. See Part One, "Planning and Research," for a discussion of outlining and organizing.

DEVELOPING THE PARAGRAPH TOPIC

Good paragraphs have an appropriate pattern or order. You must decide how you will go about developing your topic. Will you begin with your topic sentence? pose a question? contrast or define? As you gain experience in writing effective paragraphs, you will find that even though paragraphs can follow certain patterns, they don't necessarily follow them in exactly the same way. You may use a topic sentence to start your paragraph, end it, or carry the middle. In fact, your paragraph may express one topic and still not have one particular sentence embody that topic.

Common paragraph patterns include examples, general-to-specific, specific-to-general, question-to-answer, definition, description, narration, comparison, contrast, analogy, classification, enumeration, and cause-and-effect. An example paragraph makes a statement and supports it with specific examples, such as this paragraph. General-to-specific paragraphs begin with a broad statement of the paragraph topic and progress to more and more specific points to support the topic. Specific-to-general paragraphs build the topic from the ground up, going from detail to detail until the whole idea is complete. Question-to-answer paragraphs pose a question, sometimes rhetorically or theoretically, sometimes specifically or realistically, then proceed to provide the answer. A definition paragraph defines a term or concept. A description paragraph offers a vivid impression of a person, place, or object. A narration paragraph relates an event. A comparison paragraph compares topic A with topic B. A contrast paragraph contrasts topic A with topic B. An analogy paragraph compares an unfamiliar topic to a familiar topic. A classification paragraph divides a topic into categories. An enumerative paragraph numbers the points it will cover as first, second, and third. A cause-and-effect paragraph discusses the effects of a particular cause.

ACHIEVING AN EFFECTIVE FLOW

The sentences in an effective paragraph should flow smoothly from the first to the last. This smooth flow is achieved mainly by the principles of cohesion and coherence. You create cohesion from sentence to sentence in a paragraph by including information that is old and information that is new in each sentence. Consider this example that incorporates cohesion:

> You create a presentation using a flowline metaphor. The flowline contains icons that represent the various steps in your procedure. The steps can be multimedia elements or interactive points. Multimedia and interaction can be combined to create an effective learning module.

You create coherence in a paragraph by using transitional words and phrases, repeating key words, and repeating key ideas.

PROVIDING ADEQUATE DETAILS

Controlling the level of detail in a technical document can be a challenge even to the experienced writer. The difficulty lies in finding the correct level without either confusing your readers by vagueness or drowning them in information they do not need. Naturally, the appropriate level of detail will depend on your purpose and scope. If you are preparing assembly instructions for a piece of machinery, for example, appropriate detail would include explaining what to do with all the various parts. You wouldn't want to include details of the design of the individual circuit boards or the composition of the metallic alloys used in the device. On the other hand, an instruction manual on using the device would probably not cover details necessary for its assembly if that were not a usual activity your readers would expect to perform.

Aside from purpose and scope, you must consider your audience when planning out detail. Are your readers engineers who are very familiar with the constituent parts of the item being described? If so, you can refer to the individual parts by their standard names and expect your audience to follow you. If your audience is inexperienced with this material, you may have to increase the level of detail you supply to ensure that your descriptions and explanations do not assume too much prior knowledge. Or you may want to reduce the level of detail for your novice audience and present only the material that it specifically needs to know.

Consider your genre when you plan your details. What is the appropriate level of detail in a job application cover letter, for example? Should you include as much detail as you would in a résumé? Too much detail can be as ineffective as too little.

Larger Segments

You need to give as much attention to the larger segments of your technical documents as you do to the smaller segments of paragraphs, sentences, and words. Unity, flow, order, and depth of detail—qualities that are essential to good paragraphs—are equally essential for effective larger segments. Each segment must be on one topic, the paragraphs within the larger segment must flow smoothly from one to the other, there must be a logical order to the paragraphs within the segment, and each segment must be adequately detailed.

Effective organization is one key to creating larger segments. For example, whenever possible you should place the most important information first within each segment. Also, avoid too much complexity. Present the information you are discussing in small, simple chunks and short sections. See Part Three, "Design and Illustration," for ways to use headings and lists to help your readers recognize the organization of your documents.

Revising

Effective writing is achieved only by revising and then revising some more. You're not likely to make a technical document, whether it's a memo or a technical report, just exactly what it needs to be on your first, second, or even third try. These early drafts are rough drafts, and to achieve your final draft, you will have to do a great deal of polishing. After finishing a first draft, good writers will labor over every word choice, every phrase and clause, every sentence, every paragraph, and every section in the next draft and in the draft after that.

The first draft of any document may be detailed, but it's seldom adequately detailed. The style may be okay, but it's rarely completely suited to the audience. The grammar and spelling errors may be few, but there may still be errors. The design may appear to be fine, but it usually won't be as good as it can be. Illustrations may appear to be well suited to the text, but they often aren't. If you are fortunate enough to have someone else thoroughly edit your work, you will typically be surprised at the many areas for improvement. You'll have to continue revising until you finally create a document that is strong in all areas. Of course, even after you do all of this revising, your document may still not be perfect, but at least it will be far more effective than it would have been.

In some respects, revising occurs throughout the documentation process. As you plan your document, each time you make changes in your subject, purpose, or scope, you are revising. Each time you take a different approach to your subject after brainstorming, mind mapping, or freewriting, you are revising. Each time you make changes in your outline or other organizational plan, you are revising. And, of course, as you write every word, sentence, and paragraph and make even slight changes, you are revising.

However, more formally defined, revising is what you do *after* you have planned, researched, and written a first draft of your document. Now that you have a version of your ideas on paper or a computer screen, you'll need to go back through your document and make changes as necessary.

Revising is not the same as reviewing or evaluating. Revising occurs when you make stylistic, grammatical, or other kinds of changes in your document or someone else's document. Reviewing occurs when you submit your document to someone else to critique. Evaluating occurs when you or someone else tests the appropriateness or usability of what you have written or created. See Part Four, "Editing."

TIPS: Revising Your Work

- After you finish a draft, leave the draft alone for some time, preferably a day or two.
- For paper documents, revise on paper rather than online, at least during early drafts.
- Examine each section to make sure you have followed your outline or storyboard.
- Check your level of detail. Is it consistent and appropriate for your audience, scope, and purpose?
- Are your paragraphs organized logically? Is the information presented clearly?
- Have you covered all your material? Have you eliminated unnecessarily repeated material?
- For online material, does each screen contain as much material as your readers can reasonably be expected to handle at one time and no more?
- Are your illustrations placed appropriately? Should you add nontext elements, such as figures or tables to support your text?

RESOURCES

Andrews, Deborah, ed. *International Dimensions of Technical Communication.* Arlington: Society for Technical Communication, 1996.

Cook, Claire Kehrwald. *Line by Line: How to Improve Your Own Writing.* Boston: Houghton Mifflin, 1985.

De Vries, Mary A. *Internationally Yours: Writing and Communicating Successfully in Today's Global Marketplace*. New York: Houghton Mifflin, 1994.

Ewing, David. *Writing for Results in Business, Government, the Sciences, and the Professions*. 2nd ed. New York: Wiley, 1985.

Fahey, Tom. *The Joys of Jargon*. New York: Barron's, 1990.

Frank, Francine Wattman, and Paula A Treichler. *Language, Gender, and Professional Writing: Theoretical Approaches and Guidelines for Nonsexist Usage*. New York: Modern Language Association, 1989.

Hoft, Nancy L. *International Technical Communication: How to Export Information About High Technology*. New York: Wiley, 1995.

Jones, Dan. *Technical Writing Style*. Boston: Allyn & Bacon, 1998.

Lanham, Richard A. *Analyzing Prose*. New York: Scribner's, 1983.

———. *Revising Prose*. 3rd ed. New York: Macmillan, 1992.

Lutz, William. *Doublespeak: From "Revenue Enhancement" to "Terminal Living": How Government, Business, Advertisers, and Others Use Language to Deceive You*. New York: Harper, 1989.

———. *The New Doublespeak: Why No One Knows What Anyone's Saying Anymore*. New York: HarperCollins, 1996.

Miller, Casey, and Kate Swift. *The Handbook of Nonsexist Writing*. 2nd ed. New York: Harper, 1988.

Rubens, Philip, ed. *Science and Technical Writing: A Manual of Style*. New York: Holt, 1992.

Smith, Edward L., and Stephen A. Bernhardt. *Writing at Work: Professional Writing Skills for People on the Job*. Lincolnwood: NTC, 1997.

Varner, Iris, and Linda Beamer. *Intercultural Communication in the Global Workplace*. Chicago: Irwin, 1995.

Williams, Joseph M. *Style: Ten Lessons in Clarity and Grace*. 5th ed. New York: Longman, 1997.

Design and Illustration

Illustrations

Tips

Design and Illustration

*D*esigning and illustrating technical documents—whether they are print, online, or Web documents—present many writers with special challenges. Most of us have some knowledge of the basics of effective writing, but many of us are not familiar with the basics of design or illustration. However, becoming familiar with even just a dozen or so important design principles and a dozen or so kinds of illustrations can help you to make your documents more visually attractive and convey your message more effectively.

Why are design and illustration important? Design elements and illustrations in documents are part of your message. Used well, they can help readers to find, understand, and remember the information more effectively. Used poorly, they can impede your readers' journey through the material and send them away without delivering your message. Effective design attracts readers to the content. It makes people want to read what you have written and makes your documents easier to use.

DESIGN

If you are a college student, most of your written assignments are simply a matter of gathering information and then formatting it according to the documentation style (for example, MLA, APA, CBE) of your discipline. You will typically use white paper that is 8½" by 11", double-space the text, use a ten-point or larger fairly traditional font, print on one side only, and provide a cover. However, increasingly, college students are challenged to provide more creative designs for documents in many kinds of courses. One simple and common example that you are sure to encounter is designing an effective résumé. (See Part Eight, "Correspondence and the Job Search," for a discussion of the content of a good résumé.) Your résumé is a perfect example of a document for which being accurate and complete is just not enough. An otherwise impressive résumé can fail completely because of poor design. And, as anyone who has created a résumé knows, finding the best design or designs for your résumé will require considerable time and effort. The principles covered in this section will help you to design your résumé and many other documents effectively.

Basic Principles of Design

Design concerns the appearance and placement of text, text art or line art, illustrations, and white (or blank) space on a page or screen. And design also concerns some of the production choices you make for a technical document—whether it's printed or online, for example, what kind of binding and paper you will use for a print document, and what kind of screens and resolutions you have available for an online document. (See Part Five, "Production," for a more detailed discussion of production issues.) Knowing how to combine all of these elements in a variety of effective ways requires considerable knowledge and skill.

Design elements have many purposes: to organize information for easier use, capture attention, guide eye movement, cue similarities or dissimilarities in meaning, clump information, signal visually the organization of a document, demonstrate hierarchies and distinguish the important from the less important, identify with corporate identity or a document suite, establish a tone, persuade the reader that the content is of value and manageable, and appeal to a particular audience.

Although there are many ways to consider design and many systems of grouping design elements, you will find it helpful to think of design as a visual organization of information. The basic elements of design are then selection, value, and placement. While these words have general, conversational meanings that are familiar to most of us, they have specific meanings when they're applied to design.

Selection concerns your choice of content. When you choose an illustration or decide to use a text description instead, you are making a design selection. Your choice between several different illustrations is also a design selection. In short, the items that you place on your page or screen—be they text, illustrations, rules, or even white space—are all elements in the selection process.

Value describes the attributes or qualities of the items you have selected to include in your design. Color and size, for example, are value attributes that you can vary in your design to call more or less attention to the elements on the page or screen. The decisions that you make for value determine how attention getting your elements will be to your readers and how prominent a role the elements will play in the overall design.

The third basic element in design is *placement*. How you arrange the elements you have chosen determines how those elements will contribute to the message you are trying to convey with your design. Sometimes you will want to group items near each other; other times, they will work better spread apart. Awareness of typical reading patterns and eye-scanning patterns will help you to place items where they will achieve a pleasing balance and be most effective.

You can begin planning your design by deciding whether you will use a grid to organize the placement of elements in your document. Graphic designers often use grids to balance text, headings and titles, visual aids, and white space on a page. Grids are a good way to start the planning of many kinds of designs. A grid is a division of your page or screen into a combination of vertical and horizontal lines used to guide the placement of other elements such as text, graphics, and white (or blank) space. The grid lines are not visible to your reader; they are lines that you use during the design process to arrange elements on the page or screen. The grid will indicate top and bottom margins, side margins, paragraph indention, and any other layout elements you need to plan for. With a grid you can determine the best relationship among your text, headings and titles, and visual aids. You can line up these separate elements along the same grid line to show a visual connection. More complex grids involving columns give you even more creative flexibility. You can experiment with headings and titles, text, visual aids, and white space on various grids to determine the design that you think will emphasize your message and appeal to your readers. (See Figure 3.1 on page 110 for a few sample grids.)

Several basic guidelines will help you to make design decisions about the elements of selection, value, and placement, with or without a grid. Mastering the principles of simplicity, symmetry, consistency, usability, and readability will aid you in keeping track of the most effective ways of reaching your audience and achieving your communication goals.

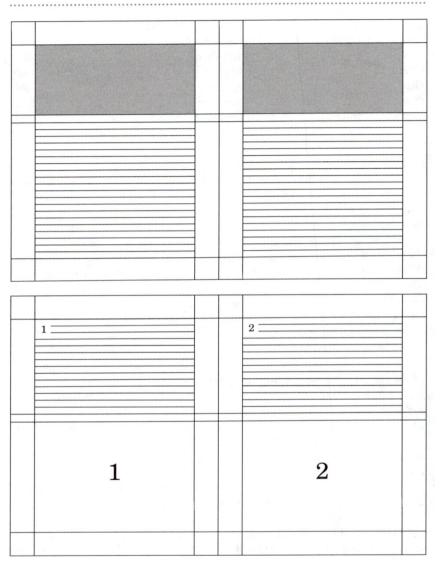

FIGURE 3.1
Sample Grids

SIMPLICITY

Simplicity is a matter of restraint. Keep combinations on the page to a minimum. Don't provide too many combinations of text, text art, and graphics. Use blank space to punctuate your design and separate items of text and graphics. Plan the way the user will access the material. How does it flow? What path does the eye follow through the page or screen? What type of motion are you setting up for the reader? Your document will look better and function more effectively if you present a few elements in harmony rather than many elements in discord. If there are too many elements on a page, they compete for attention.

SYMMETRY

Symmetry concerns how you balance or contrast design elements on a page. Keep elements balanced visually on the page. Try to arrange items on the page so that busier areas have a counterbalancing section of similar weight opposing them. Step back from the page and view it as an arrangement of objects rather than as words and pictures. Do all the elements fall on the same half of the page—upper, lower, left, or right? Shuffle them around to see whether another arrangement would be more pleasing. Not all design has to be symmetrical around a center line, and in fact it is often more effective and visually pleasing not to have complete symmetry. Sometimes you may want to organize your text and graphical elements around a guide. You can line up several elements flush to a vertical line, either present or implied, and provide a pleasing effect of alignment. If you create an asymmetrical design, try to provide for balance. The effect you are striving for is the one that uses the tension between opposing elements in a positive way.

CONSISTENCY

Consistency concerns how you use repetition effectively in your design choices. Use design elements with consistency on each page and from page to page within a document. A ruled line at the top or bottom of each page can unify the look of your document. Consistent headings, fonts, margins, white space, and other elements all contribute to an effective design. Color is another strongly unifying element when used carefully. And of course your margins and paragraph styles lend a consistent look to your page.

READABILITY

The elements on the page or screen should not be difficult to read. More than that, readability means that your page or screen is appealing to read. Text should be an appropriate size and style (see

"Typography," page 113), and all graphics should be easy to read. Your text should not be written so densely as to be incomprehensible. Group related ideas or concepts together on the page or screen to aid the reader in following connections. In designing a résumé, for example, you should group your name and contact information (address, telephone number) together, and group career information under major headings (Objective, Education, Work Experience, and so on).

USABILITY

The key information should be easily identifiable with headings, subheadings, and rules, if appropriate. Test the usability of your document in conditions that are as similar as possible to those your readers will encounter. For example, if they will be using the document in dim light, the print will have to be bigger, and you will need to provide even more white space than usual.

Planning the Design

Just as you must devote a great deal of your time to focusing your topic and outlining before you write, you must spend some time planning how your document or screen will look. You need to consider many factors to help you make the best design decisions—the purpose of your document, the audience, the context, your budget, your time, the kind of document. Poorly designed documents are not going to deliver your message as effectively as well-designed ones. Whether you are planning a brochure or a book, a video or a video game, a printed computer manual or online help, a technical report or assembly instructions, you will have to make many design decisions. You pay for poor design in loss of sales, increased calls to technical support, and missed opportunities for employment or research funding. A well-designed document attracts readers; a poorly designed one sends them away.

Consider a Web site for a small business, for example. A great deal of planning is necessary for even a relatively modest site. Before creating a business Web site, a company needs to consider whether it wants to sell products or services directly to the customer, distribute information about a product or service to existing and potential customers, gather sales leads, or establish a Web presence to remain competitive. The company should clearly understand which of these goals best applies to the Web project before it begins building it. With every feature added to the site—from images to video—the company should ask whether the feature will help it reach its goals.

The company will also have to plan ahead for many other variables: the cost of maintaining the site, how much the company can afford to spend on page design, whether custom illustrations or a domain name is affordable, and so on. Once these decisions and many others are made, then the company is ready to begin creating the Web site.

Of course, the design of many other kinds of documents requires considerable planning as well. Brochures, business cards, flyers, and posters, for example, may appear to be relatively easy to design, but designing them according to the best principles and pleasing the people who have asked you to design them present many challenges. In sum, to plan your design, you must plan according to design principles and according to audience or client preferences.

Typography

Typography, the visual image of your text, is an important factor in the production of your document. Typography is what your reader sees. Depending on your purpose, you may want the typography to go relatively unnoticed by your readers, or you may want the typography to be a significant design feature of your text. Some people believe that the more unobtrusive your typography is, the better. Others, such as the editors of certain magazines, want the typography to convey an attitude. Typography can be very distracting to the reader and subtract from the effectiveness of your prose, or it can enhance your document's effectiveness and subtly reinforce your message. For example, your choice of typeface can brand your business as solid and conservative or trendy and iconoclastic. How you handle typography depends on what you want the typography to accomplish. An inappropriate choice can undermine your message and even convey the opposite of what you mean to present.

When you make decisions about typographic elements in your text, you are concerned with the appearance of the letters, numbers, and symbols and also with their arrangement. Legibility concerns the presence of easily recognizable letter forms; readability refers to the easy accessibility of the message. Unless you are aiming to make your writing difficult or daunting to read, as in some legal contracts, your goal should be to obtain the highest legibility and readability possible through your typographic choices.

TYPEFACES AND FONTS

Font—or, more correctly, typeface—refers to the characteristics of the letterforms themselves. The term *font* actually means the spe-

This typeface has serifs.

FIGURE 3.2
Serif Typeface

cific combination of typeface and size that together form a unit. For convenience, typefaces are grouped into various categories according to their appearance and typical use. You will find it helpful to familiarize yourself with the general characteristics and uses of these categories so that you can make informed decisions about which ones you will use and which ones can be reasonably substituted for one another.

Typefaces developed from the handwriting styles that were in use before the invention of the printing press. Because they derive from penned letterforms, many typefaces retain some of the features of handwritten lettering. You will often see little bars on the ends of letters, particularly in the body text of books. These terminating lines, called serifs, hearken back to the time when scribes used to end their letters and words with these finishing strokes. Even texts chiseled in stone often show these serifs. (See Figure 3.2.) Times New Roman and Garamond are typefaces with serifs. Serif typefaces are generally thought to be easy to read, perhaps because the combined effect of the serifs along the base of the written line guides the reader's eyes forward.

Another major category of typefaces is the sans serif typeface. *Sans* means "without," and sans serif typefaces, of course, have no serifs. Arial and Helvetica are popular sans serif typefaces. Very often, you will see sans serif typefaces in headlines, on computer screens, and on road signs. Sans serif typefaces present a very clean, uncluttered appearance. (See Figure 3.3.) They are thought to be more difficult to read in body text than serif typefaces, but some studies suggest that in fact the easiest typeface to read is the one you have become accustomed to reading, be it serif or sans serif.

Within each category of typeface, there are various families of type. A type family comprises all the sizes of a particular typeface,

This typeface has no serifs.

FIGURE 3.3
Sans Serif Typeface

along with their associated italic, bold, and specialty forms. Specialty forms may include ligatures, that is, pairs of letters that are specially modified to fit together well, such as "fi" or "fl"; extended character sets, such as accented letters that are useful in non-English prose; and mathematical symbols.

Broadly, serif typefaces fall into categories such as Old Style, Modern, and Slab Serif, based on differences in the style of the serifs, the angle at which the serifs terminate, and the relative thinness or thickness of the letter strokes. Examples of some popular and useful serif typefaces are Times New Roman, Caslon, Palatino, Baskerville, and Garamond. The sans serif typeface families are not normally divided into groups, since they have no serifs and, for the most part, they do not feature varying stroke widths. Some well-known examples of sans serif typefaces are Helvetica, Univers, Gill Sans, and Frutiger.

In addition to the categories of serif and sans serif, you will see a great many novelty typefaces. Of great interest to would-be desktop publishers, novelty typefaces have become the toy of the home and small-business computer owner. These novelty typefaces must be used sparingly, if at all, for technical prose. They can add a touch of humor to an otherwise dry document, but they are visually distracting and can easily detract from the effectiveness of your document.

Still another category of typeface, one that has attained great popularity in recent years, is the symbols typeface. Wingdings, dingbats, musical symbols, funny faces, and hundreds of other nonalphabetic typefaces help the text designer to bring a graphical presence to a document as easily as typing words. Putting clip art and other illustrations on your page or screen can be as easy as clicking your mouse and typing a few keystrokes.

SIZE

Since normally your document must fit into certain page or screen size limits, you will naturally have to consider how large your type will be. Type height is measured in points (pt), with 72 points to an inch. So a font that is 12 pt is one-sixth of an inch in height. The height itself refers to the size of the body of metal the type would have been cast onto back before type became software. The term is still used, and it is a convenient, although approximate, guide to how big your letters are likely to be. Standard body text normally falls in the range of 10 to 12 points. Headlines can be 18 point, 24 point, 36 point, or larger. Keep in mind that, depending on its design, the let-

Serif Typefaces

Times Roman
Lorem ipsum dolor sit amet, consectetuer adipiscing elit, sed diam nonummy nibh euismod tincidunt ut laoreet dolore congue nihil imperdiet doming id quod mazim placerat facer possim assum.

Century
Lorem ipsum dolor sit amet, consectetuer adipiscing elit, sed diam nonummy nibh euismod tincidunt ut laoreet dolore congue nihil imperdiet doming id quod mazim placerat facer possim assum.

Baskerville
Lorem ipsum dolor sit amet, consectetuer adipiscing elit, sed diam nonummy nibh euismod tincidunt ut laoreet dolore congue nihil imperdiet doming id quod mazim placerat facer possim assum.

Sans Serif Typefaces

Helvetica
Lorem ipsum dolor sit amet, consectetuer adipiscing elit, sed diam nonummy nibh euismod tincidunt ut laoreet dolore congue nihil imperdiet doming id quod mazim placerat facer possim assum.

Frutiger
Lorem ipsum dolor sit amet, consectetuer adipiscing elit, sed diam nonummy nibh euismod tincidunt ut laoreet dolore congue nihil imperdiet doming id quod mazim placerat facer possim assum.

Gill Sans
Lorem ipsum dolor sit amet, consectetuer adipiscing elit, sed diam nonummy nibh euismod tincidunt ut laoreet dolore congue nihil imperdiet doming id quod mazim placerat facer possim assum.

FIGURE 3.4
Some Common Typefaces

ters for a particular point size can vary quite a bit from one typeface to the next. (See Figure 3.4 for examples of common typefaces.)

ALIGNMENT/JUSTIFICATION

When you design how your text will look on the page or screen, one of the important decisions you must make is how to align and justify your text. If your company or school follows a style guide, text alignment and justification will undoubtedly be one of the topics it covers.

Alignment describes how the text fits within its space. Text can be left aligned, center aligned (or centered), or right aligned. Justifica-

Left-Justified (or Ragged Right) Text

Lorem ipsum dolor sit amet, consectetuer adipiscing elit, sed diam nonummy nibh euismod tincidunt ut laoreet dolore congue nihil imperdiet doming id quod mazim placerat facer possim assum.

Right-Justified (or Ragged Left) Text

Lorem ipsum dolor sit amet, consectetuer adipiscing elit, sed diam nonummy nibh euismod tincidunt ut laoreet dolore congue nihil imperdiet doming id quod mazim placerat facer possim assum.

Full-Justified Text

Lorem ipsum dolor sit amet, consectetuer adipiscing elit, sed diam nonummy nibh euismod tincidunt ut laoreet dolore congue nihil imperdiet doming id quod mazim placerat facer possim assum.

Centered Text

Lorem ipsum dolor sit amet, consectetuer adipiscing elit, sed diam nonummy nibh euismod tincidunt ut laoreet dolore congue nihil imperdiet doming id quod mazim placerat facer possim assum.

FIGURE 3.5
Justification in Text

tion describes whether text is smoothed out at the margins. Edges are either justified or ragged. So text can be *left justified* (also called *ragged right*), *right justified* (also called *ragged left*), or *full justified*. Full-justified text differs from centered text in that centered text generally has both ragged right and ragged left margins, while fully justified text has its right and left sides smoothly lined up at the margins. (See Figure 3.5.)

WEIGHT/COLOR

Weight and *color* are terms that pertain to the effect of type on the page or screen. The *weight* of a typeface refers to the heaviness of its letterforms. Try to select a typeface that will do the job without weighing down the page with its darkness or failing to carry its message because of its lightness. These discrepancies of weight can sometimes be adjusted by changing the size of your font and its spacing, but selecting the right font in the first place will save you much trouble later on. *Color* is the impression the text makes on the reader with its overall density. You will want to have your text display an evenness of color by selecting appropriate typefaces, using

Century lorem ipsum dolor sit amet, consectetuer adipiscing elit, sed diam nonummy nibh euismod tincidunt ut laoreet dolore magna aliquam erat volutpat. Ut wisi enim ad minim veniam, quis nostrud exerci tation ullamcorper suscipit lobortis nisl ut aliquip ex ea commodo consequat. Duis autem vel eum iriure dolor in hendrerit in vulputate velit esse molestie consequat, vel illum dolore eu feugiat nulla facilisis at vero eros et accumsan et iusto odio dignissim qui blandit praesent luptatum zzril delenit augue duis dolore te feugait nulla facilisi.

Helvetica lorem ipsum dolor sit amet, consectetuer adipiscing elit, sed diam nonummy nibh euismod tincidunt ut laoreet dolore magna aliquam erat volutpat. Ut wisi enim ad minim veniam, quis nostrud exerci tation ullamcorper suscipit lobortis nisl ut aliquip ex ea commodo consequat. Duis autem vel eum iriure dolor in hendrerit in vulputate velit esse molestie consequat, vel illum dolore eu feugiat nulla facilisis at vero eros et accumsan et iusto odio dignissim qui blandit praesent luptatum zzril delenit augue duis dolore te feugait nulla facilisi.

Frutiger lorem ipsum dolor sit amet, consectetuer adipiscing elit, sed diam nonummy nibh euismod tincidunt ut laoreet dolore magna aliquam erat volutpat. Ut wisi enim ad minim veniam, quis nostrud exerci tation ullamcorper suscipit lobortis nisl ut aliquip ex ea commodo consequat. Duis autem vel eum iriure dolor in hendrerit in vulputate velit esse molestie consequat, vel illum dolore eu feugiat nulla facilisis at vero eros et accumsan et iusto odio dignissim qui blandit praesent luptatum zzril delenit augue duis dolore te feugait nulla facilisi.

Baskerville lorem ipsum dolor sit amet, consectetuer adipiscing elit, sed diam nonummy nibh euismod tincidunt ut laoreet dolore magna aliquam erat volutpat. Ut wisi enim ad minim veniam, quis nostrud exerci tation ullamcorper suscipit lobortis nisl ut aliquip ex ea commodo consequat. Duis autem vel eum iriure dolor in hendrerit in vulputate velit esse molestie consequat, vel illum dolore eu feugiat nulla facilisis at vero eros et accumsan et iusto odio dignissim qui blandit praesent luptatum zzril delenit augue duis dolore te feugait nulla facilisi.

FIGURE 3.6
Text with Different Weights

capital letters sparingly, and controlling the spaces between letters, words, and lines. (See Figure 3.6.)

KERNING/TRACKING/LEADING/WORD SPACING

Most word-processing programs and page-layout programs give you great flexibility in how you use the typeface you have selected. Although you will probably not need to concern yourself with some of the more arcane adjustments that are available to you, sometimes quite minor tweaking of your typeface can dramatically increase its usefulness.

Kerning refers to the adjustment of the spaces between adjacent pairs of letters. All well-designed typefaces incorporate some kerning to alter the default distance between letters. Certain letters, when they appear next to each other, create an awkward appearance because of the amount of white space between them. Such combinations as "To" and "VA" need to be shifted closer together. Others—like italic *f*—benefit from being given a little extra room. Even though you will normally not have to kern your type, occasionally, because of design or space considerations or because you are using an unusually large or small font, you may have to make minor adjustments to the spacing between specific letter pairs.

Tracking refers to the looseness or tightness of your font. Unlike kerning, which deals with specific letter pairs, tracking affects the entire alphabet. If you track your text tightly, you squeeze your letters closer together. Tight tracking is often used for headlines and other display types that spread out too much in large sizes. If you track your text loosely, you spread your letters out. Loose tracking is useful when you use small fonts and when you are displaying text on a low-resolution medium such as a computer screen. By spreading small letters out, you make them easier to read.

Leading (pronounced "ledding") gives you another way to control the white space in your document by adjusting the distance between lines. The vertical distance between the baseline of one line of text and the baseline of the following line is referred to as the leading. Leading is measured in points, just as font sizes are. The leading measure consists of two numbers: the size in points of the typeface and the distance in points between baselines. So a typical leading for text set in 12 point type might be 12/14 (pronounced "twelve over fourteen.") Leading adjustments, like tracking adjustments, are useful when you have very large or very small fonts. Large fonts, such as those found in headlines, are frequently set with closely spaced lines or tight leading. A typical headline might be 24/24. Naturally, if you use such a tight leading, you will need to make sure the headline or other text is easy to read without too much effort.

Word spacing, just as the name implies, concerns the distance between words. You can set your word spacing as a percentage of the default spacing used by your word processor or layout program. If you want your text to be pushed closer together, as you might do if you were short on room, you could specify a 90 percent or even an 80 percent word spacing. If, on the other hand, you want your text to be more spread out, as you would if you were creating justified left and right margins, you would need to add space between words. Your word processor or page layout program will automatically add word spaces when you specify full (left and right) justification of margins.

FONT CREATION AND MANAGEMENT

With the ready availability of software tools for typeface creation, new fonts are being created every day, not only by professional typographers but increasingly by font hobbyists. Special purpose typefaces, created for specific purposes or even just for the novelty, have sprung up and can easily be purchased, borrowed, or shared from a variety of sources. Usenet newsgroups devoted to fonts have an enthusiastic following, many Web sites offer free fonts, and type foundries are doing a booming business in digitized font sales.

If you decide you want to create a typeface, either because you can't find just what you need on your computer or in a catalog or because you have the urge to create a typeface that is uniquely your own, you can use various techniques to input letterforms into your computer. The most popular method is to use a scanner to copy an existing typeface. You can also use a font you already have as the basis for your new font. Another method is to use your mouse or a stylus to draw the letterforms by hand.

Once you have entered the letterforms into your computer, you can modify them or replace them easily using a font creation program, such as Fontographer <http://www.macromedia.com/software/fontographer/>. Once you are satisfied with the form of your letters, you will want to work on creating kerning pairs for specific letter combinations.

Since at various sizes, letters need to have their line thickness (called *stem weight*) altered for greater readability, typographic programs enable you to create multiple instances of each letter, depending on the letter sizes you are creating.

Once your letterforms are finished, you must instruct your software to create fonts in True Type or PostScript form.

For most people, the challenge is not creating new fonts, but managing the fonts they already have. Fonts are so easy to acquire—many word processing and design programs install their own on your computer whether you want them or not—that the typical student or employee soon has more fonts than can be managed easily through regular system software. Many people have found it more useful to rely on font management software. These font managers help you to organize your fonts by enabling you to inventory, catalog, and view your installed fonts. To free up space on your computer, you can uninstall and reinstall the fonts you need for particular projects while ensuring that your basic work fonts are easily accessible.

TIPS: Typography

- Select typefaces that are easy to read in preference to novelty typefaces that will distract from your content.
- Stay with one type style for body text and one coordinating style for display or headline text. Generally, a serif typeface for body text and a sans serif typeface for headers work well together. Avoid mixing many typefaces on the same page, because the conflict of type designs causes visual clutter, distracts from your message, and can make your text harder to read as users shift mental gears to decode the text.
- Use left justification or full justification in your document. Centered text and right-justified text are more difficult to read than left-justified text. Full-justified text is usually as easy to read as left-justified text, according to studies conducted with experienced readers (see Colin Wheildon's *Type & Layout*, for example). Left justification is the recommended alignment for readers who ordinarily have difficulty reading.
- Line widths of approximately two and a half alphabets worth of letters and spaces generally make for easily readable text.
- Avoid using all capital letters. Fully capitalized headlines are more difficult to read than headlines in which only the first letter of each word is capitalized.
- Avoid extensive use of italic or boldface type. After a few words or at most a sentence or two, italic and bold text are fatiguing to read.

Print Design

Despite the prevalence of online documents, from help screens to Web sites, most of us still spend most of our time reading and responding to printed documents of one kind or another. Historically, print has been with us long enough to have attracted a substantial body of study and attention. When you are faced with designing a print document, you may visualize the page as an artist's canvas, ready to accept whatever design your imagination conjures up. While there is great leeway in the choices you make to create your document, a thorough familiarity with established patterns and conventions will make your reports, correspondence, and other documents as professional looking as possible. Other sections of this handbook discuss specific design features of letters and memos, and résumés, for example. (See Part Eight, "Correspondence and the Job Search.")

You can make several major decisions to improve the design of your print documents immediately. First, break your topic into chunks and use headings frequently. Long paragraphs create too much work for readers. Second, provide ample white space on every page. White space makes each page less intimidating for readers and gives them a resting spot as they work through your text. Third, use headings in a size that is appropriate for the body text. Don't use headings that are much too large for the body text. Fourth, avoid excess punctuation wherever possible. You don't need to use periods or colons after headings. The fewer punctuation marks you use, the better.

Online Design

Computer-based training and online help are two typical online areas for which you may be developing screen designs.

COMPUTER-BASED TRAINING (CBT)

Computer-based training (CBT) modules are usually online tutorials that teach people how to perform a task or use an application in a series of steps. In addition to tutorials, CBT uses other techniques such as demonstration (for example, to introduce new information), drill and practice (to help users master a new skill or information after an initial introduction), training games (to help motivate users to learn a new skill or information in an entertaining fashion), simulation (to help users practice a skill that would otherwise be too expensive or dangerous), and problem solving (to help users develop skills in logic, solving problems, and following directions). For example, with CBT, companies can offer employees instant feedback on what and how they are learning. CBT also helps companies in placement by matching a user with the type of training the user needs. And CBT, with its abilities to use text, graphics, video, and sound, can help students or employees to retain more of what they learn.

ONLINE HELP

Online help provides answers to users' questions about features, procedures, shortcuts, troubleshooting, and so on for a particular program or application. Other kinds of online documents include kiosks

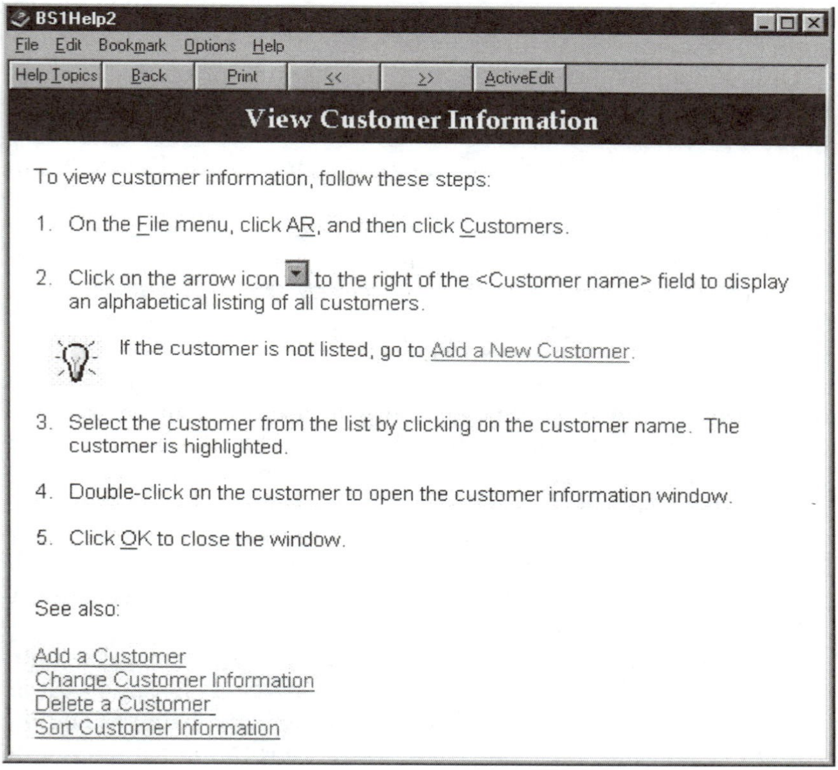

FIGURE 3.7
Online Help Screen

that provide reference information or demonstrate the features of a product, CD-ROMs of various sorts, wizards that help users perform tasks related to a software program, and Web pages. (See "Web Page Design," page 126, for an overview of the elements of Web pages and tips for creating Web pages.)

As an example of online design, consider online help. (See Figure 3.7.) With online help becoming more and more the norm, you will probably find yourself writing documentation that either supplements printed documentation or, more likely, replaces it. While many of the same principles apply to each form, creating online information presents certain challenges that don't apply in the print realm.

The first principle to remember is that online help is not printed help that happens to appear on a computer screen. Both conceptually and visually, online help is designed to be very different from print documentation. The same information may very well appear in both, but the approach will likely be quite different.

Online help consists of screens of information presented to users in one or more windows. It is normally available along with the software it is documenting, so users can switch back and forth from working with their application to getting information about using the software. Online help can be invoked by pressing a key (for example, F1 on the keyboard), clicking on a query button, passing the mouse cursor over an area, clicking on an entry in a help table of contents, typing in a question, or selecting from a menu of topics. Online help provides user-directed navigation in the form of buttons, hyperlinks, or typed text. The help topic itself can be text, illustrations, animations, sounds, videos, or any combination of these. And online help can be created in any of a number of formats, depending on the platform, the help software, and the location of the information (database, intranet, extranet, Internet).

Regardless of these variables, all online help, to be effective, has to be readily accessible to its users. If users can't easily find the information they need, they will not use the online help, and if they do not use the online help, they will either avoid your product or inundate your telephone support lines with calls for assistance.

Before you begin creating online help, thoroughly familiarize yourself with the product you are documenting. If you do not know all its features, you cannot present those features effectively in your online help. Also, don't assume that your users know as much as you do about the product or even about using a computer. Write online help that novice users can understand, but provide shortcuts that experienced users can follow to avoid the more elementary explanations.

TIPS: Designing Online Help

- Break up information into packets or small digestible pieces. Keep each screen simple. Address one topic per screen, with perhaps links to other related topics provided in a standard format to which users can go if they need further information.

- Keep your sentences short and simple. The tendency in online writing is for the style to be even more informal and personal than that for printed matter.

- Aid your users in finding out where they are, how they can go back to a more basic topic, where they can go for more detailed information, and how to leave the help window. Provide buttons, tabs, arrows, or text boxes as navigational tools. Present context labeling such as title bars to provide cues to the users as they navigate through successive help screens so that they always know where they are. Failure to manage navigation is the biggest obstacle to successful online help. If your software permits it, provide a bookmarking feature so that users can return to a particular topic later without having to search for it again.

- Always provide some way for the users to backtrack and review previous material.

- Use icons, buttons, tabs, labels, and other design elements consistently throughout your online help. Once users learn the conventions of your online help presentation, they should be able to find their way around each screen automatically and easily.

- Add troubleshooting sections to your online help where appropriate. Try to anticipate where users may run into trouble with your application, and design the troubleshooting section to address problems they may encounter.

- Provide a Frequently Asked Questions (FAQ) section. This can be a big help, since many users will readily turn to this section. Try to include all questions users might have about using various features and windows within the program. Don't provide the answer to the questions in this section. Instead, merely pose the question and link it to another window that will answer or assist the user.

- Provide a table of contents to the online help. Organize your topics into a hierarchy, and preserve that hierarchy when you create your help table of contents. Users can follow links in the table of contents to acquire an overview of the topics that interest them.

- Create an index from entry points into the help system. Remember, these are the terms users will enter when they want to find information, so make the index as comprehensive as possible. Use like terminology or topics for similar fields, and use multiple index topics where necessary. Anticipate how users will ask for help. Remember, new users may not know the exact name or topic they are looking for, so make the index as accommodating as possible. If necessary, provide different approaches into the same material to account for the various ways in which users will discover a need for the information.

- Think about any additional topics novice users may need, such as how to use your help system or how to navigate in help windows. Often, a glossary is helpful.

- Keep in mind that your users may have a variety of computers running your online help. Your help system needs to support the lowest system or video level, which can limit the type of help system you develop or the number of colors available for your illustrations and help windows.

(continued)

Designing Online Help *(continued)*

In addition, if you create help systems for a variety of platforms, for example, Windows 3.1, Windows 95, Windows 98, Windows NT, OS/2, and Macintosh OS, you have to make sure you are not using features that are unsupported in a particular target operating system.

■ Choose fonts, colors, and illustrations carefully. Test your choices on a variety of monitors. Make sure that your typeface and font size can be easily read with the low resolution of a computer screen. Be careful that your colors work well together and that together they don't cause eyestrain for the viewer. Select few colors rather than many. Use special effects (blinking, animation) sparingly if at all. They can distract users from the information you are trying to convey. If your online help will be transmitted over an intranet, extranet, or the Internet, choose illustrations that download quickly.

■ Make sure that the help window is displayed on top of other windows. This feature ensures that clicking outside the help window does not cause the help window to disappear. Also, make it easy for users to close or hide the help window when they are finished with it.

■ Consider creating your online help from screen shots of the application. Users can click on elements in the workspace for context-sensitive help.

TOOLS FOR ONLINE HELP

Many tools are available for developing online help. Popular online help programs that work with Windows help include RoboHELP, ForeHelp, Doc-to-Help, Help Breeze, Help Scribble, Help Magician, and NetHelp. RoboHELP is one of the most widely used and most respected programs, but increasingly, help tools are moving toward creating HTML-based help. RoboHTML is just one of many products available for creating HTML-based help. HTML (Hypertext Markup Language), the language that is used to create Web pages, greatly increases the possible interactivity of the online help. With HTML and its associated features, online help can incorporate forms, questionnaires, Shockwave modules, and VRML (Virtual Reality Modeling Language). Additionally, HTML-based help can handle multimedia and Web links quite naturally. HTML works across platforms, so you don't have to create separate online help for each operating system. And when software is revised, HTML-based help is easy to update.

Web Page Design

Web pages are, of course, online documents, but designing and illustrating Web pages present challenges beyond those you will en-

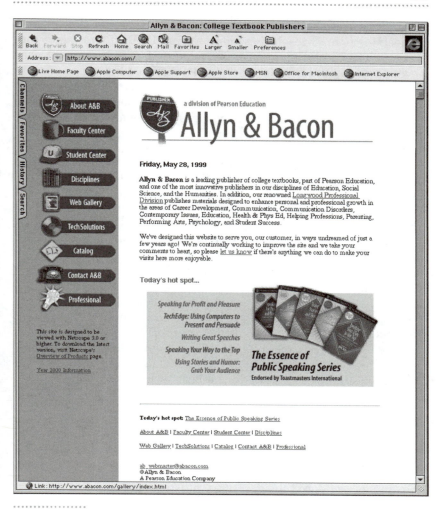

FIGURE 3.8
A Web Page

counter in creating online help screens or online tutorials, for example. (See Figure 3.8.) For one thing, you are designing your site for an unknown audience that is bound to be more varied than the audience for your print and online help work. After all, the World Wide Web is, as its title suggests, accessible to anyone with a Web browser. So you will have to accommodate your design to a possibly multicultural audience. In addition, your readers have a range of hardware capabilities, ranging from low-resolution monitors and slow modem connections to high-resolution monitors and ISDN, T1, or faster connections. The resolution and connection speeds will limit what your viewers will see and be able to access in a reasonable time, so you will want to constrain your graphics sizes, video files, color depth, and so on. Another limitation is browser compatibility. Some site designers make an effort to ensure that their pages can be viewed equally well on all browsers, but most do not, leading to a variety of unforeseen effects for users. Finally, many effects depend on the reader's having particular plug-ins, or programs that work along with browsers to read and display certain types of files. Without these plug-ins, the embedded files cannot be displayed. These considerations differentiate Web pages from other online design models. Effective Web page design is a complex topic, and these concerns are only highlighted here. Some of the best resources you'll find on Web design and content are available on the Web itself. In fact, there are countless Web sites offering all kinds of information, both basic and advanced, for creating effective Web sites. A few of the better print sources on the topic are listed in the Resources list at the end of this section.

Whether you are producing printed or online information, there are three aspects to consider: design and technical issues, content, and prose style. Web pages offer special challenges in all three areas. These three areas are briefly discussed below, along with detailed tips that will help you to create the best Web site for your subject, purpose, and audience.

DESIGN AND TECHNICAL ISSUES

Some typical design issues for Web pages are basic. These include, for example, audience considerations, illustrations, simplicity, chunking, styles, and consistency. People accessing Web sites typically are in a hurry or impatient, so time is an important issue in Web site design. Users quickly skip from link to link and from page to page looking for what they need or are interested in and skimming through or skipping over what they don't care about. Your design decisions must keep in mind this typical user behavior. Yet, even though most users fit this description, they represent a wide variety of backgrounds, education, needs, and familiarity with your subject or topic. That is why you have to provide simple introductory material that is suitable for a novice, taking into account a range of computer knowledge and equipment while still providing the more experienced user a way to access your information without passing through the hand-holding sections.

Illustrations must be used with care because of many technical considerations but also because your graphics convey so much in so little space. Graphics are easily placed in Web pages, so it's tempting to provide too many of them or to provide graphics that are too large. Here again, time becomes an important consideration. Slow-loading graphics are probably the greatest deterrent to staying with a site. Different readers have different access times to your site. Some have a slow connection and others have a fast connection, so what seems to you to be a quick-to-load graphic may in fact be an impediment to your readers' use of your site.

Determining the best position for graphics on the screen presents other challenges. What your readers can see depends on the settings on their computer, so you have to plan accordingly. For example, graphics may appear in different locations, depending on the user's screen resolution and size, creating unforeseen effects. And large de-signs may not be easily viewable on smaller monitors or monitors that are set at lower resolutions. Your readers will have all sizes of monitors, ranging from a small laptop screen to a 21-inch or larger screen. You can probably count on your readers having monitors that support resolutions of at least 640 by 480 pixels but perhaps no greater, so develop your Web page with that in mind.

Navigation presents another design problem. Your readers need to know where they are in your Web site and how to get from one part of the site to another quickly and easily. Make it easy for your readers to move between sections of your site. Try to have a navigation bar or other navigation aid displayed on each screen in the same place for consistency.

Simplicity in Web site design is essential. Screens are small and easily take on a cluttered look. *Chunking,* the grouping of information into discrete parts and clearly labeling each part, helps readers to see the structure in your information.

There must be a great deal of consistency in the overall design of the site. The design elements should be consistent from node to node (any basic element on a Web site) and from page to page. Use styles to achieve consistency in your Web page design. Styles are recurring characteristics that can provide your site with a pleasing rhythm as users see repeating features connecting similar elements. Typical style features are bold, italics, underlining, paragraph alignment, color, and typefaces. Styles can help or hinder your site design, de-pending on their use. Bold, for example, can often replace headings, but using bold too often or mixing too many different styles can be distracting and confusing.

Other Web design issues are more advanced, concerning multimedia, scrolling, linking, and color. Multimedia (video and audio clips) can help a Web site by capturing viewers' attention, but multimedia can easily be overused. Not all browsers support playing audio and video, and much audio and video must still be downloaded before readers can access them. To overcome this restriction, many browsers now support plug-ins that can play audio and video as they are being downloaded (a process known as *streaming*), thus eliminating the long wait that is typical for multimedia files.

You should also limit the amount of scrolling a reader must do. Many readers will scroll through several windows of information, but most are not willing to scroll through more than that. Linking is necessary when you can't cover a topic in just a few pages of information. You need to be consistent in your linking so that your readers will see why you are linking. The danger in excessive linking is that your readers will lose track of where they have come from and will not want to return to their starting point to follow other topics.

Color is a major design concern for Web pages. Color on Web pages varies according to settings on a particular computer, the quality of the monitor, and the browser. Because color is very easy to include on Web pages and difficult and expensive to achieve on paper, online designers are excited about the new possibilities that are open to them. But unlike color on paper, color on screen is not well standardized, so take care to use colors that will show well on a variety of systems and monitors. Use the browser-safe palette of 216 colors described in many Web design books to ensure that your readers see the same colors that you want them to see on your Web pages.

TIPS: Designing a Web Page

- *Learn from others.* Become familiar with the Web by browsing a variety of sites to see which techniques contribute to a good site and which ones hinder the usability of a site.
- *Planning.* Begin by planning the organizational framework of your site. The site will be most consistent and easy to navigate if it is well structured from the beginning. A physical drawing that shows which pages

will link to one another and roughly how each page will be laid out can help you start to envision the site. Storyboarding software (described in Part One, "Planning and Research") may help you with this step.

- *Ways of viewing.* Consider all the ways in which the user may want to view your page. This process will help you to determine which links to provide. Links should go between text and text, between illustrations and illustrations, between text and illustrations, and between page and page.

- *Links.* Make sure your links are indicated by underlines and perhaps color coded. If you use color alone and no underlines, users who are color-blind or have monochrome monitors will have difficulty finding the links. Choose different colors for visited and unvisited links, so that your readers can identify where they have already been. Choose colors that will not fade into the background and render the link labels unintelligible.

- *Logic.* Break pages up into logical categories or associated topics. For proper navigation, it is helpful to list links to specific topics at the top of the page, thus preventing users from having to search through an entire page.

- *Scaling.* Scale text to all page elements such as heading, illustrations, and paragraphs. Choose standard styles and sizes, avoiding stylized or cursive typefaces and condensed or stretched typefaces. Standard font sizes are considered 10–12 points. Avoid using too many styles on one page. Be consistent.

- *Legibility.* Make paragraphs legible. Use high contrast between text and backgrounds. Do not use all caps, all bold, or all italicized text. Keep lines short, and left justify text instead of fully justifying it.

- *Hierarchies.* Create visual hierarchies. Help the reader to discern the relationships between topic headings and lower-level material. Use lists, indentations, and paragraph spacing to establish how your information fits together. Use contrasts of weight, color, shape, and size to emphasize important text and other elements.

- *Color.* Add color. Adding color makes a more inviting document but it should highlight the content, not detract from it. Too much color can be distracting, especially in backgrounds. Contrast between text, pictures, and backgrounds is also important, which is why more people are choosing white or very pale pastels as their background color. Limit your backgrounds and illustrations to the 216-color palette that most Web browsers support. If you use more colors, users will experience unpleasant color effects and graininess of the display.

- *Typography.* Use typographic effects with caution. Use mixed upper and lower case for headings (don't use all capitals). If you have white letters on a dark background, and if someone wants to print out this page, the

(continued)

Designing a Web Page *(continued)*

white letters will not likely be viewable when printed out on white paper. Background colors or illustrations are not printed out on the printer. Choose a type and size, and be consistent with their use. As constricting as it may sound, be modest. And realize that whatever type-face you specify must already be installed on your users' computers. If you select esoteric fonts and those fonts are not present, the users' browsers will display a default font, making your page appear very differ-ent from what you probably intended.

■ *Visual contrast.* Visual contrast is a basic requirement for attractively de-signed Web pages: break up text walls with headings, lists, blank space, lines, illustration dingbats, inline pictures, bold text, and italic text. Tables are a good way to break up the screen into areas and control the place-ment of text and graphics. As with any design element, try not to have too much novelty, which can be fatiguing, nor too much sameness, which can be boring. Seek a balance between contrasting elements and re-peated elements.

■ *Navigation.* Don't use "Click here" or "click me" to cue a navigational link. Always embed the link in the text for easier reading and printouts that make sense. There should be a way back to the main page. Don't trap readers in dead ends. And while most browsers have a "back" but-ton to retrace steps, readers who want to skip directly back to the main page should be able to do it in one jump. For complex sites, consider putting "home," "top of page," or images that consistently link to main pages of the site to help users move easily from page to page. Be careful with the number of links you use. Too many links will confuse users who are trying to navigate through your site. Also, don't include outside links on your page or site so soon that people surf away before you have de-livered your message.

■ *Page length.* Long pages are appropriate for documents that users are likely to print out and keep for future reference. However, short pages are the most effective way to focus on a subject.

■ *Unity.* Use an illustration or navigation bar at the top or along the side of each page. A small illustration can also be a good way to identify each of your pages. Try to maintain a consistent look and feel for a set of re-lated pages.

■ *Background consistency.* Be consistent with your background choices. Keep in mind that some monitors and browsers may not support some backgrounds. If you specify an image for your background, make sure you have also specified a solid color as an alternative for those browsers that don't display background images. Check to make sure that your text color is viewable equally well against your background image and your alternative background color.

CONTENT ISSUES

Typical content issues concern content quality, document identification, complete headings, redundancy, and brevity. Too many sites offer lots of "bells and whistles" but have little or no valuable content. Whether you're creating a personal home page or a sophisticated commercial site for a company or organization, you need to provide content that is valuable for your intended audience, and you need to update your site regularly to make sure the content provides something new and interesting to your audience. Also, too many sites lack proper identifying information. Your readers need to know what they are reading, so identify your document clearly by providing an appropriate title and introduction to the site. And you need to provide the date (usually the date of the last time the site was revised), a contact name, an e-mail address, copyright information (if applicable), and other essential information. Your headings need to be complete, parallel, and as descriptive as possible. Redundancy can easily be avoided on a Web site by linking to other sites that provide the same information. But remember that once users leave your site, they may forget to come back. The size of the Web site, of course, depends a great deal on the subject, purpose, scope, and intended audience of the site, but you should aim at keeping every node as brief as possible.

TIPS: Contents of a Web Page

- *Title.* The title will appear in the top line of the Web browser screen. This is the title that will, most likely, show up in search engines. Make sure the title is descriptive, using keywords only.

- *Purpose.* Clearly identify the purpose of the site on the main page. If it's your personal home page, say so. If it's the opening screen for your company's site, say that.

- *Illustrations.* Provide visual appeal. Use an appropriate number of illustrations and an appropriate size and resolution for your illustrations. (See "Illustrating for Online and the Web," page 142.)

- *Main illustration.* Use a main illustration on the main or home page. This can be a company logo or other graphic that identifies your page in some characteristic way.

- *Main menu.* Provide a menu on the main page showing the key links of the site. This is the screen from which readers will choose where they want to go, and they will probably come back to this screen many times. Make this screen attractive, easy to use, and quick to load.

(continued)

Contents of a Web Page *(continued)*

■ *Authorship or contact information.* Provide the Web site author's name or the name of a contact person. It is just general courtesy to list the author, even yourself, so that people can get in contact with you. Provide an e-mail address. Allowing people to contact you keeps you up-to-date on what they want to know. And if there is a problem with your site, such as a broken or outdated link, including contact information enables your readers to let you know about the problems.

■ *Dates.* Provide a date of page creation or last revision. List the last update of the page so that users know how current your information is. Keep your pages current. Check your links monthly. Change and update your information regularly. A page that never changes doesn't encourage revisits.

■ *Copyright.* Provide a copyright statement, typically at the bottom of the main page and often at the bottom of every page in the Web document. You or your company owns the copyright for the original aspects of your Web page. You cannot claim the copyright for material that you use but that belongs to others.

■ *Address.* Consider placing the Internet address (URL) at the top or the bottom of the page so that printed pages will have an address on them. Although some browsers will do this automatically, others will not.

■ *Index.* Consider providing an index or site map for large sites. A site map will help users to find the information they are looking for quickly, without having to navigate through several levels of links.

■ *Meta tags.* Take advantage of meta tags. Meta tags are keyword tags you add to your Web pages that allow search engines to publicize your site and attract users.

■ *Self-contained pages.* Don't assume that readers have seen previous pages when you plan the content of a page. Make sure that each page can function by itself. If a search engine picks up information from several pages on your site, users may enter the site from any page. Once you have attracted users, you want them to be able to find out more about the rest of your site. Don't turn them away by not allowing them access to other pages. Give them navigation options for each page.

PROSE STYLE ISSUES

Typical prose style issues concern the audience, scannable text, and link wording. Your readers will have all kinds of backgrounds and computer skill levels, so you'll have to plan the prose style of your site accordingly. Adopt a prose style and tone that are easily accessible to the least sophisticated of your readers without offending or boring your more experienced readers. You'll also have to present text in highlighted keywords, headings and subheadings, bulleted lists, short paragraphs, and so on in such a way that your readers can easily scan the text. The wording of each and every link is also an important consideration. Make the link names short and descriptive; don't make the reader guess. The reader should know what to expect before clicking on each link.

TIPS: Prose Style of a Web Page

- *Focus.* On the Web, it is possible to communicate with people from all over the world, but it is easier to conceptualize your site if you narrow your audience down a bit. Consider what you have to offer users. Give your site a clear purpose. Then ask yourself which users will be interested in what you have to offer. The usual audience considerations for print apply on the Web, but in addition you must consider the computer and Web surfing experience of your potential audience and make style decisions accordingly.

- *Inverted pyramid.* Users read quickly on the Web. They scan pages, looking for information. To keep their attention, adopt the inverted pyramid style of journalism. Start by giving readers the most important information first. Begin with the conclusion. Then provide supporting and background information. Readers who are really interested in the information will continue reading. But if you delay the important details until the end, users may become frustrated and confused about what the page's purpose is.

- *Meaningful headings.* Provide headings and subheadings that allow readers to find information quickly and easily. Avoid using clever wording in your headings. Tell readers exactly what they should expect to find under each heading.

- *Keywords.* Choose important words that will need special emphasis, and highlight them with bold or colored text to help readers scan the text. Hypertext is inherently highlighted. Because it is underlined and colored, the text draws readers' attention. Make sure highlighted text is distinguishable from linked text.

- *Chunk information.* Break information into small, digestible chunks. If possible, limit each paragraph to one subject. Because users scan so quickly, they may miss a second point that is presented late in the paragraph.

- *Bulleted lists.* Bulleted lists are easier to read on the screen than long paragraphs are. Lists help to organize information in a logical format and are a good format for presenting links. Lists are most effective on the screen when each list item is short.

(continued)

Prose Style of a Web Page *(continued)*

- *Informal prose.* The Web is an informal medium. A personal tone is inviting to users who are sitting in front of impersonal computer terminals. Don't be afraid to experiment with a conversational tone and use modern language. Use simple sentences that are easy for readers to follow. Web readers are notorious for their sense of humor. Use humor to communicate with readers on a personal level and make your site more attractive.

- *Correct mechanics and grammar.* Check for errors in mechanics and grammar before you upload your pages. Even though the Web is an informal medium, readers look for credibility in sites. A site that looks well edited and well thought out will appear more credible than one that is sloppily thrown together without the author's paying attention to the basics of clear communication.

- *Objective tone.* Web writers can improve their credibility by offering their readers good writing with an objective tone. Because readers are often unsure of who has written the information at a site and how reliable their facts are, they search for sites that seem objective. Exaggerated language may seem dishonest and turn potential readers off.

- *Jargon.* Remember that your audience will have varying knowledge of the Web and computer usage. To accommodate the largest audience possible, avoid jargon and other language that could exclude potential users.

- *Links.* Don't use "Click here" or "click me" to cue a navigational link. Always embed the link in the text for easier reading and so that printouts will make sense. Write links within a context that tells readers what information they will find if they follow the link. Don't highlight whole sentences with link text. It is distracting. Choose key words that indicate why this link is different from others.

- *International audience.* If you plan to take advantage of the Web to communicate with users from other countries, you must tailor your document to meet their needs. Avoid jargon, colloquialisms, and acronyms that are unexplained. Write simple sentences. Consider including a glossary, metric measurements, and dates that are formatted in the conventional style of users from your intended audience.

TIPS: What to Avoid When Designing a Web Page

- *Blinking elements.* Be careful not to put too many distracting objects on your Web pages. The classic example is the <BLINK> ... </BLINK> HTML tag, which causes the text between the tags to blink on and off.

Avoid providing anything that blinks. Because the blinking can be so distracting to the user, the <BLINK> tag has been called the most reviled tag in HTML.

- *Animation.* Use animation sparingly. The human eye is attracted to motion, so animated GIF illustrations can be effective attention getters. For the same reason, animations tend to make it difficult to concentrate on the other content of the site. And since animated GIF files contain more image information, they take longer to download than standard images.

- *Frames.* Avoid using frames unless you know what you are doing. They are another potentially deadly element in the hands of the untrained because they make readers scroll more and they break up the viewing area. Although frames can occasionally be used effectively, most often they are not. Often, tables are a better way to divide your screen into different content areas.

- *Image maps.* An image map is an easy way to let a user choose an element of an image to click on to follow a related link. However, this good idea can easily be turned into a bad one when done incorrectly. Without accompanying text as an alternative choice, viewers will have no idea what the page is or where they are linking to if the image fails to load properly or the viewer is not using a graphical browser. Even in the best case, viewers will have to wait for enough of the image to load to get an idea of what the links would be. Alternative text links should always be provided, since the potential for problems is great and, in many cases, people won't wait for large images to load. In addition, text links can show, through color change, which links have already been visited; image maps cannot.

- *Unnecessary words.* Avoid stating the obvious. Many people use the words "click here" on their Web pages. A link exists to be clicked, and to say so in the link text isn't necessary.

- *Careless links.* It's a shame to waste the power of links by creating them randomly. Consider their cumulative effect on your online readers and avoid these pitfalls: links that point to a page that sends the reader to yet another page, links that circulate through parts of the very same page, links that do nothing more than make a smart remark and send the reader back, and links that go somewhere irrelevant to the topic.

- *Excessive use of links.* When links appear in almost every line of a document, viewers are less likely to follow any link at all. By eliminating links that are irrelevant, extraneous, or redundant, you do a great favor for impatient users. Reference lists and search engines exist for anyone who wants to find every site that is related to a given topic.

- *Gratuitous use of external links.* You can fill a document with links that go to Web sites all around the world, but keep in mind that a person who leaves your site will probably never return. The more you send them

(continued)

What to Avoid When Designing a Web Page *(continued)*

away, the greater the impression that your own site has little to offer. Provide links to good external sites as appropriate, especially when they add useful information that your site can't provide. But don't send people away until you are finished with them.

■ *A dearth of link options.* If a Web page has only one link, it may as well be printed on paper. Never try to force a single path on the reader. Give your reader choices.

■ *Unnecessary scrolling.* Try not to require too large an area to display your page content in either the horizontal or vertical direction. If you do, a user with a small monitor may have to do a lot of scrolling to see everything. Avoid making readers scroll sideways to read each line of text. Pay attention to what is above the vertical scroll line as well. Readers' initial impressions of each page come from the view they get when the page first loads in the window.

ILLUSTRATION

You may have heard people saying that they think visually or that to understand or remember something, they need to see it, not just hear it. This very common preference reflects the impact that pictorial information has on us. We are accustomed to "seeing the big picture," having a picture drawn for us, picturing something we are asked to imagine. And isn't a picture worth a thousand words? All of these ideas follow from the fact that pictorial representations are rich in information. So much can be said by even a simple diagram that it would be foolish to ignore this very powerful tool. But like any tool, it can be used well or poorly, effectively or ineffectively, efficiently or wastefully. Your goal should be to use illustrations to serve your purpose; in other words, make the illustrations work for you. And to do that, you have to be aware of some basic principles about how illustrations and people interact.

Your illustrations should have a specific purpose. They are used primarily to illustrate concepts, but they can also be used to help readers navigate through the text or clarify, simplify, emphasize, summarize, reinforce, attract, impress, show a relationship, and even save space. Illustrations can be used to show what something looks like or how it works, clarify information, distinguish between two or more objects, and get quicker and more accurate responses from readers. Illustrations and text are not mutually exclusive. Ideally,

ILLUSTRATION   **139**

they will be used to enhance each other's effectiveness. Illustrations and text should be well balanced and complement each other. Highly professional, formal text with cartoons may look out of place and pull the reader out of the text, just as comic books with flowcharts and photographs would seem inappropriate.

Illustrations are just one of the major elements of effective document design, but handling them appropriately and effectively is so important that the topic requires discussion separately from design. Illustrations aid job performance, help to make documents internationally accessible, assist nonreaders, bring in wary readers, add credibility to work, and promote creative thinking and efficient reading.

Sometimes space in a report or other technical document is limited. Use illustrations in place of words for these situations: if the concepts in the text use numbers, symbols, or measures; if the ideas being presented are structural or pictorial; if the readers are visually perceptive rather than word perceptive; if the subject is too complex to explain in words; or if your illustrative skills are stronger than your writing skills.

Illustrations serve three roles when combined with prose in a technical document: They get the reader's attention and create interest, they explain information, and they help the reader to retain the text material. Illustrations attract attention and generate interest by keeping the reader interested, breaking up long passages, emphasizing specific data, making the reading enjoyable, and providing quick references. They explain information by simplifying complex subjects, helping the reader to absorb many facts and figures, condensing detailed information, showing relationships between things, getting across concepts that words cannot convey, and comparing and contrasting diverse subjects. They help to retain information by taking advantage of the fact that illustrations can be remembered more easily than text and the reader may be able to retrieve words from memory that are associated with the pictures.

Tables and figures are the two major kinds of illustrations. Tables present data in rows and columns and may be informal or formal. Figures include everything from bar graphs to drawings to photographs.

Illustrating for Print

People who must communicate technical information are challenged to make the information they portray as clear and concise as possible. Many times, to illustrate a point or make the text clearer, an illustration is inserted in a document. An illustration is a visual aid that helps the reader to access the information faster.

You should follow some basic guidelines when planning to use illustrations in a document. Consider first how people take in information and how best to display the information you have. Motivate your readers by getting their attention and relating the material to them.

TIPS: Displaying Illustrations

- Illustrations must be appropriately and effectively displayed if you want them to have the intended effect on your audience. All of us have seen textbooks, reports, manuals, and many other kinds of documents containing illustrations that seem out of place or that are placed far from where the discussion (if there is any discussion) of the illustrations occurs. It's not unusual to see a report, for example, in which all of the illustrations are placed at the end of the report or perhaps in an appendix.

- Refer to illustrations before they are displayed. You may refer to an illustration or a table in several different ways. For example, "As Figure 1.1 illustrates, you remove the disk from the drive by pressing the button on the lower right corner." Or simply, "See Figure 1.1." or "(Figure 1.1)." Sometimes informal illustrations (for example, cartoons or informal tables) are neither discussed nor referred to in the text. Just make sure these informal illustrations are necessary and add something to the discussion.

- Number all figures consecutively as they appear in the document (Figure 1, Figure 2, and so on). Number all tables consecutively, too, separately from the figures (Table 1, Table 2, and so on).

- Illustrations should be properly labeled. Every illustration should have a number, a title or caption, and, if necessary, a citation of its source. Use callouts (numbers and letters) to label and identify parts that are listed near the figure. Use a legend or key to identify symbols, variations, or the size scale. Do not directly label parts in an illustration if there is any possibility that it will be reused in another context. Callouts and a legend make the illustration more generic for reuse later. Use standard symbols and abbreviations for any text that is used. Be consistent with terminology in the text and on the illustration. (See Figure 3.9 for a properly labeled figure and a properly formatted table.)

- Illustrations should not be too close to the surrounding text on the page. Allow for plenty of white space on all sides of the illustration.

- The illustration should be effectively displayed in the top third, middle third, or bottom third of the page. Use a balance of text and illustrations on any pages where illustrations occur. If an illustration occupies an entire page, make sure it is adequately sized to the page. It should not be too large or too small for the page. All captions, callouts, and legends should be clearly readable.

ILLUSTRATION **141**

- Illustrations printed in landscape format should face out from the binding of the document if the document is bound.
- Foldout illustrations are permissible in many kinds of documents but require special handling.
- All illustrations should have a professional appearance. Poor photocopies or copies printed with low-quality printers are not acceptable except in draft versions of the document.
- Don't glue, paste, or tape illustrations onto the page of the final copy of a document. Doing so for photocopying purposes may be acceptable.
- Properly indicate the source of any illustration borrowed from or adapted from another source in the lower left-hand corner of the illustration. Make sure you have permission to adapt or borrow the source illustration, unless you have reason to believe your use qualifies as fair use under the copyright law.
- Consider providing a list of illustrations as a separate table of contents. This list of illustrations can be an additional reader aid.
- Provide ethical and responsible illustrations. Make illustrations truthful and never try to mislead the reader. Compare like items and avoid deceptive comparisons. Convey information simply and directly. Label and annotate illustrations to avoid ambiguity. Make horizontal and vertical scales comparable. If you have a good reason for not doing so, explain it.
- If any illustrations are not essential to the text of the document but you still want to include them, place them in an appendix for the document.

Display figure number and caption above, next to, or below figure. Provide source information below.

Place table number and caption above table. Provide source information below.

Figure 1.1. Data on Three Regions.
Source: Author, Book Title (City: Publisher, year), page number.

Table 1.1.
Four Most Populous U.S. States

State	Population
California	32,000,000
Texas	19,000,000
New York	18,000,000
Florida	15,000,000

Source: Author, Book Title (City: Publisher, year), page number.

FIGURE 3.9
Properly Labeled Figures and Tables

Illustrating for Online and the Web

Many of the basic principles of print illustrations apply to illustrations for online and the Web. Because online illustrations and illustrations for Web pages must adhere to the same principles, both are covered in this section.

Much of the difficulty in planning online illustration stems from the wide variety of screen sizes and resolution. These differences mean that you must know exactly when and where to use an illustration. Try out your illustrations on a variety of monitors, platforms, and browsers if possible. Always remember to design online pages to be pleasing to the eye.

TIPS: Illustrating for Online and the Web

- Select the information that you want to display and decide whether the illustration will ask a question, provide an answer, or create an emotional response. Determine a specific objective including who your audience will be, what decisions they will have to make, and how well they will be able to understand the illustration.

- Consider the medium that will be used to display an illustration. For example, if an illustration is to be used in a book, is it necessary to include the same illustration online?

- Choose your illustrations according to image quality, size of image (large images take a long time to download), need for transparency, progressive rendering or streaming, and possible necessity to pay licensing fees.

- Select the format in which you want to present the illustration. Purchase and learn to use software that might assist you. Often, software is available in shareware versions or demo versions, so you can try before you buy. Use the JPG (pronounced "jay-peg") illustration format for photos and GIF illustration format for solid color drawings or clip art. GIF should be used for any illustration in which a portion should be transparent against a background, since GIF supports transparency. GIF animations are an easy way to add motion to your pages if that is desired. Provide a text-only option for readers who either won't or can't display illustrations as they browse.

- Decrease the size of your illustration files by saving them in a low-resolution format. Remember, computer monitors are low-resolution devices, so high-resolution graphics with 24-bit color are wasted in this medium. Worse, the size of the illustrations will make the loading time

ILLUSTRATION **143**

prohibitively long. Smaller files will look just as good and will download in a more reasonable length of time. File download time often makes the difference between a site that users like to visit and browse and one that is glanced at and abandoned. Count on screen sizes of 640 by 480 pixels, even though many monitors will do better than that. Most browsers default to 500 pixels wide. Headers should be less than 480 pixels wide so that viewers do not have to scroll across. Issue size warnings for larger illustrations or sound and movie files.

- Choose small illustrations with few colors, since these are quicker to download than larger ones and those with more colors. Consider using interlaced illustrations (illustrations that originally appear in very low resolution and refresh at higher resolution) to download, since viewers are more willing to wait through a long download if they can see the picture becoming sharper and sharper than if they have to look at a blank screen slowly filling in from the top.

- Size your illustrations appropriately. The size of the illustration signals a certain degree of importance. Make the illustration appropriate to its context. If your illustration is too large to fit on the screen, eliminate unnecessary information, consider making the illustration smaller, break the illustration up into smaller parts, and provide scroll bars.

- Illustrations should be close to their related text. Consider anchoring your illustration to the related text. *Anchoring* means that if the text moves, the illustration will move as well. If you have several illustrations, they can be clustered together or arranged in a vertical or horizontal arrangement. Your material can have different interpretations based on the position of the illustrations, so make sure you have placed the illustrations just where you mean them to be.

- The shape and orientation of an illustration can determine the attitude the reader will have toward it. For important illustrations, leave the shape and orientation basic so that users can recognize them easily. Textures and patterns can determine the sharpness of an illustration and make the illustration more recognizable. Size will affect the recognizability of illustrations as well.

- When applying text to your images, apply the text last and don't resize the image once the text is in place. Use antialiasing to make curves in text seem less jagged. *Antialiasing* is a technique that decreases the effect of low-resolution display on curved images such as letterforms. Instead of using stair-step jagged lines for diagonals and curves, antialiasing blurs those lines with the background pixels. The effect, rather than seeming more blurry, is to have the lines appear smoother. Ensure that the background and text contrast adequately by adjusting the color combinations, line weight, or stroke width.

(continued)

Illustrating for Online and the Web *(continued)*

■ Adjust the contrast between the illustration and the background to make the illustration easier for the reader to interpret.

■ Use color carefully. Color can enhance illustrations but may be difficult for some users who have monochromatic monitors or who are color-blind. Also variations in certain colors make illustrations difficult to distinguish. Mix color values (as few or as many as you have) to create more color options unless you are designing for the Web. Web browsers display colors differently; therefore, confine your Web page colors to the 216-color browser-safe palette.

■ Allow viewers to zoom in on the illustration if possible. One way is to provide a low-resolution thumbnail that links to the full-sized illustration when the user clicks on the smaller image.

■ Dynamics or movement of illustrations should be kept to a minimum. Marquee, blinking, or vibrating illustrations are distracting.

■ Evaluate illustrations and test them with users. Ensure that the message you want to deliver is being conveyed. Plan for change. If your illustrations do not achieve your purpose, modify them until they do.

Tables

Illustrations are generally divided into tables and figures. Tables show comparisons of figures by displaying them in rows and columns. Tables are convenient for displaying several factors at once. A reader can quickly and easily compare data in one column with data in another.

Typical parts of a table include a table number, a table title, column headings, row headings, body, rules, source line, and footnotes. (See Figure 3.10.)

Table Number. Table Title.

Stubhead	Column Head	Column Head	Column Head
Row Head	Table Body	Table Body	Table Body
Row Head	Table Body	Table Body	Table Body

Source: Author, Title of Work (City: Publisher, year), page number.

FIGURE 3.10
Parts of a Table

ILLUSTRATION   145

Average Annual Temperatures in Florida

	Low	*High*
January	30	70
February	40	72
March	42	75
April	50	76
May	55	80
June	70	90
July	72	95
August	72	95
September	72	95
October	55	75
November	55	75
December	52	70

Tables are useful for presenting either simple information or technical data. A decision must be made to determine whether an informal table will provide sufficient clarity to the subject or whether a formal table will be necessary to illustrate complex information.

INFORMAL TABLES

Informal tables differ from formal tables in these ways: they are brief and simple, they are not identified by table numbers and frequently have no title, they have no ruled frame around them, and they have only the necessary minimum of internal ruled lines. Ruled frames are omitted so that informal tables are viewed as a continuation of text. Informal tables are also not recorded in the list of tables at the front of the report in the table of contents. (See Figure 3.11.)

FORMAL TABLES

Formal tables are used more often than informal tables in reports. Also, they are presented as separate items rather than as ordinary paragraphs of text. Formal tables have framed borders, descriptive titles, and table numbers. As a result, formal tables are included in the list of illustrations. Tables are best used for displaying numbers and units of measurement that must be illustrated precisely. For example, there is an obvious difference between 2.0123 and 2.0150 when they are presented side-by-side in a table, but if they are shown as contrasting heights on a graph, the difference would not be so obvious. (See Figure 3.12.)

Resident Population of the United States: Estimates, by Age and Sex
(Numbers in thousands. Consistent with the 1990 census, as enumerated.)

	July 1, 1990	July 1, 1992	July 1, 1994	July 1, 1996	July 1, 1998	Nov. 1, 1998
BOTH SEXES						
Population, all ages	249,440	255,002	260,292	265,179	270,029	270,933
Summary indicators						
Median age	32.8	33.4	34.0	34.6	35.2	35.3
Mean age	35.2	35.4	35.7	35.9	36.2	36.3
Five-year age groups						
Under 5 years	18,851	19,489	19,694	19,324	19,020	18,974
5 to 9 years	18,058	18,285	18,742	19,425	19,912	19,931
10 to 14 years	17,191	18,065	18,666	18,949	19,184	19,291
15 to 19 years	17,763	17,170	17,707	18,644	19,460	19,554
20 to 24 years	19,137	19,085	18,451	17,562	17,685	17,796
25 to 29 years	21,233	20,152	19,142	18,993	18,621	18,513
30 to 34 years	21,909	22,237	22,141	21,328	20,163	19,965
35 to 39 years	19,980	21,092	21,973	22,550	22,600	22,589
40 to 44 years	17,793	18,806	19,714	20,809	21,875	22,014
45 to 49 years	13,823	15,362	16,685	18,438	18,850	19,007
50 to 54 years	11,370	12,059	13,199	13,931	15,727	15,973
55 to 59 years	10,474	10,487	10,937	11,362	12,408	12,631
60 to 64 years	10,619	10,440	10,079	9,997	10,256	10,358
65 to 69 years	10,076	9,973	9,963	9,895	9,575	9,515
70 to 74 years	8,022	8,467	8,733	8,778	8,781	8,780
75 to 79 years	6,146	6,392	6,575	6,873	7,195	7,238
80 to 84 years	3,934	4,135	4,350	4,559	4,712	4,748
85 to 89 years	2,050	2,170	2,287	2,395	2,533	2,560
90 to 94 years	765	860	956	1,024	1,094	1,108
95 to 99 years	206	231	249	287	317	324
100 years and over	37	44	50	57	63	62
Special age categories						
16 years and over	192,000	195,711	199,590	203,699	208,039	208,901
18 years and over	185,276	188,928	192,489	196,157	200,224	201,096
15 to 44 years	117,816	118,543	119,127	119,886	120,403	120,432
65 years and over	31,237	32,273	33,164	33,867	34,269	34,336
85 years and over	3,059	3,306	3,542	3,762	4,006	4,055

Source: U.S. Bureau of the Census
Internet Release Date: December 28, 1998

FIGURE 3.12
Formal Table

ILLUSTRATION   147

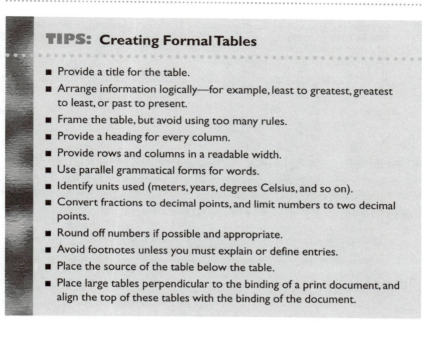

TIPS: Creating Formal Tables

- Provide a title for the table.
- Arrange information logically—for example, least to greatest, greatest to least, or past to present.
- Frame the table, but avoid using too many rules.
- Provide a heading for every column.
- Provide rows and columns in a readable width.
- Use parallel grammatical forms for words.
- Identify units used (meters, years, degrees Celsius, and so on).
- Convert fractions to decimal points, and limit numbers to two decimal points.
- Round off numbers if possible and appropriate.
- Avoid footnotes unless you must explain or define entries.
- Place the source of the table below the table.
- Place large tables perpendicular to the binding of a print document, and align the top of these tables with the binding of the document.

There are many other kinds of figures in addition to the one covered here, but these are some of the most commonly used ones.

Graphs

Graphs (sometimes also called charts) are designed by plotting a set of points on a coordinate system. One advantage of using graphs over tables is that measurements of trends, movements, distributions, and cycles can be presented more clearly and effectively with graphs. In using graphs to illustrate data, the essential meaning of large amounts of information can be visualized quickly. Also, graphs provide a more complete interpretation of data by offering a comprehensive view of the problem.

There are several basic types of graphs that can be used to illustrate statistical data. Some of the most notable kinds are bar graphs, line graphs, circle or pie graphs, and pictographs.

BAR GRAPHS

Bar graphs use bars to compare specific measurements. Bar graphs consist of horizontal or vertical bars of equal width that are scaled in

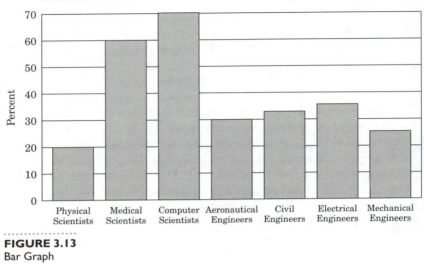

Projected growth in employment of U.S. engineers and scientists from 1990 to 2005

FIGURE 3.13
Bar Graph

length to represent some quantity. When bars are presented horizontally, the figure is referred to as a bar graph. If bars are shown in a vertical position, the figure can also be identified as a column graph. These two kinds of graphs may also be combined to illustrate statistical data.

The simple bar graph is one of the most useful and widely used formal graph aids. It is used to make comparisons between two or more coordinate items, and comparisons are based on linear values. The lengths of the bars are determined by the value of each category.

Bar graphs may be used to display a change in quantity over time, but their intention is to display individual data. The length of each bar represents a specific measurement. (See Figure 3.13.)

TIPS: Creating Bar Graphs

- Make each bar the same width, and clearly label each one.
- Make the space between bars approximately one-half the bar width.
- Group bars according to categories when appropriate.

ILLUSTRATION   149

- Use a legend or a key, if appropriate, to distinguish different bars, and vary the pattern or color in multiple bar graphs.
- Try to make the vertical or Y scale at least 75 percent as long as the horizontal scale.
- Make the longest bar extend nearly to the end of its parallel axis.

LINE GRAPHS

Line graphs are useful for showing how quantities change over time. Money, time, and temperature are common subjects for line graphs. The line graph can be used to plot the movement of two variables: the independent variable and dependent variable. The independent variable is usually plotted horizontally. The dependent variable is plotted vertically. A normal practice in presenting statistical data is to allow the independent variable to denote time and the dependent variable to denote the amount. A point marks the recorded data and gives readers reference points for the graph as a whole. (See Figure 3.14.)

Sales Trends: 1998

FIGURE 3.14
Line Graph

TIPS: Creating Line Graphs

- Place the dependent variable on the Y (vertical) axis and the independent variable on the X (horizontal) axis.
- Keep vertical and horizontal axes proportionate.
- Make sure the slope or steepness of the line accurately indicates the trend indicated by the data.
- Place no more than three lines on each graph. If the lines fall very near each other, consider changing the scale of the graph to make the lines more distinguishable.
- Use colors and symbols (dots, dashes, x's) to distinguish lines. Color alone will not be useful if your graph is photocopied or viewed by a color-blind person.
- Label each line or use a legend if necessary to avoid clutter.
- When showing positive and negative Y values, use the X axis to represent zero.

PIE GRAPHS

The pie graph or circle graph is another type of graph that is used to present statistical information. This graphic aid provides for easy and accurate interpretation of data. The primary advantage of the pie graph is that it clearly depicts proportions of a whole. This feature makes the pie graph a valuable tool for executives, for example, to use in financial reports. Another advantage of the pie graph is that data can be arranged in an orderly manner. (See Figure 3.15.)

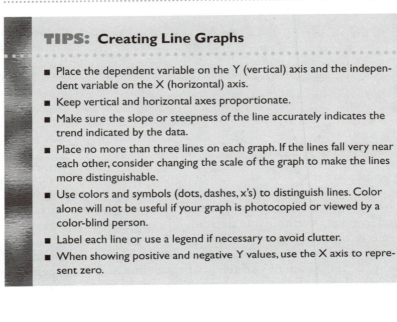

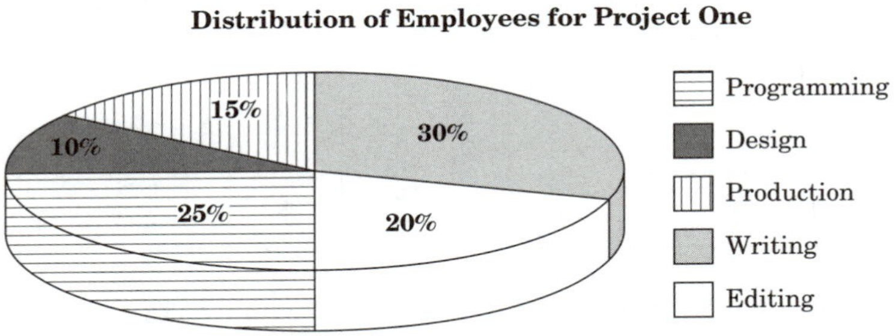
Distribution of Employees for Project One
15%
30%
10%
25%
20%
Programming
Design
Production
Writing
Editing

FIGURE 3.15
Pie Graph

ILLUSTRATION   151

TIPS: Creating Pie Graphs

- As a rule, the largest segment of data should always begin at the twelve o'clock position.

- Percentage numbers and any description may be located inside the graph when space allows. But if your graph will be used in different cultural and linguistic contexts, descriptions should be outside the pie graph in a legend or in labeled callouts.

- Use wedges to represent approximate, relative amounts of a whole. Some pieces of the graph can be exploded or separated out from the whole. Separate at most one wedge when you are using the separation for emphasis.

- Combine very small wedges together into a larger category or under "other" if feasible. Keep in mind that too many wedges are difficult to see because each wedge is so small.

- Do not use a pie graph if the purpose of the graph is to show precise amounts.

- Limit your pie graph to eight wedges or fewer.

PICTOGRAPHS

A pictograph (or pictogram, visual table, or picture graph) is a graphic form that uses pictures to represent data. The advantage of using a pictograph to portray data is that it is an effective, interest-getting device and can be useful for a lay audience. The pictograph is essentially a pared-down graph that uses the objects displayed themselves—stick figures, cows, or whatever else might be related in some way to the object being quantified—to represent the actual data. Unlike a bar graph, a pictograph uses no axes to show numeric value or scale, so you have to show any relevant information through the number of the objects displayed. (See Figure 3.16 on page 152.)

TIPS: Creating Pictographs

- Keep the pictograph simple and clear.

- In drawing pictographs, make sure the design is displayed so that it can be easily seen in relation to the desired units.

(continued)

FIGURE 3.16
Pictograph

Number of Chickens

Creating Pictographs *(continued)*

- Choose symbols that closely represent your topic, value, or idea.
- Choose symbols that are readily understood and are simple and recognizable when reduced or enlarged.
- Space all symbols equally.
- When displaying different quantities, change the number of symbols, not the size or height.
- If possible, round off numbers to eliminate fractions.

Charts

Charts show relationships instead of amounts. Flowcharts, organizational charts, and schedule charts are examples of charts.

FLOWCHARTS

A flowchart provides an overview of a process. A flowchart is an illustration that involves a sequence of stages that are displayed from

ILLUSTRATION   153

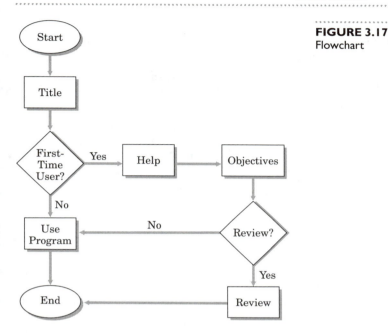

FIGURE 3.17
Flowchart

beginning to end. The items in the chart should be connected according to the sequence in which the steps of the process occur. The main advantage of using a flowchart to illustrate information is that it displays an overview of a process, enabling the reader to comprehend the important steps of the process quickly and easily.

There are several varieties of flowcharts that can be used to signify the steps of a process. These forms include labeled blocks, a pictorial representation, or standardized symbols. The items of any flowchart should always be connected according to the order in which the steps occur. Flowcharts usually have left to right or top to bottom movement. Should the movement be otherwise, always use arrows to show the proper movement. (See Figure 3.17.)

TIPS: Creating Flowcharts

- Break down each step into the smallest unit appropriate for your purpose and scope. A broad overview will have a few very general steps. A detailed programming module may have many specific steps.

(continued)

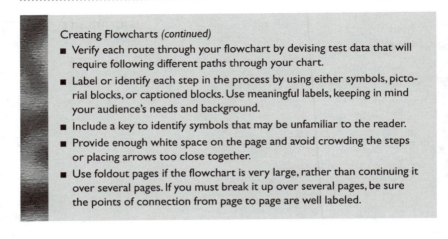

Creating Flowcharts *(continued)*

■ Verify each route through your flowchart by devising test data that will require following different paths through your chart.

■ Label or identify each step in the process by using either symbols, pictorial blocks, or captioned blocks. Use meaningful labels, keeping in mind your audience's needs and background.

■ Include a key to identify symbols that may be unfamiliar to the reader.

■ Provide enough white space on the page and avoid crowding the steps or placing arrows too close together.

■ Use foldout pages if the flowchart is very large, rather than continuing it over several pages. If you must break it up over several pages, be sure the points of connection from page to page are well labeled.

ORGANIZATIONAL CHARTS

Organizational charts demonstrate the components of an organization and how they relate to each other. They emphasize the hierarchy of positions within a company or organization. These charts are effective for providing an overview of an organization to readers or to present the lines of authority within the organization. (See Figure 3.18.)

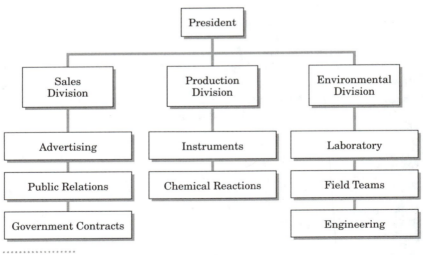

FIGURE 3.18
Organizational Chart

ILLUSTRATION   **155**

TIPS: Creating Organizational Charts

- Specify each component of the chart in a single box.
- Boxes on the same horizontal level should represent similar levels of authority.
- Clearly show how each position relates to those above and below as well as to the sides.
- Link the boxes appropriately to a central authority.
- Include, if necessary, the names of personnel who fill each position in the box.

SCHEDULE OR GANTT CHARTS

Schedule charts (also called Gantt charts) identify the major steps in a project and tell when they will be performed. A schedule chart enables readers to see what will be done, when each activity will start and end, and when activities will overlap. Schedule charts are often used in proposals to show the projected plan of work. You can also use them in progress reports to show what you have accomplished and what you still have to do. One of the principal considerations in creating a schedule chart is deciding how much detail to include. The scope or detail is determined by the purpose of the chart and the audience. (See Figure 3.19 on page 156.)

TIPS: Creating Schedule Charts

- When planning the schedule chart, decide on the appropriate time units for your chart: days, weeks, months, or longer.
- Label all activities clearly. The chart should be easy to read in its normal location (on a wall, in a report, under shop conditions, and so on).
- Time is normally shown along the horizontal axis.
- For each type of activity, designate a distinguishing color. For example, planning stages might be in blue, writing stages in green, review cycles in yellow, and evaluation or maintenance periods in red.
- Don't forget to allow for holidays, vacations, and schedule slippage from unanticipated causes.

(continued)

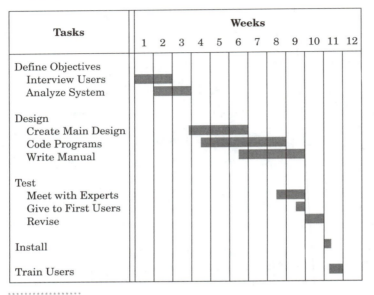

FIGURE 3.19
Schedule Chart

Creating Schedule Charts *(continued)*

■ If some of the steps depend on activities of other divisions or organizations (such as a printing service bureau), those steps will have to be placed on the chart on the basis of that entity's time constraints.

■ Be prepared to modify the chart if deadlines are not being achieved. It is better to have a realistic schedule chart that functions as a helpful tool than one that is ideal but unusable.

Drawings, Diagrams, and Maps

Drawings are illustrations that are designed manually by professional artists or laypeople to portray a subject or object. Sometimes they are produced by mechanical means. Drawings have many advantages over photographs. They can convey ideas that are difficult to explain in prose. Also, they can include as much or as little information as is needed, according to the purpose of the drawing. For example, drawings can highlight details or relationships that photographs cannot convey. Or they can display the main components of an object, eliminate clutter, enlarge important areas of a picture, cut

ILLUSTRATION   **157**

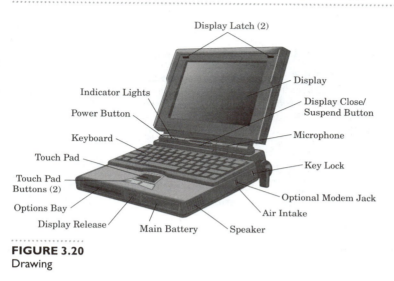

FIGURE 3.20
Drawing

away the outside view of equipment so that readers can see the insides, and offer an exploded view, in which the parts of the object are spread out so that the reader can see their relationship and how they fit when assembled. Some commonly used drawings are artist's conceptions, block diagrams, cartoons, cutaway views, engineering drawings, exploded views, line drawings, logic drawings, perspective drawings, and schematics (for example, for wiring circuits). (See Figure 3.20 for an example of a drawing.)

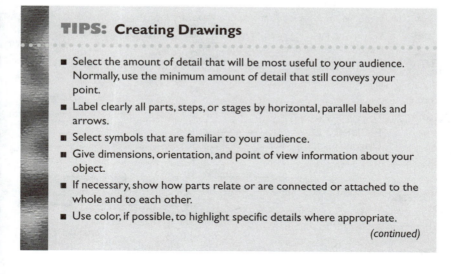

TIPS: Creating Drawings

- Select the amount of detail that will be most useful to your audience. Normally, use the minimum amount of detail that still conveys your point.
- Label clearly all parts, steps, or stages by horizontal, parallel labels and arrows.
- Select symbols that are familiar to your audience.
- Give dimensions, orientation, and point of view information about your object.
- If necessary, show how parts relate or are connected or attached to the whole and to each other.
- Use color, if possible, to highlight specific details where appropriate.

(continued)

Creating Drawings *(continued)*

■ Place drawings close to text that refers to them.

■ Draw information to scale.

■ Provide an inset schematic showing where the area in the drawing is located with respect to the larger object. If you are showing a blowup of a part of a larger unit, provide an outline sketch that shows its context in the bigger picture.

Diagrams are pictures of objects or events that use conventionally defined symbols to convey information. Diagrams may show, for example, the wiring of a room, the assembly of a computer, or the changes in a weather pattern. Diagrams combine literal elements (drawings of parts) and symbolic ones (arrows to show movement, direction, or associations; shading to show curvature). (See Figure 3.21.)

TIPS: Creating Diagrams

■ In an exploded diagram, show parts near their actual locations on or in the object depicted.

■ Highlight the most important material and de-emphasize or delete unimportant material.

■ Use the standard symbols of a discipline or profession where appropriate. Electronics, genetics, and linguistics, for example, all have their own standard symbols for representing various concepts.

■ Clarify the meaning of any symbol through a caption if you think your audience may be unclear about a symbol.

■ Provide context for the diagram by showing where it fits in the larger picture.

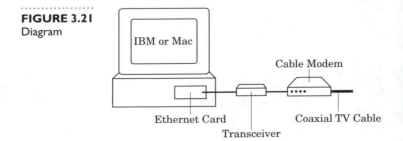

FIGURE 3.21
Diagram

ILLUSTRATION   **159**

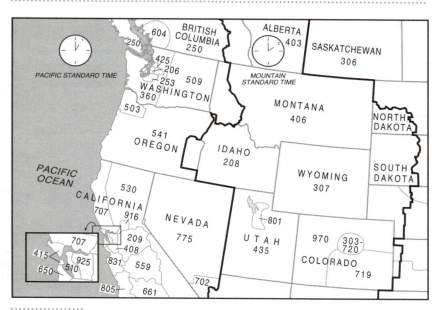

FIGURE 3.22
Map

Maps are drawings that provide a physical layout. Like graphs, they can show quantitative information such as average temperatures for a region, voting patterns, and population distribution. Maps are used to portray geographic characteristics such as roads, mountains, or rivers. They are valuable for displaying geographic distribution such as housing and industrial centers. Also, simple maps are used to explain transit and highway systems. (See Figure 3.22.)

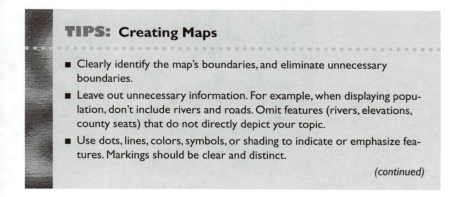

TIPS: Creating Maps

■ Clearly identify the map's boundaries, and eliminate unnecessary boundaries.

■ Leave out unnecessary information. For example, when displaying population, don't include rivers and roads. Omit features (rivers, elevations, county seats) that do not directly depict your topic.

■ Use dots, lines, colors, symbols, or shading to indicate or emphasize features. Markings should be clear and distinct.

(continued)

Creating Maps *(continued)*

- If necessary, include a legend, or map key, providing an explanation of dotted lines, colors, shading, or symbols.
- Indicate direction. Conventionally, maps show north, often by including an arrow and the letter N.
- Include a scale of miles or feet for proportions.
- Present proportions with as much accuracy as possible. If you are using a projection that has appreciable distortion, add a note to inform your readers of this fact.
- Acknowledge your source if you did not construct the map yourself.

Photographs and Clip Art

Photographs are preferred over drawings when a precise replica is needed. Photographs are useful to help technicians understand complex equipment. They are one of the best ways to show results. Photographs can be visually appealing; however, a drawing of the same subject may provide more information. Consider using good-quality black-and-white photographs for reports when quick pictures are needed. Usually, photographs to be printed with text must be halftones (represented by a system of dots). Include photographs in your document by scanning a conventional photograph or using a photograph taken by a digital camera. Photographs with good contrast can be photocopied. (See Figure 3.23.)

FIGURE 3.23
Photograph

ILLUSTRATION   **161**

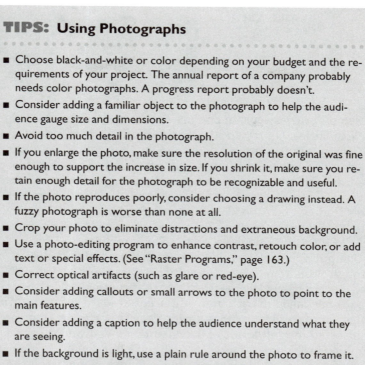

TIPS: **Using Photographs**

- Choose black-and-white or color depending on your budget and the requirements of your project. The annual report of a company probably needs color photographs. A progress report probably doesn't.

- Consider adding a familiar object to the photograph to help the audience gauge size and dimensions.

- Avoid too much detail in the photograph.

- If you enlarge the photo, make sure the resolution of the original was fine enough to support the increase in size. If you shrink it, make sure you retain enough detail for the photograph to be recognizable and useful.

- If the photo reproduces poorly, consider choosing a drawing instead. A fuzzy photograph is worse than none at all.

- Crop your photo to eliminate distractions and extraneous background.

- Use a photo-editing program to enhance contrast, retouch color, or add text or special effects. (See "Raster Programs," page 163.)

- Correct optical artifacts (such as glare or red-eye).

- Consider adding callouts or small arrows to the photo to point to the main features.

- Consider adding a caption to help the audience understand what they are seeing.

- If the background is light, use a plain rule around the photo to frame it.

- Avoid using paper clips or staples on photographs.

- Avoid folding or creasing photographs.

- Avoid writing on either the front or back of photographs.

Clip art consists of commercially produced drawings, symbols, and icons, licensed to be used in a wide variety of situations. Pictures in many categories—office, people, home—are available for use in your documents to provide instant visual interest. An immediate advantage of clip art is that most of it may be used without specific permission once you have purchased a license for the collection. You may insert, photocopy, or redraw and publish clip art without any attribution. When you purchase a license to use a clip-art collection, you purchase the right to use it in your own work without paying further royalties or licensing fees. Some clip art carries restrictions about multiple uses within a given project, prohibitions against online or Web-page use, or limitations on resale. But in general, you may use clip art freely in your work. (See Figure 3.24 on page 162.)

FIGURE 3.24
Clip Art

Some libraries contain clip-art books, such as the *Dover Pictorial Archive* series. Up to ten illustrations may be reproduced from this source on any one project or in any single publication without permission or cost. In addition to clip art that you may photocopy and use from sources such as this one, clip art is traditionally available as transfers (thin sheets with plastic symbols, letters, numbers, or pictures that can be rubbed onto paper), templates (plastic sheets with cutouts of shapes such as furniture or computer terminals; math, logic, or electronic symbols; or numbers and letters).

Computer clip art and photo images are also widely available through numerous computer software programs and via the Internet. You can easily locate CD-ROMs that contain only clip art or find clip art packaged in with many different kinds of software. Microsoft Office, for example, provides a wide variety of clip art that can easily be inserted into all kinds of documents. As you would expect, many graphics programs come with a clip-art library. And some include fairly extensive collections of photo images. Sometimes the use of clip art and photo-image libraries is restricted. Corel Draw, for example, provides many kinds of clip art and photo images for you to use but also restricts how you may use them. Restrictions involve a variety of issues; you may not redistribute or sell the images, link any images related to individuals to any products or services, or create obscene or offensive works using the images. You need to review carefully any restrictions specified by the company that licenses the clip art or the manufacturer of the graphics program you use.

Of course, you can find numerous clip art sites on the Internet as well as many sites providing all kinds of photo images. You will have to be sure which sites are providing clip art that is clearly in the public domain and which sites have certain restrictions. You must be especially careful in using graphics and photo images that you find on the World Wide Web and other Internet locations. Photo and graphic images are copyrighted and may be used only with the copyright owner's explicit permission. (See "Copyright," page 41, for a more detailed discussion of this important issue.) Very often, the graphics that are displayed on a Web page do not belong to the person

ILLUSTRATION **163**

who put them there. Unfortunately, the ease of copying graphics and photos has led to an overly casual concept of image ownership. Just because someone puts a picture or other graphic on a site does not mean that the image belongs to that person. If you use a graphic that turns out to have been taken from the actual owner's site, you can be held liable for copyright infringement. Sometimes it takes real detective work to find out who actually owns an image.

Computer Illustrations

Computers have made it easier than ever before to create your own professional-looking illustrations in just a few minutes. With many computer programs, once you have entered the essential information, you can switch almost instantly to viewing the data as a bar graph, a pie graph, or a variety of other illustrations. Then you can choose the illustration that you think best represents the data for your particular audience. Computer programs may save you a lot of time and work in creating illustrations, but you will still need to learn the essentials of effective illustrations to get the most benefit from them. Because of the wide array of illustration tools available, it's possible even for design novices to create professional-looking designs and illustrations. Nevertheless, it's important to remember that these are tools, and no matter how sophisticated they are, they cannot produce excellent designs in the absence of design skills judiciously applied.

There are three basic types of graphics programs that are essential for anyone who wants to create computer illustrations: vector, raster, and three-dimensional. In addition, a wide range of useful image utilities round out the tools that are helpful to students and employees. Powerful computer programs can help you to do all kinds of sophisticated photo editing, drawing, painting, screen captures, and illustration management. Some of the better-known tools are discussed briefly below. Of course, some of them are effective for several of the categories discussed here. You will find much more detailed information at each of the Web sites provided. It's also a good idea to read at least several independent reviews of any of these products. Stroud's is one good source for reviews at <http://www.stroud.com>.

RASTER PROGRAMS

Raster programs are ones with which you create or edit images made up of tiny squares of color called pixels (picture elements). Each screen pixel has only one color, which is described by its red, green, and blue (RGB) component values. Once all these colored pixels are side by side, they form an image. You will also see these images referred to as *bitmaps* or *bitmapped images*. All images on the Web (GIF, JPG, and others) are bitmaps. Raster programs are useful when you want to edit scanned images or create your own images from scratch. They enable you to perform sophisticated transformations on the various colors and positions of items in a graphic image. The more powerful programs, such as Adobe PhotoShop and Jasc Paint Shop Pro, have many useful options that mimic the effects of artists' tools. Photo editors and paint programs are examples of raster programs.

Photo Editors.   Photo and other graphics editors have become very popular as they became easier to use and people's taste in graphics and graphical effects became more developed. As readers became accustomed to seeing shadow effects behind floating titles and strange and intriguing photo montages, many companies came out with photo editors and other graphics programs to create these effects. Probably the leader in the photo-editing field is Adobe PhotoShop <http://www.adobe.com>. PhotoShop provides a host of options— custom brush strokes, retouching tools, masks, channels, layers, text effects, filters, output formats—with a correspondingly steep learning curve. Still, it is a program worth learning, as it is very widely used in industry. Jasc Paint Shop Pro <http://www.jasc.com/> is another very widely used program. It features a screen capture capability, and it supports a wide range of bitmap file formats. Its popularity stems from its versatility and relatively low price. Since Paint Shop Pro is shareware, you can try it for a limited time before you buy it. LView Pro <http://www.lview.com/> is another shareware graphics program that enables you to create many effects that are found in the more expensive commercial programs. Another program, Deneba Canvas <http://ww2.deneba.com/>, includes layout capabilities and a clip-art library along with its graphics features. Alchemy Mindworks Graphic Workshop <http://www.mindworkshop .com/alchemy/gww.html> is another relatively inexpensive graphics editor that enables you to manipulate, convert, and package your graphics files. Macromedia xRes <http://www.macromedia.com/ software/xres/> is a commercial application that you can use to publish your work on the Web, create and modify graphics using a wide variety of brush and pastel effects, and retouch photographs. Unlike most other photograph editors, xRes does not require a lot of computer memory to perform its work.

ILLUSTRATION   **165**

Paint Programs.   Very often, you will not need the full range of paint-ing features that the high-end programs provide, and a simple (or not so simple!) paint program will suffice. Most computers have a paint program bundled with them, and many of the shareware drawing programs found on the Internet are raster based. These programs, which usually have the word *paint* somewhere in their name, provide many useful features to create and modify bitmapped images. Use a paint program to color, enlarge, retouch, and crop your images. Once you have learned to use one of these programs, the oth-ers are fairly easy to figure out, since they all do more or less the same things. Some of the more popular paint programs are Micro-soft Paint <http://www.microsoft.com>, which comes bundled with the various Windows operating systems, Corel Photo-Paint <http://www.corel.com>, which adds the possibility of layering to in-crease the flexibility of the program, and MacPaint and Painting <http://www.interlog.com/~sarwat/painting/> for the Macintosh.

VECTOR PROGRAMS

Vector programs use mathematical calculations to draw shapes. By layering these shapes, you get the appearance of an image. These programs are generally used for creating original artwork, logos, or special text effects that you can't create with a raster (pixel-based) program. Vector drawings, unlike their raster counterparts, can be scaled up or down in size without losing sharpness or introducing stair-step effects, called "jaggies." Once you create an image in this type of program, you can import it into a raster paint program to add more color or depth to the image. With vector programs, you do not save the image as a GIF or JPG image; each program saves images in its own proprietary format. Use vector programs for a variety of drawing and illustrating needs and when you want to create certain specialized kinds of forms such as flowcharts.

Drawing and Illustrating.   Even though many of the photo-editing pro-grams also enable you to create art and other graphics from scratch, there are programs that are specifically designed to do that and do it well. From a drawing and illustrating program, you should expect tools that enable you to draw complex shapes and fill them with a va-riety of patterns, fills, gradients, and textures. You can move shapes around, rotate and reflect them, distort them, group them, resize them, sharpen and unsharpen their edges, manipulate contrast, and apply a variety of text effects. Drawing programs can be used as page-layout programs, although they usually do not afford the same breadth of features as programs dedicated to that purpose.

Corel Draw <http://www.corel.com> is one of the most established drawing programs on the market and still one of the best. It can be used to scan, trace, and manage illustrations. Corel Draw has editing utilities including textures, 3-D, sticky notes, and screen captures. Adobe Illustrator <http://www.adobe.com/> provides numerous drawing and painting features along with sophisticated typographic controls. Illustrator enables you to prepare color separations, work in layers, and import and export many file formats. DrawPlus <http://www.spco.com/products/dp3main.htm> is a drawing program that is especially suited to the less-experienced graphics software user. It provides wizards (helper applications) to lead you through the creation of interesting effects; libraries of clip art, photos, and fonts that you can use in your designs; and predefined "envelopes" to convert text and graphics into a variety of forms. Macromedia Freehand <http://www.macromedia.com/software/freehand/> is particularly helpful in creating Web animation, although it is also a very well regarded print application. It features a customizable user interface, innovative lens and transparency effects, and a wide range of import and export file formats.

Microsoft Draw <http://www.microsoft.com> is an add-on program for Microsoft Office. The program functions as an integral part of Microsoft Word. Microsoft Draw allows you to create many kinds of illustrations. In addition to its drawing features, it allows you to make captions, callouts, and flowcharts. The program also includes many ready-made shapes called AutoShapes. As with many other drawing programs, you can resize, rotate, or flip illustrations; fill them with colors, patterns, textures, or pictures; enhance them with shadow or three-dimensional effects; and arrange illustrations in many ways in relation to each other on the same page.

Specialty Drawing Programs. SmartDraw <http://www.smartdraw.com>, unlike the other programs in this section, is an example of a special purpose drawing program. It is used to create flowcharts, diagrams, floor plans, storyboards, time lines, banners, forms, and maps. It can even be used to create flyers and posters. SmartDraw is easy to use because of its drag-and-drop interface and its many formatting options. Visio Corporation <http://www.visio.com/> provides four widely used specialty drawing programs with Visio Standard, Visio Professional, Visio Enterprise, and Visio Technical. Visio Stan-

ILLUSTRATION   167

dard, for example, enables you to create many kinds of flowcharts, diagrams, organizational charts, project time lines, maps, and much more. The three other programs offer other advanced features.

THREE-DIMENSIONAL PROGRAMS

Three-dimensional graphics programs offer sophisticated, easy-to-learn features that help Web illustrations attract attention. They are designed to provide software rendering of three-dimensional objects, taking advantage of the powerful multimedia hardware that is now available on home computers and individual workstations. With three-dimensional (3D) programs, you take a simple shape and stretch, bend, and twist it into 3D models. You can create anything from a rotating square to a realistic dinosaur with this type of program. These programs can be used to animate logos or create other simple animations, but they are capable of much more. Macromedia Extreme 3D <http://www.macromedia.com/> will help you to create animations and graphics for a variety of print, Web-based, or CD-ROM-based applications. Its many choices of output file format enable you to transport the resulting image files to a variety of other applications. 3D Studio Max <http://www.ktx.com> was developed and designed specifically for the Windows NT platform and therefore takes full advantage of the 32-bit environment for three-dimensional rendering and animation. Other companies provide plug-ins to increase the functionality of this powerful graphics application. Another well-regarded 3D graphics application, LightWave 3D <http://www.newtek.com/products/frameset_lightwave.html> is available for virtually all popular platforms: Windows 95/98 and NT, Power Macintosh, DEC Alpha, Silicon Graphics, and Sun Microsystems.

GRAPHICS UTILITIES

Screen-Capture Programs.   In software documentation, screen captures are commonly used to show the various screens—main menus, dialog boxes, toolbars—of a software program. Using screen capture software, you can easily show your readers exactly what they are going to see when they use a particular feature of the software you are describing. You can show toolbars, buttons, cursors, menu items, dialog boxes, error messages, and a variety of realistic examples in as much detail as you require. Some programs even enable you to show animations in your screen captures, so readers can better understand the processes you are describing. Screen-capture programs have a variety of options, enabling you to select which portion of the screen you would like to capture and the form in which you would like to save your resulting image. HyperSnap-DX <http://www.hyperionics.com>, for example, offers many features to edit your screen captures. HyperSnap can capture whole windows, parts of windows, or screen regions, with or without the cursor. Screen captures are copied to the Windows clipboard and can be saved in a variety of formats. SnagIt <http://www.techsmith.com> is a versatile screen-capture program that could also be listed along with the graphics-viewing and management programs, since it incorporates a thumbnail catalog browser and multimedia viewer. In addition to capturing static screen images, SnagIt can also copy desktop animation sequences, including video and user-interface activities. Once you have created an image using SnagIt, you can modify it in various ways—change the colors; convert color images to 256 colors, 16 colors, monochrome, or grayscale; and replace any color in the captured image.

Graphics Viewers and Managers.   Many kinds of commercial, shareware, and freeware graphics programs are available to help you organize, manage, and view the illustrations in your directories and files. In addition to the programs listed here, many of the graphics programs listed in other categories will perform some of these same functions for you. Thumbs Plus <http://www.thumbsplus.com/> can take your images and present them as a slide show, print them as contact sheets, create thumbnails from all or part of the image, crop, filter, retouch, adjust colors, rotate and resize, and do all of this on files in a variety of formats. PolyView <http://www.polybytes.com/> can manipulate images in many file formats and can create animated GIFs. PolyView also works well with images from scanners and digital cameras. It can organize your images and help manage your image collection, but it can also perform a variety of image-manipulation functions. Hijaak Pro <http://www.imsisoft.com/> is a well-regarded, general purpose image manager. It can handle and convert over eighty-five file formats, including 3D and full postscript. It creates thumbnails for two- and three-dimensional images. Hijaak Pro can convert raster images into vector images, as well as perform screen captures and manipulate the size, color, and orientation of images. It can print thumbnails, organize images, create catalogs of images, manage fonts, and generally serve as an all-purpose image viewer.

RESOURCES

Barker, Thomas. *Writing Software Documentation: A Task-Oriented Approach*. Boston: Allyn & Bacon, 1998.

Binns, Betty. *Better Type*. New York: Watson-Guptill, 1989.

Bringhurst, Robert. *The Elements of Typographic Style*. Vancouver: Hartley & Marks, 1992.

Brockmann, R. John. *Writing Better Computer User Documentation: From Paper to Hypertext, Version 2.0*. New York: Wiley, 1990.

Dinucci, Darcy, Maria Giudici, and Lynn Stiles. *Elements of Web Design: The Designer's Guide to a New Medium*. 2nd ed. Berkeley: Peachpit, 1998.

Fernandez, Judi N. *GIFs, JPEGs, and BMPs: Handling Internet Graphics*. New York: MIS, 1997.

Fleming, Jennifer. *Web Navigation: Designing the User Experience*. Sebastopol: O'Reilly, 1998.

Hackos, JoAnn T., and Dawn M. Stevens. *Standards for Online Communication*. New York: Wiley, 1997.

Horton, William. *Designing and Writing Online Documentation*. 2nd ed. New York: Wiley, 1995.

————. *The Icon Book: Visual Symbols for Computer Systems and Documentation*. New York: Wiley, 1994.

————. *Illustrating Computer Documentation: The Art of Presenting Information Graphically on Paper and Online*. New York: Wiley, 1991.

Kosslyn, Stephen M. *Elements of Graph Design*. New York: Freeman, 1994.

Kostelnick, Charles, and David D. Roberts. *Designing Visual Language: Strategies for Professional Writing*. Boston: Allyn & Bacon, 1998.

McLean, Ruari. *The Thames and Hudson Manual of Typography*. London: Thames and Hudson, 1980. Rpt. 1992.

Parker, Roger C. *Roger C. Parker's Guide to Web Content and Design: Eight Steps to Web Site Success*. New York: MIS, 1997.

Parker, Roger C., and Patrick Berry. *Looking Good in Print*. 4th ed. Scottsdale: Coriolis, 1998.

Pirouz, Raymond. *Click Here: Web Communication Design*. Indianapolis: New Riders, 1997.

Price, Jonathan, and Henry Korman. *How to Communicate Technical Information: A Handbook of Software and Hardware Documentation*. Redwood City: Benjamin/Cummings, 1993.

Schriver, Karen. *Dynamics in Document Design: Creating Texts for Readers*. New York: Wiley, 1997.

Siegel, David. *Creating Killer Web Sites: The Art of Third-Generation Site Design*. 2nd ed. Indianapolis: Hayden, 1997.

Tufte, Edward R. *Envisioning Information*. Cheshire: Graphics, 1990.

————. *The Visual Display of Quantitative Information*. Cheshire: Graphics, 1993.

————. *Visual Explanations: Images and Quantities, Evidence and Narrative*. Cheshire: Graphics, 1997.

Waters, Crystal. *Universal Web Design*. Indianapolis: New Riders, 1997.

————. *Web Concept and Design: A Comprehensive Guide for Creating Effective Web Sites*. Indianapolis: New Riders, 1996.

Wheildon, Colin. *Type & Layout: How Typography and Design Can Get Your Message Across—or Get in the Way*. 1984. Berkeley: Strathmoor, 1995.

Williams, Robin. *The Non-Designer's Design Book: Design and Typographic Principles for the Visual Novice*. Berkeley: Peachpit, 1994.

Williams, Robin, and John Tollett. *The Non-Designer's Web Book: An Easy Guide to Creating, Designing, and Posting Your Own Web Site*. Berkeley: Peachpit, 1998.

Woolever, Kristin R., and Helen M. Loeb. *Writing for the Computer Industry*. Englewood Cliffs: Prentice Hall, 1994.

Editing

(continued)

● ●

Tips

PART 4

Editing

*E*diting, reviewing, and evaluating are all aspects of the same process, a process that refines, smoothes, and polishes a work until it is as perfect as it can be, whatever its constraints. Editing, reviewing, and evaluating are essential to ensure the quality of your documents. In fact, typically, you will be constantly writing, editing, and having others review and evaluate your work. These activities go hand in hand, one following the other, and in recurring fashion. Editing is performed at many stages and by many people. You may, of course, edit your own work, and you should edit while you are writing and after you revise. More commonly, others will edit your work. If you are fortunate, a professional editor will examine every word, phrase, clause, sentence, paragraph, and larger segment. The review process occurs when others read your work and accept it, suggest modifications, or reject it. As a college student, you may be involved in a peer review in which other students look over your work according to certain criteria provided by your professor. In industry, a review may involve many phases of review from peer to technical to legal. Finally, when the work has gone through its edits and reviews, it can be evaluated and tested to see whether the documents are appropriately written for their audience and whether they function as they are required to do.

In some ways, approaches to editing vary almost as much as editors. Editing is a task that probably no two people do the same way. Some people edit their work page by page, simply looking at each word, sentence, paragraph, and larger section, making changes at both the micro level and the macro level at the same time. In short, they edit their work piecemeal and randomly. They don't have a system. Other people systematically use an approach called bottom-up editing. They begin editing at the sentence level and work their way to the larger elements of a document. Still other people prefer what is called top-down editing. They start with the larger features of the document—including content, organization, and design—and then work their way down to the word and sentence level of the document. People using this top-down approach feel that they will save time over using the bottom-up approach. Using the top-down approach, editors will not have wasted time editing a subsection at the sentence level if they later decide to delete an entire subsection. On the other hand, those who follow the bottom-up method find it distracting to read past word and sentence errors and awkwardness and believe that they can follow the motion of ideas through a text better when their view is not obstructed by lower-level errors.

Whatever system you use, the important thing is to have some kind of system. More important, it's necessary to understand that editing is a complicated process that comprises the important phases of developmental editing, substantive editing, copyediting, and proofreading.

DEVELOPMENTAL EDITING

Editing is a deceptively short word for an extensive process that begins before you write your first word and ends when your work goes out of print. In between, you subject your projects to scrutiny at every level of organization and at every stage of preparation and production.

The first step in that extensive process is the developmental edit. During developmental editing, you evaluate your design, the organization of your major areas, the genre of your deliverables, and many other aspects of your project. During developmental editing, you look at your documents, Web pages, online help files, graphics, and other output from the broad perspective of the user's needs and the producer's resources. Unlike the editing that you perform at the paragraph, sentence, or word level later on, developmental editing seeks to uncover fundamental weaknesses in organization, arrangement, or premise.

Developmental editing continues through the earlier stages of your project, during planning and early drafts. You are performing a developmental edit when you evaluate a project proposal. You are performing a developmental edit when you develop and critique a project outline. You are performing a developmental edit when you scrutinize models and flowcharts for appropriateness and logic. You are performing a developmental edit when you redesign the visuals in your project. Perhaps you will test your concept of the project by interviewing users of your product. You may decide that your chosen genre is not the most appropriate one for your purpose. Revisions based on your interviews are part of a developmental edit. During a developmental edit you check your storyboards for appropriate continuity and technical feasibility and evaluate your graphics for usefulness and clarity. Changes you make during developmental editing will save you time and money later, since you will catch problems before they have been elaborated on and while they can easily be undone.

The stage of developmental editing is not the one in which you correct grammar, spelling, or punctuation. In fact, paying attention to those details during a developmental editing is a waste of time, since those very words you correct now may not even make it into the next draft once you have performed a developmental edit on the document.

TIPS: Developmental Editing

- Don't proofread. Don't copyedit. Developmental editing is a planning and evaluating tool, not a spelling checker.
- Put yourself in your reader's shoes. Will your document or Web site effectively convey its message? Could it be reorganized to do a better job?
- Ask yourself whether information is presented in a logical order. Does it build from general to specific, from specific to general, chronologically, in size order, in order of complexity, or in some other order?
- Check that similar items and subjects are treated similarly.
- Verify that the parts form a coherent whole.
- Make sure that the information is complete but not redundant.
- Consider whether the genre is appropriate for the content. Would the text be more effective as a visual? with a visual? For an online project, would multimedia aids add to or subtract from the impact of the text?

(continued)

Developmental Editing *(continued)*

- Use an appropriate level of theory. Is the text too theoretical? Should you have more concrete examples? Is it too concrete? Do you need to include more theoretical underpinning for your assertions?
- Take care of any legal issues that you must consider in presenting this content. Are copyright and trademark issues being respected? Are permissions needed?

SUBSTANTIVE EDITING

In this phase of editing, the editor thoroughly checks the content and meaning of the technical document. The editor checks, for example, the purpose, scope, and accuracy of the information, and the organization, design and layout, style, and logic of the document. Substantive editing may occur with each successive draft of a document. While developmental editing will begin as the document is still being planned and substantive editing reviews the document after it has been drafted, both types of edits treat similar concerns. They both seek to ensure that the document is as effective as possible.

TIPS: Substantive Editing

- Check that material is presented in a logical order and that items are grouped logically.
- Verify that information is complete but not redundant.
- Establish that the information is presented appropriately for the audience, scope, and purpose.
- Verify that the level of detail is appropriate for the audience, scope, and purpose.
- Work through the design of the document. Are the headings, headers, footers, and other design elements appropriate and helpful?
- Ensure that the tone is consistent and appropriate for the genre.
- Examine text to determine if visuals would be useful.
- If there are visuals, check that they are necessary, that they are clear and easy to understand, and that they display information appropriately and accurately.

COPYEDITING

Copyediting occurs after developmental and substantive editing. The copyeditor checks for punctuation and spelling errors, correctness, consistency, accuracy, and completeness.

TIPS: Copyediting

- Evaluate text for flow. Can the reader follow the argument from one sentence to the next? from one paragraph to the next? from one section to the next?
- Correct faulty grammar, spelling, punctuation, capitalization.
- Check that text references match graphics, figures, tables.
- Verify cross references to other sections of the work.
- Verify the consistent use of headings, terminology, spelling, and punctuation.
- Ensure that acronyms, initialisms, and abbreviations are explained when they are first mentioned unless they are in such common use that the reader can be expected to know them.
- Check for internal consistency throughout the document.
- Ensure that the information that is provided is correct and complete.
- Check that figures, tables, and illustrations are marked for insertion in the most appropriate location.
- Check the quality of graphics for accuracy and readability.
- Ensure that the text conforms to the chosen style guide and house style.

PROOFREADING

Proofreading occurs at the end of the process. It is sometimes confused with copyediting, but the two have distinct differences. Copyediting occurs earlier in the process. It essentially helps you to make the document ready to be printed. Proofreading occurs after the document has been printed and verifies that the text has been printed according to your requirements. Proofreading marks are made chiefly in the margins of the printed copy. Copyediting marks and comments are typically made both in the margins and within the lines of text in a manuscript. Many dictionaries and some style manuals list the standard proofreading marks. They can be found online as well, for example, at <http://www.eeicom.com/staffing/marks.html>.

TIPS: Proofreading

- Check spelling, punctuation, and capitalization.
- Verify format consistency.
- Check line breaks and word breaks for correctness.
- Eliminate repeated lines; replace missing lines.
- Review the placement of graphics for correctness.
- Ensure the consistent use of heading styles.
- Verify the alignment of text and figures.
- Review indentations, margins, and alignments.
- Verify page number accuracy.
- Check callouts for accuracy.
- Compare captions with text descriptions.
- Ensure that house style is followed throughout.

EDITING FOR STYLE

Editing for style, your manner of expression in your prose, presents some difficult challenges because so many style choices are personal choices. Often, writers have an array of choices for an effective or appropriate word, they can select among many ways to write an effective sentence, and they can craft paragraphs and larger segments in countless ways. Editing for style in your own writing presents difficulties because often you are too familiar with what you have written. After writing and revising a number of drafts, you will find it more difficult to see whether you are choosing the most appropriate or best words, expressing your thoughts in the most effective sentences, and ordering your paragraphs or larger segments in ways most suited for your audience. If you are revising your work well, you are continually editing for style because you are continually trying to make every word count, every sentence concise, and every paragraph necessary.

Editing the style of other writers presents additional challenges. You will be constantly tempted to change another writer's style to make it more like your own. Particularly when you are used to editing your own work, you may be in the habit of editing what you read into a certain style. Now is the time to resist the impulse to impose your own style on another person's writing. Naturally, you will edit to conform to your professor's requirements or your employer's style guide. But be aware that what you are reading may be perfectly fine as it is, even if it is not phrased exactly as you would have done it yourself. The same thought can often be phrased many different ways; it is up to you to make sure that your suggested changes are actually improvements.

TIPS: Editing for Style

- Check the discourse community. The expectations of audiences can vary widely within a discourse community as well as differ greatly in different discourse communities. Different discourse communities have different styles, and to write successfully in these discourse communities, you must know what these different styles are and how to use them to your advantage. Check to see whether the document assumes any basic knowledge that in fact the audience may not have. Check to see whether there are special preferences in the discourse community to which this document should give particular attention.

- Check the appropriateness of the style. Check to see whether the style is too simple or too complex. Check to see whether the audience will be familiar and comfortable with the style used in the document.

- Check the persuasiveness of the prose. Check to see whether you have presented your argument in the most convincing way for your particular audience.

- Check the diction. Are you using the most appropriate words for your purpose? Are any of your words unclear, or could any of them be clearer? Are you using concrete and specific words and avoiding, wherever possible, unnecessary abstract words? Are you consistent in your terminology throughout the document? Do you avoid pretentious words and use simpler words wherever possible? Are you using words that express your sincerity?

- Check the technical terms and jargon. Will your audience be familiar with all of the technical terms you use? Are you clearly defining any terms that may be unfamiliar to your audience? Where necessary, are you defining terms the first time you use them? Are you also providing a glossary?

- Check the sentences. Is every sentence you have written necessary? May any be omitted? Are any sentences wordy or redundant? Are any sentences overly complex? Does every sentence state your meaning as clearly as possible?

(continued)

Editing for Style *(continued)*

■ Check the paragraphs and other segments. Is every paragraph unified, complete, ordered, and coherent? Are any paragraphs merely transitional paragraphs, and do they work effectively as transitional paragraphs? Are any paragraphs too long?

■ Check the tone (including bias). Have you given careful consideration to the attitudes and feelings of your audience? Have you been sensitive to the way your readers see the subject or issue? Have you treated your audience courteously, seriously, and professionally?

■ Check for ethics. Have you omitted any key facts or evidence that may weaken your argument but that may be essential for your audience to understand your point of view? Have you been fair to all sides of the issue if you are discussing an issue? Have you misrepresented or distorted any facts or opinions? Have you used your sources reliably and documented them fully?

Accuracy

Accuracy—freedom from error—is important in many types of writing, but it is an absolute necessity in most kinds of technical writing. When you write technical prose, you provide your readers with information that they will depend on, and its accuracy ensures that they can understand, evaluate, and use the information you have conveyed. Consider the following terms, for example: *DNA–chromosome* and *inapt–inept*. To a layperson there may not be much difference between each word in these word pairs. To a specialist, however, there are essential differences. For example, most people recognize DNA as the genetic material of living organisms, whereas a biologist would realize that DNA is a collection of nucleic acids that code for specific genes in chromosomes. Also, words such as *inapt* and *inept* sound similar but have different meanings: The former means "inappropriate," and the latter signifies "incompetent." It's essential that these terms and countless others be used accurately and not interchangeably.

When you are writing about technical subjects, you have to understand your subject well to express it clearly to your readers. If you don't understand the technical details, how will you explain them in your prose? Even if you are quite knowledgeable about your subject, you may find it useful to have your work reviewed for technical accuracy. Whether your subject is highly technical, moderately technical,

slightly technical, or not technical at all, accuracy is a basic requirement to create functional, effective documents.

Active Verbs and Active Voice

Verbs express the action in a sentence. When you use passive voice
and weak, colorless verbs, you are depriving your prose of a simple
yet powerful tool for carrying your message effectively. Choose verbs
that express action and use the active voice. Often writers bury the
action of a sentence in nouns, adjectives, or other parts of speech.
Writers create sentences such as the following: *The chemist completed the dilution of the solution and performed the transference of
the diluted solution to the beaker.* In this sentence the nouns—*dilution, transference*—are trying to carry the sentence. With a little revision, the verbs could do most of the work: *The chemist diluted the
solution and transferred it to the beaker.* This revision is shorter and
more direct.

Good writers make abundant use of action verbs. Action verbs are
specific verbs that descriptively tell the audience what was or will be
done. They make your writing more direct and forceful. Action verbs
in an employment letter or résumé may include *designed, collabo-
rated, edited, wrote, accomplished, organized, directed, polished, re-
worked, networked, pioneered, supervised,* and *advanced.* The list
of possible action verbs for a résumé or employment letter is almost
endless.

In instructions, simple imperative statements use action verbs
to tell readers exactly what needs to be done: *Insert the cable into
socket A; press the Enter key; right click on the desktop and select
Properties from the menu;* or *Cut the blue wire ¹/₄ inch from its base.*

To place the action in the verbs of your sentences, ask yourself
what the main action of the sentence is, identify the actor, and try to
begin your sentences with the actor followed by the main action. For
example, consider the following sentence: *The rats were fed protein
pellets by the researcher in Phase II of the experiment.* The main action concerns feeding the rats. The actor is the researcher. A possible
revision is *The researcher fed the rats protein pellets in Phase II of
the experiment.*

Often, the actor is not specified, because the information is either
irrelevant or unknown. In the above example, the original sentence
might have read *The rats were fed protein pellets in Phase II of the
experiment.* Now you have a choice when editing this sentence. If the

writer is describing the stages in a procedure, it may not be important to know who is doing the action, only that it was accomplished. Sometimes, you will have to guess who the actor is and verify your guess with the author.

Active voice is another strategy for helping your reader recognize the agent or doer of an action, and, like action verbs, active voice helps to make your writing more direct and forceful. Your readers are told quite clearly who is doing what to whom. To write in the active voice, place the actor in the subject position of the sentence. For example, write *The systems administrator set up the Novell Netware network in the lab so twenty workstations could share programs and printers*. Or write *This program installs itself after you double-click the executable file*. In the first sentence, you know who set up the network, and in the second sentence you know that the program installs itself. Active voice helps readers to recognize more quickly who is doing what action, helps to make sentences shorter, and helps to avoid the ambiguity created by using passive voice.

Passive voice makes the agent of the action less clear. Sometimes passive voice weakens a sentence because it increases the ambiguity. *Every evening the lab must be cleaned and the computers checked for viruses*. In this sentence it is not clear who must perform the actions. In the revision, the actor is clearer: *Every evening you must clean the lab and check the computers for viruses*.

Depending on your purpose, you may need to use passive voice at times. If you want to deemphasize or obscure the agent of the action, for example, you may use passive voice. Since passive voice weakens the impact of your verb, you will want to ensure that when you use it, you are using it for a good reason.

Appropriateness

The appropriateness of a document refers to a writer's degree of success in aptly choosing language that is especially suitable or compatible for a given subject, purpose, scope, and audience. When you choose a word, a phrase, or even a tone, your choice should be suitable for the situation. By using inapt language, you distract from your message and sabotage your document's effectiveness. When you choose words, their associations can make the difference between appropriate and inappropriate usage.

Consider carefully the message your words carry beyond their dictionary denotation. Words bring connotations that help to determine their suitability in a given context. Every word you choose helps to shape your readers' perceptions of your subject matter. Consider, for example, the word *inspect*. The word means to examine carefully, especially for flaws. Synonyms are *investigate, probe,* or *scan*. Telling a client that you are going to *inspect* a product report has a different connotation from suggesting that you are going to *scan* the product report. *Inspect* is a more appropriate diction choice and suggests that you are going to examine the report very carefully, looking for flaws. *Scan* is a less intensive word and signifies that you are going to examine the report more superficially. Your word choices strongly influence the effectiveness of your prose and how your audience views your own seriousness, credibility, and persuasiveness.

Clarity

Normally, you will want your readers to understand your prose without having to struggle over your meaning. One important way to achieve this effect is to strive for clarity in your writing. Clarity, quite simply, is the state of being clear. Writing clearly means using words in the right way and focusing on achieving a single meaning for your words, phrases, and sentences. So you can think of clarity as the absence of ambiguity. Far from being the exception in technical prose, ambiguity is all too often the rule. Unclear prose shows up in many technical documents, and ambiguity may occur at the word, phrase, clause, sentence, paragraph, or other segment level. If what you write can easily be misinterpreted by your readers, you have not been clear. If your readers misconstrue your message and interpret it in a different way from what you intended, you have not been clear. Sometimes, it is difficult to imagine how our words can be misconstrued, since for us they seem so unambiguous. Here is where an outside reviewer can be helpful in identifying possible sources of confusion in your writing.

Conciseness

When you write with conciseness, you capture your meaning in exactly the length and depth required, without stinting or overdoing. Your treatment is neither incomplete nor superfluous. It is actually more difficult to write fewer words that express your meaning than more words. Concise writing often requires lots of rewriting to eliminate unnecessary words from early drafts. But when you cut back to achieve conciseness, take care that you are not falling into the trap of writing cryptically. True conciseness consists of achieving a balance between lean prose and sufficient detail.

Two weaknesses to look out for are unnecessary words and long words. Consider, for example, the unnecessary words in the following sentence: *Lift the cover, and after lifting the cover, place the test tube inside the tray of the spectrophotometer and then close the spectrophotometer cover.* Instead try: *Lift the spectrophotometer's cover, place the test tube inside the tray, and then close the cover.*

Concreteness

Concreteness refers to language that describes tangible things. Its opposite is abstractness. Concrete words are words that refer to specific objects, persons, places, and acts, things that can be detected by the senses. *Apple Computer Corporation* is more concrete than *the computer industry; Gatorland* is more concrete than *zoological park; amylase* is more specific than *enzyme;* a *1997 Toyota Corolla* is more specific than *car;* and *using the Infoseek search engine to perform a Boolean search on frogs* is more specific than *using a search engine.*

Abstractness, in contrast, refers to general ideas, acts, conditions, relationships, qualities, or anything that cannot be perceived by the five senses: *courage, evil, goodness, truth, freedom,* and *justice.*

Concreteness and abstraction are rarely encountered as absolutes; they exist along a range. Choose words suitable for your audience. You will use both types of words in your writing, but an overabundance of abstract words will mask your meaning and bore readers. Use concrete details in conjunction with abstract words as well as examples to explain concepts. Many abstract words present readers with too many possibilities for interpretation. In technical prose, abstract words are often necessary, but there are times when a more concrete word is a better choice.

Consistency

Consistency refers to your treatment of repeated elements in your document. You achieve consistency when you stay at the same level of formality throughout your document. You achieve consistency when you control your vocabulary and don't overuse synonyms. You achieve consistency when your punctuation and spelling of technical terms remains constant throughout your document.

You must be careful about spelling or punctuating certain words consistently. You need to decide, for example, whether you will refer to your audience members as *readers* or *reader(s)* or *users* or *user(s)*. If you are referring to the document you are writing, will you use the term *book, guide, manual, instructions,* or *document*? Usually, you will want to make sure you are using terms consistently. You may have been taught to vary your writing by using synonyms or rephrasings to avoid excessive repetition. Although this advice often holds for non-technical prose, in technical writing you may have to forgo this nicety of expression in the interest of effectiveness. Readers assume that different technical terms refer to different items, and if you sometimes use *fixture* and other times use *light,* your audience may very well believe that you are talking about two different items even when you're not. So in general, be consistent and use the same term throughout the document. Of course, you must make decisions about capitalization and usage as well. To achieve consistency, your company may have developed a style guide you must follow; if not, you can create your own to ensure consistency within your own documents.

Simplicity

Most often, a simple or plain style should be your objective in writing technical documents. The strategy of simplicity does not mean to patronize your audience or simplify your prose unnecessarily. After all, you are often writing about a complex subject. However, it does mean that as much as possible and as often as it suits your purpose, you should prefer the familiar word to the unfamiliar word, the simpler word to the more complex word, and a plain style to an affected or pretentious style. Many writers of technical prose use language that is more complex than it needs to be. And many needlessly use jargon, in the sense of unnecessarily technical language, that is totally unsuited to their audience.

Sincerity

Sincerity is not a feature you would expect to be concerned about for a technical document. After all, doesn't it go without saying that your user guide, for example, would present its instructions honestly? Why would technical writers intentionally try to deceive their readers? Although sincerity may not be a quality you strive for explicitly, it underlies all your writing. Yet even ethical writers grapple with these issues in technical prose. Lapses in sincerity may occur when software bugs, often euphemistically lumped together with "undocumented features," go discreetly unmentioned in the reference manual or instruction sheet. And sometimes, insincerity is more a question of information misleadingly omitted than information falsified.

To create a relationship with your readers, you will want to establish that you are a trustworthy source of information. If your readers don't believe in your sincerity, your text will not effectively convey its message. You may have valid reasons for withholding or distorting information, but sincerity is a powerful tool to achieve believability. Good writers establish a bond with their readers, and sincerity is one very effective way to maintain that bond.

In general, write truthfully and honestly, and unless using humor suits your purpose, you should also treat your subject and your readers seriously. Choose words that help you to convey your tone, establish your credibility, and build trust with your audience. Pay attention to those words, phrases, clauses, sentences, and larger segments that can undermine the relationship that you are carefully crafting with your reader.

EDITING FOR GRAMMAR

To be an effective editor of your own work or someone else's, you must have a thorough understanding of the rules of grammar, and, of course, grammar is a complex subject. A few major concerns are highlighted here to help you with some of the most basic points.

Parts of Speech

NOUN

A noun is the name of a person, place, thing, or abstraction. Examples: *PC, pointer, placement*. Nouns function as the subject of a sentence, as the complement of a verb, as an appositive, and as a term of direct address.

PRONOUN

A pronoun substitutes for a noun. Pronouns can be personal *(you, me, ours, mine)*, interrogative *(who, what, which, whom)*, indefinite *(one, some, each, all)*, relative *(who, whose, which, that)*, demonstrative *(this, that, these, those)*, reciprocal *(one another, each other)*, reflexive *(himself, myself)*, or intensive *(myself, herself)*. The same pronoun can function differently, depending on the context.

VERB

Verbs perform an action or describe a state of being. Action verbs express a physical action (She *pressed* the enter key; He *shut* down the computer). Other verbs express a mental action (He *thought* he *understood* the computer operating system). Verbs can appear in many forms. They change tense to indicate time *(walk, walked);* voice to move from actor to acted upon *(sing, is sung);* mood to show action, command, or wish *(he is staying; halt! so be it)*. Verbs, unlike most other parts of speech, can carry a sentence by themselves.

ADJECTIVE

An adjective describes or modifies a noun or pronoun. Adjectives may modify nouns or pronouns in one of several ways. An adjective may tell what kind (a *large* laptop; a *fast* computer processor); may point out which one (*this* technical manual; *that* document plan); or may tell how many (*several* volumes; *nine* computers).

ADVERB

An adverb modifies a verb, adjective, other adverb, or sentence. An adverb most commonly modifies a verb. Modifying a verb, an adverb tells how (he worked *quickly*), when (he called *immediately*), where (the team worked *there*), or to what extent (the team worked *hard*). Adverbs may also modify an adjective (she is *extremely* knowledgeable). And adverbs may modify another adverb (the program worked *very* slowly).

PREPOSITION

A preposition describes the relation of a noun or verb to another word. Prepositions form phrases along with their noun or pronoun objects and modifiers. The company *across* the street went out of business; Our meeting is *after* lunch; Everyone *in* the workroom escaped injury.

CONJUNCTION

A conjunction joins words, phrases, and clauses. Conjunctions can be coordinate conjunctions, which join words, phrases, or clauses of equal importance *(and, but, or, nor, for, so, yet)*; conjunctive adverbs, which join independent clauses *(accordingly, afterwards, also, besides, consequently, furthermore, hence, however, later, moreover, otherwise, still, then, therefore)*; or subordinate conjunctions *(until, since, if, because, after, before, although, that, as if, so that, though, unless, while, when, where)*.

INTERJECTION

An interjection is a word that expresses emotion or surprise *(Wow! Hey! No way!)*. Interjections have no grammatical relationship to other words in a sentence.

Problematic Grammar Issues

We have all been taught grammar since grade school, but several issues seem to provide problems for many people. To be effective, your writing should be as free from error as possible. If you haven't mastered the use of your language and its grammar, those in your audience will question the accuracy of the information you are trying to convey. Correct grammar and usage are the prerequisites to believable prose.

FRAGMENTS

Fragments are incomplete sentences. Of course, many professional writers use fragments for a deliberate effect, but many inexperienced writers create fragments when they think they are writing complete sentences.

Incorrect:	Waiting for the computer to boot up.
Correct:	Wait for the computer to boot up.
Incorrect:	After recording the test results.
Correct:	After recording the test results, we calculated the distributions.

RUN-ON SENTENCES

Run-on sentences are unnecessarily long or wordy sentences:

Incorrect:	After connecting all of the necessary wires to the CPU, printer, speakers, and scanner, we turned on the computer to see if it would work properly only to

discover that something was not properly connected because the printer was not working at all and we knew the problem was not a software problem.

Correct: After connecting all of the necessary wires to the CPU, printer, speakers, and scanner, we turned on the computer to see if it would work properly. That is when we discovered that something was not properly connected. The printer was not working at all and we knew the problem was not a software problem.

COMMA SPLICES

Comma splices occur when you combine two independent clauses by splicing them together with a comma. Typically, you can correct a comma splice by creating two separate sentences, by adding a coordinating conjunction, or by separating the two independent clauses with a semicolon.

Incorrect: We soon had to admit that we had failed to find the problem, whatever the problem was we were not going to discover it today.

Correct: We soon had to admit that we had failed to find the problem. Whatever the problem was, we were not going to discover it today.

DANGLING OR MISPLACED MODIFIERS

Dangling modifiers occur when the sentence is worded in such a way that the modifying phrase or word does not modify the subject:

Incorrect: Opening the software package, the disk was not enclosed.

Correct: Opening the software package, we realized that the disk was not enclosed.

FAULTY PARALLELISM

Faulty parallelism occurs in sentences or lists whose component parts do not agree in form.

Incorrect: Connect the cable, install the printer driver, and the light will blink to show that the device is working.

Correct:	Connect the cable, install the printer driver, and check that the light is blinking to show that the device is working.

SUBJECT–VERB AGREEMENT

Subjects and verbs must work in pairs, agreeing in number. Singular subjects take singular verbs. Plural subjects take plural verbs. It is not always simple to determine whether a subject is singular or plural, however.

Incorrect:	Either my CD-ROM drive or my scanner are causing the problem.
Correct:	Either my CD-ROM drive or my scanner is causing the problem.

PRONOUN–ANTECEDENT AGREEMENT

Pronouns stand in for nouns or other pronouns. The nouns and pronouns that they stand for are their antecedents. A plural pronoun should stand for a plural antecedent.

Incorrect:	Tell whoever wants to come that they can.
Correct:	Tell those who want to come that they can.

PRONOUN CASE

Pronouns have case just as nouns do, and the correct case to use depends on their function in the sentence. When pronouns are used as the subject of the sentence or clause, they are in the nominative case *(I, we, they)*. When they function as an object of a verb or a preposition, they are in the objective case *(me, us, them)*.

TENSE, MOOD, VOICE

Verbs have tense, mood, and voice. Tense tells when an action occurred—in the past, present, or future. There are six tenses. Consider the verb *to go*: present tense *(I go)*, past tense *(I went)*, future tense *(I shall go)*, present perfect tense *(I have gone)*, past perfect tense *(I had gone)*, and future perfect tense *(I shall have gone)*.

The present tense expresses action (or state of being) occurring now; past tense expresses action (or state of being) that occurred in the past but did not continue into the present; the future tense expresses action (or state of being) at some time in the future (using *shall* or *will*); the present perfect tense expresses action (or state of being) occurring at no definite time in the past (formed with *have* or *has*); the past perfect tense expresses action (or state of being) completed in the past before some other past action or event (formed with *had*); and the future perfect tense expresses action (or state of being) that will be completed in the future before some other future action or event (formed with *shall have* or *will have*). Use tenses consistently and in context.

Mood refers to whether the verb is stating something (indicative: *The printer is making a strange noise*), requesting something (imperative: *Then turn it off!*), or speculating about something contrary to fact or to be desired (subjunctive: *I wish it were working now*). Use the subjunctive mood in contrary-to-fact statements (after *if* or *as though:* If I *were* you, I'd buy this computer) and in statements expressing a wish (I wish it *were* true).

Voice refers to whether the subject of the verb is the agent of the action or the recipient of the action. A verb is in the active voice when it expresses an action performed by its subject *(The stylus tore the paper)*. A verb is in the passive voice when it expresses an action performed upon its subject *(The paper was torn by the stylus)* or when the subject is the result of the action *(The paper was bought by my room-mate)*. Both active and passive voice may be used in many instances. However, the passive voice may produce awkwardness in writing, and, in general, a passive verb is weaker in its effect than an active verb. (See "Active Verb and Active Voice," on page 179.)

EDITING FOR MECHANICS

Effective editing requires a thorough understanding of mechanics: spelling, punctuation, bullets, abbreviations, acronyms, initialisms, numbers, and capitalization, among other topics.

Spelling

Many people are poor spellers. Even many good writers are poor spellers. Of course, a full-featured word-processing program has a spelling checker feature, but even a first-rate spelling checker won't catch certain kinds of spelling errors. It certainly won't catch words that are confused with similar words *(here, hear; their, there; your, you're)*. There are no easy shortcuts to improving your ability to spell correctly. Reading constantly and widely can certainly help. Sounding out words can help, too. And caring about spelling correctly is a factor as well. Make a list of words that you commonly misspell. (You'll find that many people misspell many of the same words.) Also, be aware that many words may be spelled in more than one way (for example, *employee, employe; judgement, judgment; online, on-line; disk, disc*). In these cases you should use the preferred spelling or the more common spelling. Your employer or professor may require you follow a specific style guide on these and other issues.

Punctuation

Punctuation is the use of standard marks, such as commas and periods, to separate sentences, change tone, or indicate pauses. The purpose of punctuation is to increase clarity and readability of text. Use punctuation consistently; that is, punctuate similar situations in the same way.

PERIOD

The principal use of a period (.) is to indicate the end of a sentence. It is also used in abbreviations, after the last item in a list of incomplete sentences, and in Internet addresses to separate elements of domain names.

QUESTION MARK

The question mark (?) is used at the end of a question sentence. It can be used to separate smaller question elements within a sentence.

Was it at the library? the laboratory? the operating room?

EXCLAMATION POINT

The exclamation point (!) is used to indicate surprise, irony, or strong emotion.

Never close your file without first saving it!

COMMA

The comma (,) is a mark of separation and enclosure that indicates a grouping of words in a sentence. A comma is used to add emphasis or to indicate a pause. A comma helps your audience to read the text with ease. Use a comma to set off an introductory clause, phrase, or word.

After running the setup program, look at the Readme file.

A series is composed of three or more words, phrases, or clauses of equal grammatical rank. Use a comma to separate items in a series, including before the conjunction *(and, or)*.

You will learn to use the menus, toolbar, and keyboard macros.

The old computer was slow, dirty, small, and unusable. (four adjectives)

The software package contained five disks, two compact disks, one user guide, and a manual. (four nouns)

She started the computer, inserted the disk, and ran the program. (phrases)

Larry installed the floppy drive, Mort wired the wall, and I bought the software. (independent clauses)

Use a comma in dates, before abbreviations after names, and in addresses.

Monday, February 11, 1991   Sam Scientist, Jr.   New York, NY

Use a pair of commas to enclose parenthetical interrupters.

A mouse, while not a necessity, is certainly a convenience.

Commas are used to separate parts of a sentence to make the sentence clear and direct. Use a comma before *and, but, for, or, nor,* and *yet* when they join the clauses of a compound sentence.

The printer light blinked, but the document didn't print.

I knew the problem was fixed, for the document printed.

Use a comma between coordinate adjectives preceding a noun. A comma separating two adjectives indicates that the two adjectives are equal in their modifying force.

He bought a fast, powerful computer.

SEMICOLON

A semicolon (;) marks a major break in a sentence and indicates a pause. The semicolon shows that two statements within one sentence are related. A semicolon is also used to separate major sentence elements of equal grammatical weight. It signals a greater break in thought than a comma. Use a semicolon to separate independent clauses when there is no connective word.

> The programmer completed the revisions; the system rebooted successfully.

Use a semicolon to separate items in a series when the items contain internal punctuation.

> WordPerfect, version 5.1; Microsoft Word, version 4.0; and Lotus, version 2.0 were installed onto the computers in the lab.

COLON

A colon (:) indicates that what follows is an example, explanation, or extended quotation. It is also used to introduce a list. The first letter after the colon may be either a capital or lowercase letter. The usage must be consistent throughout the document. Use colons for a list.

> Each person must bring the following to computer class: blank floppy disk, mouse pad, and lab manual.

Use colons in bibliographies before page numbers of journal articles. Use colons in an informal table or unnumbered figure.

> Table 1.1: Sample Spreadsheet   Figure: Start-up screen

Use colons in a memorandum after identification words:

> TO: FROM: DATE: SUBJECT:

Use colons as a separator.

> Hours and minutes: 14:50

Use colons in a ratio.

> The ratio of computers to personnel is 1:4.

Use colons to introduce a quotation.

> In my computer class, my professor says about the Internet: "The Internet is an indispensable tool for the completion of this course."

Do not use a colon between a verb and its complement or object.

> **Incorrect:**   The winners were: Pat, Lydia, and Jack.
>
> **Correct:**   The winners were Pat, Lydia, and Jack.
>
> **Correct:**   There were three winners: Pat, Lydia, and Jack.

Do not use a colon between a preposition and its object, or after *such as.*

Incorrect:	Many people do not take advantage of computer capabilities, such as: software, e-mail, games, and desktop publishing.
Correct:	Many people do not take advantage of computer capabilities, such as software, e-mail, games, and desktop publishing.

DASH

There are two kinds of dashes, and they differ in both length and function. The em dash (—), so called because it is the approximate width of a capital M, is used to show an abrupt change of thought in a sentence. It must be used sparingly and never as a substitute for other punctuation marks.

Properly loaded paper—notice the word *properly*—will not jam in the printer.

The en dash (–), which is approximately the width of a capital N, is used to indicate a range, such as of pages, dates, or times: *Lab hours are 11 a.m.–7 p.m.* It can also be used to separate the two numbers in a sports score: *The Yankees won, 7–2.* Use an en dash to join equal partners in an adjective pair, such as, *He never involved himself in PC–Mac debates.* In addition, en dashes are used in place of hyphens when an element of the hyphenated expression consists of more than one word, for example, *The company developed Year 2000–compliant software.*

HYPHEN

A hyphen (-) is used to connect words to make a compound word. The hyphenated word is read as a single unit, such as an *up-to-date* system. Be consistent throughout the document with compound words that are hyphenated. Use a hyphen for compound numbers from twenty-one to ninety-nine. Use a hyphen between the number and the unit of measure when they modify a noun.

110-lb weight

Use a hyphen to join words used as a single adjective before a noun.

full-time job   well-written manual

Omit the hyphen when a compound adjective occurs after the noun it modifies.

> Her job was full time.

> The manual is well written.

SLASH

Slashes (/) are short diagonal lines sometimes used between two words to show that either word applies in the sentence, such as *and/or*. Avoid this usage in technical prose. Slashes are also used in dates and fractions. Slashes indicate the word *per*. For example, feet/second means "feet per second."

APOSTROPHE

The apostrophe (') has three uses: to form the possessive case of nouns and indefinite pronouns, to indicate contractions, and sometimes to form the plurals of numbers, letters, and symbols.

Apostrophes are used to stand for the omitted material in contractions.

doesn't	does not
can't	cannot
o'clock	of the clock

Learn to distinguish between the following pairs of contractions and possessives:

it's (it is)	its
there's (there is)	theirs
they're (they are)	their
who's (who is)	whose
you're (you are)	your

Apostrophes indicate contractions and sometimes possessives. Contractions are a shortened forms of two words. For example, the contraction for the words *do* and *not* is *don't*. Do not use contractions in formal writing. When writing less formal documents, such as letters of complaint or praise, you may use contractions to give your writing a conversational tone. Form the possessive of a noun that does not end in *s* by adding an apostrophe and an *s*.

Programmer's code

Form the possessive of a noun that ends in *s* by adding an apostrophe after the *s*.

Congress' recess

Use an apostrophe at the end of the word to form the possessive of a plural noun that already ends in *s*. Plurals that do not end in s form their possessives by adding *'s*.

All of the programmers' codes are easy to use.

Paint the men's locker room door.

Use apostrophes to form the plurals of letters, but do not use them in plurals of abbreviations or numbers.

p's and q's   VDTs   4680s series

Exception: Sometimes you have to mix plural forms to avoid a more awkward formation.

do's and don'ts

QUOTATION MARKS AND ELLIPSES

Quotation marks (" ") are used to enclose quoted material and words that are used in some special way. Quotation marks always come in pairs. The marks show the beginning and the end of a speech, whether it is part of a sentence, one sentence, or several sentences. If a speech is interrupted by material showing who said it, quotation marks set off the quoted material from the explanatory material. Indirect quotations are not set off by quotation marks. Use quotation marks to enclose the exact words of a quoted speech.

"I am sure," said Jan, "that your answer is correct."

Peggy answered, "I worked hard on this. I checked all my data."

Peggy answered that she had worked hard and checked all her data.

Note: This is an indirect quotation. Words that are not directly quoted do not need quotation marks.

Use quotation marks to set off subdivisions of books, names of songs, and titles of units of less than book length, such as short stories, short poems, essays, and articles.

The second chapter of the manual is entitled "Installing a Hard Drive."

The first article I read when researching recommendations for computer memory was "Questions about RAM" in the consumer magazine.

Use quotation marks to set off slang words used in serious writing.

If the computer has an undetected virus, the entire system can "crash."

Do not use double punctuation when ending a sentence with a quote.

| **Incorrect:** | John asked, "Where are the extra floppy disks kept?". |
| **Correct:** | John asked, "Where are the extra floppy disks kept?" |

Follow these rules for placing quotation marks in relation to other punctuation marks:

Place commas and periods inside quotes.

"Come in," said John, "and have a seat."

"Jenny," he said, "let's have lunch."

Place semicolons and colons outside quotes.

Mr. Lowe said, "I heartily endorse this brand"; unfortunately, the customer thought he said "hardly" instead of "heartily."

In the summary of the article Judy referred to "programs"; however I remembered only one.

Place question marks and exclamation points inside if they belong to the quoted part, outside if they do not belong to the quoted part.

"Heavens!" he exclaimed. "Is this the best you can do?"

Can you believe he said "no"?

Use single quotation marks (' ') to enclose a speech within a speech or a quoted word within a speech.

"Jim said, "The tech support guys told me, 'Give it a kick!'""

Ellipses (...) are used to mark an omission from a quoted passage and to mark a reflective pause or hesitation. Use the three-dot ellipses to mark an omission from a quoted passage.

In the article about the use of color, Beth Mazur writes, "When designing GIF images ... avoid using subtle gradations of color."

Use ellipses to mark a reflective pause or hesitation.

Computers, like other technological devices, can be ... temperamental.

Use the four-dot ellipses to indicate material omitted at the end of a sentence.

The hard drive just stopped working. . . .

PARENTHESES AND BRACKETS

Brackets [ ] are used to set off interpolations in quoted material and to replace parentheses within parentheses.

The reporter wrote: "No one could have predicted that it [the computer] would have had such an impact."

Not every expert agrees. (See Colin Wheildon's [*Type & Layout.*])

In some discourse communities, such as mathematics, brackets are placed outside the parenthetical expression, with inner groupings indicated by parentheses.

$$[(a - b)x + (c - d)y]$$

Parentheses ( ) are used to enclose explanatory or supplementary information and to indicate that the enclosed words are of lesser importance. When parentheses are used at the end of a sentence, put the period after the closing parenthesis. When the entire sentence occurs within parentheses, though, punctuate the sentence as usual, within the parentheses. Use parentheses to reference other text.

Before inserting a diskette into the A drive, verify that it is not write-protected (see Figure 3.1).

Use parentheses to enclose enumerated points.

To reboot the system, (1) remove any diskettes from the A drive; (2) press and hold the ALT and CTRL keys; (3) press the DEL key; (4) release all keys.

MARKS OF EMPHASIS—BOLD, UNDERLINE, AND ITALIC

Modifying your text with bold, underline, and italic effects draws attention to the marked words. While marks of emphasis have a place in technical writing, their overuse detracts from their effectiveness. If you use them, use them for a good reason and use them consistently. Bold text should be used very sparingly for emphasis, but it can be useful in headings. Underlined text is the handwritten and typewritten equivalent of italics. If you can use italics instead, do so.

Italics are used to differentiate regular text—set in roman, or regular, style—from unusual text. Italics are used for titles of books, operas, periodicals, plays, and films; names of genus, species, and varieties; for emphasis; to refer to less commonly used foreign words; and for names of boats, trains, airplanes, and spacecraft. Words used as words, rather than for their meaning, are italicized, as in "*Supersede* is often misspelled." Unfamiliar words are often set in italics the first time they appear in text. The variables in mathematical expressions are set in italics:

$$ax - 4y = 20$$

Bullets

Bullets (•) are used before each item in a list of equally important items. Bullets are often used in lists when rank or sequence is not relevant or important. Do not use bullets for a list of fewer than three items. Bulleted items should be constructed of parallel items. For example, if some list elements are noun phrases, they all must be noun phrases. If some are complete sentences, all should be complete sentences.

Abbreviations, Acronyms, and Initialisms

Abbreviations are used to save space. They can be shortened words, contractions, acronyms, or initialisms. Some abbreviations are so well known that many people use them without thinking about what the letters actually stand for. This practice leads to the misuse of abbreviations at times. For example, if you say that you are looking for an ATM machine, in effect you are talking about an "automatic teller machine" machine. And when you finally locate this device, don't expect to be asked for your PIN number, since that would be your "personal identification number" number.

While contractions and many other abbreviations are quite well known and need no explanation, others, particularly some acronyms and initialisms, are not so familiar. Define these abbreviations the first time they appear and any other time you think your readers need the reminder. For a document that is not read sequentially, that is, cover to cover in page order, define abbreviations every time they appear in a new paragraph unless the paragraphs are short and appear near each other.

Of course, many common abbreviations are in everyday use. Abbreviate the words *avenue, boulevard,* and *street* in numbered addresses but not when they are used without numbers. The words *drive* and *road* are always spelled out in full.

You'll find the San Elijo Campus on Manchester Avenue.

Write to the San Elijo Campus at 3333 Manchester Ave., Cardiff-by-the-Sea.

When a month is used with a specific date, abbreviate only Jan., Feb., Aug., Sept., Oct., Nov., and Dec. Spell out month names when using them without a date or with a year alone. In tabular material, use these three-letter forms without a period: Jan, Feb, Mar, Apr, May, Jun, Jul, Aug, Sep, Oct, Nov, Dec.

A sale is slated for Feb. 3, 1997.

February 1996 was our biggest month ever.

Acronyms and initialisms are abbreviations formed from the first letter or letters of the words in a multiword phrase. They are a special type of abbreviation, since they are not words that have been shortened but are rather a combination of several words. Acronyms are pronounced as words; initialisms are pronounced as letters. For example, LAN is an acronym formed from Local Area Network, while FTP is an initialism standing for File Transfer Protocol. Some acronyms have become accepted as actual nouns, and they are written in lowercase letters, as, for example, *sysop* (systems operator). But in general, both acronyms and initialisms are written in all capital letters, without punctuation.

The two exceptions to this rule are geographic names, which are often (although not always) punctuated (such as U.S.A.) and academic degrees, such as Ph.D. or B.A. (By convention, the *h* in Ph.D. is written lowercase.)

In the computer industry, common usage has not definitely settled on whether certain abbreviations are acronyms (and therefore pronounced as words) or initialisms (and therefore spelled out). This topic is endlessly debated in various online forums. Whether FAQ (Frequently Asked Questions) is pronounced 'fack' or 'eff-ay-cue' may not seem to matter very much. But if you are a writer, it is exactly in these situations that you will find yourself challenged to be inventive in order to avoid problems. If FAQ is an acronym, then you would write about it as "a FAQ." If it is an initialism, then you would write "an FAQ." Sometimes you will have to spell out an abbreviation to avoid just such a problem: "a Frequently Asked Questions list."

Numbers

Write out (as a word) all numbers below ten, except units of measure, age, time, dates, page numbers, percentages, money, and proportions.

> nine floppy disks   zero quality defects

Write any number greater than nine in numerals.

> 20 times better   6,000 members   45,000 computer
>   users

When two or more numbers are presented in the same section of writing, write them as numerals.

> The upgrade system contains 32 megabytes of RAM, 3 disk drives, and
> I scanner.

If none of the numbers is greater than nine, write them all out as words.

> The upgrade system contains three disk drives, one scanner, and two
> speakers.

Spell out one of two numbers—usually the shorter or the one that is attached to a unit of measurement—that appear consecutively in a phrase.

> four 4-color   two 3-inch   three 5-person   forty-five 4,500-
> photos   disks   teams   component radar
>   systems

Do not begin a sentence with numerals.

> Two thousand test subjects participated.

> A quarter of a million dollars went into the preparation of the system.

Write large numbers in the form that is most familiar to your audience and easiest to understand.

> The company lost 2.3 million dollars.

Limit repeating a spelled-out number as a numeral to legal documents.

> Either party may terminate the agreement within a three (3) month
> period.

MEASUREMENTS

- Keep all units of measure consistent.
- Use the correct units for the system of measurement you have chosen.
- Write basic units of measure in word form, derived units of measure as symbols.

8 hours	5 seconds	9 feet
24 ft/s	45 m/s	

- Place a hyphen between a number and unit of measure when together they modify a noun (used as an adjective).

2-year-old system	15,000-volt charge
8-ounce disk	12-inch-high monitor

FRACTIONS AND DECIMALS

- Use the singular when fractions and decimals of 1 or less are used as adjectives.

0.7 pound	0.6 centimeter
0.22 cubic foot	0.3 kilometer

- Write decimals and fractions as numerals, not words.

0.68 not zero point six eight	2.312 not two point three one two
0.4 not zero point four	2/3 not two thirds

- Treat decimal representations consistently, especially when presenting them in columns, rows, or groups. Precede decimal fractions of value less than one with a leading zero before the decimal point.

0.28 not .28	0.8000 not .8000

Exception: Numbers that can never be greater than 1, such as probabilities, do not take the leading zero.

$p < .05$

- Do not inflate the degree of accuracy by writing decimals with too many digits. If you divide 2.34 by 5.55, your calculator displays 0.4216216, which should be written as

0.42

- Writing a number as a decimal implies precision to the last decimal place: *the nearest 0.25 cup* implies more precision than *the nearest quarter cup.*
- If a number is merely an approximation, then say so.

about half a glass of milk   approximately one-third less area

PERCENTAGES

Spell out the percentage in the text unless doing so makes the text less clear.

On average, 18 percent of the devices failed in the first year.

ADDRESSES

Spell out numbered streets from one to ten.

West Third Street

Write all building numbers, except for the building number one, as figures.

3804 East Broadway Street   One West Eighth Street

TIME

Use a.m. and p.m. rather than AM and PM. Avoid superfluous letters and numbers.

The show starts at 5 p.m. (not 5:00 p.m.).

The program lasts from 7 to 9 p.m. (not 7 p.m. to 9 p.m.).

DATES

Do not refer to dates entirely in numbers, unless you know that your audience follows the same conventional order as you do. The date 2/11/98 means February 11, 1998, to Americans but November 2, 1998, to Europeans. Express days of the month as cardinal numbers, not ordinals.

The show runs until Dec. 6 (*not* Dec. 6th).

Capitalization

Capitalize the first word of a sentence. Use a capital letter for titles before the names of officials. Capitalize a title, rank, and so on, followed by a proper name or an epithet used with—or in place of—a proper name.

> President Phil Smith will now address us.

Capitalize proper nouns and adjectives formed from proper nouns. Don't capitalize generic nouns unless they are part of a proper name.

> Do you know what committee she is on?
>
> She is serving on the Promotion and Tenure Committee.

Capitalize every word except conjunctions, articles, and short prepositions in the names (or derived adjectives, verbs, and so on) of organizations, businesses, institutions, agencies, geographical names, and so on. In general, do not capitalize prepositions of fewer than four letters or the articles *a* and *an* unless they are the first word of a title. The article *the* is capitalized only under special circumstances. Capitalize *north, south, east* and *west* only when they are locations or are part of a proper name. Don't capitalize them if they are directions.

> Drive east to get to the office.   They work in the East.

Capitalize titles of published works (except articles and prepositions).

> Have you read the book *Illustrating Computer Documentation?*
>
> The professional journal is called *Intercom.*

Capitalize names of important historical events, holidays, months, days of the week, but not the seasons of the year.

> Valentine's Day   Tuesday
> February 14   the Fourth of July

Nonsentence elements, such as table entries, captions, or footnotes, are often capitalized as a matter of style. These elements are capitalized in sentences: run-in headings; table subtitles, headnotes, box heads, and entries consisting of words, phrases, or sentences; footnotes to either the text or a table; and figure captions.

> Figure 1. Three-view sketch of the research aircraft. Dimensions are in inches.
>
> Figure 1. Computing scheme for algorithm.
>
> Figure 1. Concluded.

Capitalize figure labels and references to specific departments but not general references.

> English Department members are involved in computer technology.
>
> Each department is invited to submit a proposal.

Capitalize subjects such as art, geography, and accounting only if they are part of a specific department or specific course title.

> She is a journalism major and plans to enroll in Introduction to Journalism this fall.
>
> The Journalism Department will hold a faculty meeting on Wednesday.

REVIEWING

Reviewing is an essential step in creating effective technical documents. Reviewing occurs when you submit your document to someone else to edit. (Some people use the term *checking* for when you examine your document yourself.)

Reviewing Small Projects

One common mistake many writers make is spending lots of time writing a document and then submitting it without carefully editing it, proofreading it, or even simply checking it for spelling errors. Another common mistake is to send out important documents without taking the time to have others review the work. You certainly don't need to have someone review every e-mail or letter you send, but if the e-mail, letter, report, or other document is important, you need to have at least one other qualified person review your work. The qualified reviewer will spot shortcomings in your logic, your organization, your tone, your technical accuracy, and so on. And, of course, this person may spot the misspelled words that your word-processing program failed to pick up and any other typographical errors. In sum, if you want your communications to be as effective as possible, you can't rely on just your own judgment. You need the judgment and help of your peers, and, where appropriate, you need to test some documents to find out whether they are fulfilling their purpose as well as they should for their intended audience.

Peer Review Strategies

As a college student, you may find that many of your writing assignments will be peer reviewed as part of the assignment. Typically, your professor will ask you to bring in to class a draft of your document shortly before the due date, and other members of the class will be required to critique your document using a critique sheet with pertinent criteria concerning the assignment. Usually, you are asked to check various elements of the document to see how well they fulfill the requirements of the assignment and the requirements of good writing and design. Of course, this kind of peer review has to be well prepared for and controlled by your professor to be as beneficial as it should be. Still, you will often find that your document will be much improved because other readers have pointed out shortcomings you failed to see.

As a corporate employee, you need to develop your own peer network, two or three people you can count on to look over and critique any essential communications you have to write. If the network is to be helpful to you, your peers must have good communication skills (if some members' writing skills are better than yours, then that's good news for you), and they must take the peer reviewing seriously. They have to be willing to sit down and carefully read your document from beginning to end, and they have to know how to offer constructive criticism. And, of course, for this kind of peer-review network to work in a corporate setting, you have to be willing to return a favor with a favor. Be sure you devote as much time and care to reviewing the documents of others as they have devoted time and care to reviewing yours.

TIPS: Reviewing Documents

- It often suffices simply to make notes in the margins of someone's document to point out various weaknesses or to note various errors. However, often attaching a separate sheet of notes, handwritten or typed, will be more helpful to the writer of the document, especially if you have detailed suggestions for improvement.

- Make sure you know what the purpose of the document is (or is supposed to be) and who the intended audience is. Two of the most important elements you can scrutinize are how effectively the writing accom-

(continued)

Reviewing Documents *(continued)*

plishes the document's purpose and how well the document is suited for the intended reader. You should point out any problems that interfere with these basic goals.

■ If a critique sheet is available for critiquing the document, make sure you understand every item on the critique sheet and what you are required to comment on.

■ If no critique sheet is available, you'll have to rely on your own criteria concerning what makes up an effectively written, designed, and, where appropriate, illustrated document. See various sections throughout this handbook for helpful criteria. You may want to jot down your own ten- to twenty-item checklist of things to look for, ranging from clarity of purpose to effectiveness of design. Of course, invariably, you'll also be checking many other items that are not on your checklist. You should continue to refine this checklist as you review the work of others. Soon you will have a helpful guide to essential elements that will save you time and even help you in revising your own writing.

■ As a reviewer, rank your suggestions for revisions. The writer of the document may not have the time to make every change that you and others suggest, so a ranking by priority can be helpful. Also, if you are offering many suggestions for improvement, the writer may feel less overwhelmed if your major suggestions are ranked.

■ Offer constructive criticism. Be truthful about weaknesses in the document, but point out these weaknesses tactfully and respectfully. Remember that another writer's ego is at stake, and remember how you feel when others comment on shortcomings in your writing.

■ Take time to discuss your comments with the writer in a face-to-face meeting. This kind of discussion isn't always possible, but if it is, it can be very useful. The discussion may simply be a five-minute overview of some items that need improvement or, depending on the complexity of the document and its importance, a critique of its strengths and weaknesses that may occupy one or more days.

■ If you are having your writing critiqued by someone else, try to remember that the criticisms are about your document, not about you.

■ If your work has been reviewed by others, make sure you understand what their criticisms are. If you are uncertain, ask questions. If you don't make the changes that the reviewer has suggested that you make, the reviewer may be reluctant to help you the next time and perhaps even completely unwilling to help the time after that. In sum, take your reviewer's comments seriously.

■ Learn from the reviews of others. Don't continue to repeat the same errors. Learn what you are doing wrong and avoid making the same mistakes again. If your reviewer keeps calling attention to the same faults in punctuation, for example, make sure you learn how to avoid repeating those mistakes in other documents.

- Reward your reviewers. Buy them coffee, a beer, or lunch. Show them how much you appreciate their help, and, again, return their help in kind. You need these people to help you look good, and they need you. It's a mutually beneficial relationship.

Reviewing Large Projects

In some complex collaborative-writing college assignments and in major projects in a corporate setting, you will frequently encounter a much more established review procedure. In a collaborative college assignment your professor may require various kinds of extensive review of your collaborative document before you and your group members finally turn in the assignment. In a corporate setting you will often find that many different departments may have to be included in the review process because the company's reputation and profits may be at stake. The review cycle may call for many phases of review including a documentation plan review, a peer review, a technical review, a customer-service review, a marketing review, an editing review, a management review, and a legal review. Of course, if a document requires many or all of these kinds of reviews, you and others in your company will need extensive project management skills to make sure the review cycle proceeds smoothly. For an excellent discussion of these advanced skills, see JoAnn T. Hackos, *Managing Your Documentation Projects.*

EVALUATING

Evaluating occurs when you or someone else assesses the usability of your document for the intended audience. Some people refer to this process as testing, and they specify that testing occurs when members of your intended audience assess how effective your document is for them.

Simple Testing

It's a good idea to test the documents you write before sending them to your intended audience. In simple testing, you are not asking others merely to review your document for errors. You are asking them to read or even use the document from the intended audience's point of view. For example, if you are writing a memo, letter, or e-mail about a sensitive or delicate issue, you may want to have someone read your document to see if it is likely to achieve your intended purpose. Often this simple testing will reveal an inappropriate tone or missing essential information. If you are writing instructions or procedures (see Part Nine, "Technical Documents"), it's a good idea to test the instructions or procedures on several readers. If you observe others while they follow your instructions or procedures, you will often see problems that you didn't anticipate: topics that are not as clear to your readers as they are to you or sentences that could be much clearer or simpler. As much as possible, create an informal checklist of the kinds of strengths you would like your document to have, and then test the document with several readers to see which features need improvement.

Usability Testing

Usability testing is common in many industries to determine the quality of a product or service. Usability testing is widely used in the computer industry, for example, for both software and hardware, and many companies have usability labs designed solely for this purpose. In software usability testing, a multiple-step process is frequently used—a process that includes various pretest preparations, the actual testing, analysis of the test results, modifying of the software, and retesting of the product. All of these steps are complex and can be time consuming.

A well-constructed usability test will use testers, facilitators, observers, and recorders to document every phase of the process. The testers, test subjects, or test users are the people who sit at the computer or other apparatus and perform the defined task. The facilitator plays the role of the "host" of the usability test. The observer's role is simply to watch the users as they perform the defined tasks and keep notes on where they have difficulty using the system. The recorder observes the test subjects as the tasks are being performed and records anything that must be measured. For example, if one of the tests concerns the time to perform each task, the recorder uses a timer and records the times required for each task.

After the testing is performed and the results have been analyzed, further changes in the software, the documentation, or both, along with further testing, typically lead to a much improved product. Of course, a sophisticated software product will usually undergo many tests and versions before it is marketed; even after it is marketed many further improvements will be required before the product performs as well as it is intended to perform.

RESOURCES

Blake, Gary, and Robert W. Bly. *The Elements of Technical Writing*. New York: MacMillan, 1993.

The Chicago Manual of Style. 14th ed. Chicago: Chicago UP, 1993.

Coe, Marlana. *Human Factors for Technical Communicators*. New York: Wiley, 1996.

Coggin, William O., and Lynnette R. Porter. *Editing for the Technical Professions*. New York: Macmillan, 1993.

Dragga, Sam, and Gwendolyn Gong. *Editing: The Design of Rhetoric*. Amityville: Baywood, 1989.

Dumas, J. S., and J. C. Redish. *A Practical Guide to Usability Testing*. Norwood: Ablex, 1993.

Gordon, Karen Elizabeth. *The Deluxe Transitive Vampire: The Ultimate Handbook of Grammar for the Innocent, the Eager, and the Doomed*. 2nd ed. New York: Pantheon, 1993.

———. *The New Well-Tempered Sentence: A Punctuation Handbook for the Innocent, the Eager, and the Doomed*. New York: Ticknor and Fields, 1993.

———. *Torn Wings and Faux Pas: A Flashbook of Style, a Beastly Guide Through the Writer's Labyrinth*. New York: Pantheon, 1997.

Hackos, JoAnn T. *Managing Your Documentation Projects*. New York: Wiley, 1994.

Hackos, JoAnn T., and Janice C. Reddish. *User and Task Analysis for Interface Design*. New York: Wiley, 1998.

Hodges, John C., et al. *Harbrace College Handbook: With Revised 1998 MLA Style Updates*. 13th ed. New York: Harcourt Brace, 1998.

The Microsoft Manual of Style for Technical Publications. 2nd ed. Redmond: Microsoft, 1998.

Nielsen, J., and R. Mack. *Usability Inspection Methods*. New York: Wiley, 1994.

O'Conner, Patricia T. *Woe Is I: The Grammarphobe's Guide to Better English in Plain English*. New York: Putnam, 1996.

Rosen, Leonard J., and Laurence Behrens. *The Allyn & Bacon Handbook*. Boston: Allyn & Bacon, 2000.

Rubens, Philip, ed. *Science and Technical Writing: A Manual of Style*. New York: Henry Holt, 1992.

Rubin, Jeffrey. *Handbook of Usability Testing: How to Plan, Design, and Conduct Effective Tests*. New York: Wiley, 1994.

Rude, Carolyn. *Technical Editing*. 2nd ed. Boston: Allyn & Bacon, 1998.

Sun Technical Publications. *Read Me First!: A Style Guide for the Computer Industry*. New York: Prentice Hall, 1997.

Tarutz, Judith. *Technical Editing*. Boston: Addison, 1992.

Wheildon, Colin. *Type and Layout: How Typography Can Get Your Message Across—or Get in the Way*. 1984. Berkeley: Strathmoor, 1995.

Wood, Larry E., ed. *User Interface Design: Bridging the Gap from User Requirements to Design*. Boca Raton: CRC, 1997.

Words into Type. 3rd ed. Englewood Cliffs: Prentice-Hall, 1974.

PART **5**

Production

Tips

Production

*D*ocument production is the process that transforms a project into a finished product—either into print or into some other medium such as an online help file, a Web site, or a CD-ROM. You have many choices about how your documents will be produced, and you will make those choices on the basis of your needs, your audience's needs, your budget, and other factors. Both in-house and commercial printers now have more capabilities than ever before for helping you to print large-quantity, high-resolution print jobs. Often you will still have to make many decisions about the many production details. Because of desktop publishing and the ready availability of other media, many of the steps in production will be directly under your control. Some you will leave to others. Regardless of the medium you choose, high-quality document production is essential if you want your technical documents to communicate information effectively to your audience.

If you are producing your print job yourself, a personal computer and a laser printer can help you to produce professional-looking documents quickly and easily. And if you have access to a color laser or inkjet printer, your documents can look even more impressive. Most laser printers produce text and graphics that are suitable in quality to provide camera-ready copy for immediate use in a commercial printing press or to be used as an original for an in-house photocopier. Omitting many manual steps through these electronic publishing systems can cut production time, reduce errors, and save money.

Increasingly, both college students and company employees are producing documents of all kinds electronically, using online help tools for online help files and tutorials, Web editing programs for Web pages, and authoring programs for multimedia presentations and computer-based training. In fact, rapid developments in technology will inevitably continue to offer many different avenues for document production.

PLANNING TO PRODUCE YOUR PRINT DOCUMENT

Accounting for the production of your document is an essential part of the planning process for any major document. (See Part One, "Planning and Research," for a discussion of planning documents.) For smaller runs of print documents, the production considerations (paper stock, binding, printing, number of copies, costs, and so on) are minimal. For example, you may simply need to provide five copies of a twenty-page report printed on an inkjet or laser printer. Most college students won't have production problems bigger than this.

You'll still need to allow for some production time, however minimal, and consider possible delays. What should you do if the printer cartridge runs out of ink? if there's a paper jam? if the printer develops other problems? In the document plan for your project, you should allow for possible unforeseeable delays in the schedule. And how should you schedule the production of projects published in other media, such as on the Web? If your Internet Service Provider (ISP) is inaccessible because of scheduled or unscheduled downtime, what will you do? For your online help files, what if your compiler crashes or produces mysterious error messages?

In a corporate environment you can and will encounter the same technical glitches in the production of your projects. You'll need to anticipate them and to know in advance ways to work around these delays. Sometimes you may find yourself helping to plan for the production of a much larger volume of documents. You may, for example, be asked to help plan the production of 2,000 copies of a software documentation suite—a getting-started guide, a user manual, and a reference manual. For example, how do you determine the paper quality and binding to be used? How do you find out how much time the production will take and how much it will cost? What if there are errors in the printing? Will there be time to make corrections before your deadline? Who will handle storage and shipping? These are just some of the many factors you will have to account for in the production of a major project.

Many larger companies have printing or production departments that routinely handle these concerns. Others contract this work out to commercial printers. However, this fact doesn't preclude the possibility that you may at some time be asked to plan for the production of a relatively large project. If you contract with local commercial printers, you need to know the basics of production to help you determine what to specify and who is offering you the best deal.

What You Will Print and How You Will Print It

Planning your printing at the beginning of the project will help to prevent production problems. Make a detailed list of the printing you have done over the past few months and project how much printing you expect to do over the next six months. Consider many factors, including how quickly you will need some print jobs, how large some print jobs may be, how small others will be, and whether or not you will need color, illustrations, and so on. You should review each printing job to determine what needs to be printed, how the information should be printed, and what kind of materials should be used. Good planning for your printing jobs will save you time, materials, and money.

When reviewing each print job, your first priority should be to gather all the information you can about the document you plan to print. Make sure the printed document is really needed. You should ask what the consequences would be if the document were not printed, whether the document serves its stated purpose, whether the cost of printing and processing justifies the printing, and whether other related documents already exist. Sometimes you will discover that it isn't necessary to print the document after all. Once you have determined that the document should be produced, you should ask who, what, why, where, when, and how: Who will use it? What is the purpose? What quality of printing is necessary? Why is the document used? Where is it used? When is it used? How is it used? How many copies are needed?

In general, you will want to produce your document at the highest quality you can afford. But what constitutes quality varies. Quality in printing is closely tied to purpose. The quality of the document depends on how well the document fulfills its purpose, and that depends on what it will be used for. For example, if the document is to become part of a permanent record, you will want to ensure that the document is produced to be durable. Other factors determining how a document must be printed include whether users of the document will write on it, whether it is intended for interoffice use, whether it will be used many times or used only once and then filed or thrown away. Answering these questions will help you decide on the essential quality.

For many documents, photocopy quality is adequate and can be produced quickly and inexpensively. When high quality, large quantities, or multicolor printing is needed, the job may require more traditional printing methods, such as negatives and aluminum plates. (See "Traditional Printing Methods," page 218.)

Scheduling

Try to allow as much time as possible for a printing job, both to help the printer and to save money. Printers appreciate having as much lead time as possible, and generally, the more time they have, the better job they do and the less they have to charge. Consult with your printer early in the planning process to determine how much time you must allow in the schedule for production.

Despite your best planning, you will sometimes find yourself facing a deadline that gives you very little time for production. If you require same-day or overnight printing service, you can expect to pay much more than you would if you gave the printer a few days, a week, or longer to complete the project. Be careful about rush jobs. Rush jobs, while sometimes necessary, disrupt the scheduled work and increase production costs. And worse, costly mistakes are frequently made under the pressure of completing a rush job. Make sure you are aware of any potential late charges that you will incur if you find that you won't be able to keep to your projected schedule.

Budgeting

Budgeting for production is something you may never have to do or something you may have to do often. (See Part One, "Planning and Research," for a discussion of factors to consider when estimating costs in general for relatively large projects.) Often estimating costs for print projects is simply a matter of obtaining price quotes from your company's or organization's print shop or from several commercial printers and then accepting the quote that is most appropriate to your budget, schedule, and criteria for quality. However, to obtain an accurate or helpful price quote, you must be very informed about your printing requirements. You will need to determine the many issues or specifications before you request a printing estimate. (See

"Submitting Your Job to a Print Shop," page 222.) The more carefully and fully you have considered these production issues or specifications, the more accurate your estimate will be.

Your printing costs will be determined by many factors, including the number of production steps your document has to pass through before it is printed. The number of production steps depends, of course, on the printing method you choose. (See the various sections on traditional printing methods, special processes, and electronic methods below.) Normally, you will choose a printing method on the basis of the quantity and quality of the printed copies you need. If you don't need many copies, consider photocopies. If you will be ordering a large print run, tens of thousands or hundreds of thousands of copies, you will have to allow for the extra time and expense of other methods. Determine how many production stages are required to give you the finished product. Each step adds time and expense. For example, jobs requiring typesetting, printing, collating, and stitching (see "Binding," page 230) take longer to produce than jobs that are only printed and wrapped. However, if you are merely providing electronic copy to a printer for a relatively straightforward printing job, then you can expect to pay less than you would with a much more complicated production process.

Various special requirements and processes also affect the cost of printing a document. For example, if your document requires any bleeds (text, visuals, or artwork extending to the edge of the page), you can expect to pay more. Bleeds increase printing costs because they may require the use of an oversized sheet of paper that is later trimmed to size. Other special requirements might include folds, drills, or binding (see "Binding," page 230) that increase costs. Or certain special processes may be required, such as die cutting, embossing, foil stamping, and engraving (see "Special Processes," page 219).

Once this information is determined, you can work with a printer to develop an estimate of the final cost. An in-house production department or a commercial printer uses these specifications to determine optimum paper sizes and quantities, to select a printing press, and to determine how to oversee your project to improve quality and lower costs.

For many kinds of small print jobs, the costs range from none to minimal. If you own or have access to an inkjet or laser printer and need only a few copies of a report or other relatively brief document, then your costs may simply be the price of the paper. However, if you're out of town on a business trip and have to add ten more color transparencies to your presentation as well as print twenty more copies of a high-quality training document, then you'll have to settle on the best price quote you can find at the last minute.

The biggest problems for cost estimating occur when you have to calculate weeks or months in advance the production costs of a much larger project without knowing how long the final document will be or even exactly how many copies you will need. Documentation managers, for example, typically must include in their project budgets the production costs of documents that have yet to be written. Estimating costs in this kind of common scenario requires that you consider many variables: costs for color printing or other special features, paper kind and quality, binding, estimated document length, possible quantities needed, lead time (if any) required by commercial printer or in-house print shop, any possible additional charges for rush orders, costs for delivery from printer (if any), costs for delivery to customer, and so on.

The owner of the commercial print shop or the manager of the in-house print shop can help with many of these estimates but not all of them. To arrive at your best estimate, you'll more than likely have to calculate a low figure and a high figure for all of the variables mentioned above and then go with an estimate somewhere in between these two amounts. You will find that the more experience you have in seeing documents through the production process, the more accurate your cost estimates will be. Make sure you keep full and accurate records of these production details so that you can refer to them when estimating future jobs.

Printing projects have two major categories of costs: preparation of electronic files (design and desktop publishing) and actual printing and bindery costs. In general, when you request a printing estimate, you will receive information about printing and bindery costs only. If your project involves many other elements (for example, design, desktop publishing and electronic prepress, editing, copywriting, or photography services), you should obtain estimates for these services, too.

You should also note materials costs, changes to the project, and changes to the specifications when evaluating your estimate. Materials costs are constantly changing, so estimates may change depending on those costs at the time the order is initiated. Any changes to the project after final page proofs are approved, such as on blueline or on press, will result in additional charges. (See "Proofs," page 229, for more on this subject.) Changes in any of the specifications, including the complexity of the design, could result in additional charges. Print schedules are based on bluelines being approved and returned promptly (usually within forty-eight hours from when the customer receives them unless other arrangements have been made). Any delay in returning the bluelines or changes to the proofs could affect both the schedule and resulting costs.

Unlike estimates, production quotes are prepared only when final material is ready for production planning, engineering, and scheduling. A production quote is based on the job as presented. Because these quotes are based on actual materials and final specifications instead of descriptions or dummies, they are generally more accurate than estimates.

TIPS: Lowering Printing Costs

- Know your audience. Be sure to discuss with your printer who your audience is for your publication and what kind of quality level you're looking for. Premium-quality printing projects require extra paper, additional press and bindery time, and different technologies that can be more costly.

- Plan ahead. You can save money, for example, if you print multiple brochures at the same time and design them so that they use the same ink colors in slightly different combinations. This approach saves significant setup charges that would be incurred by changing ink colors or printing each brochure at a different time.

- Plan ahead for a second color if you are using one. When you know that you'll be using a second color selectively throughout your publication, your printer can help you to plan your layout so that only one side of a press sheet will require the second color. You'll save money but still have the impact of two-color printing throughout your publication.

- Review copy thoroughly. Be sure that everyone involved in your publication process has an opportunity to review the text thoroughly before you submit it to a printer. The more errors you catch before the printer sees your file, the fewer errors you will have to correct on proofs or bluelines. It's a good idea to have someone who isn't involved with the publication proof your copy.

- Choose in-stock papers so that your printer doesn't have to special order paper for you.

- Choose uncoated paper stock if possible instead of the more expensive coated stock.

- Choose lighter paper that doesn't need scoring to fold cleanly.

- Design pieces so that they can be mailed without envelopes and so that they fit U.S. Postal Service automation requirements.

(continued)

Lowering Printing Costs *(continued)*

- Know your computer. Give the printer a sample file well in advance of submitting your job to make sure there are no compatibility problems. Let your printer know all the details about your computer file; your operating system; the type and version level of the software; and the format, application, and version for any graphics, spreadsheets, or scanned artwork that you include in your file. Also include a separate copy of your graphics and your fonts, including dingbats. (See "Typography," page 113.)
- Provide detailed instructions. Your instructions must be as clear as possible. If you are providing a disk with your project already laid out, it's a good idea to include either a marked-up dummy copy of what you want your project to look like or thumbnails (reduced-size schematics) of the page layouts marked with your instructions. Be sure to mark clearly ink colors, screens, reverse type, art to be picked up from previous publications, and any other important details.
- Minimize changes.
- Allow plenty of time.

TRADITIONAL PRINTING METHODS

Despite the enormous popularity of desktop and now Web publishing, several older, more traditional methods are still used for many kinds of projects. Choose the method that best suits your needs, keeping in mind budget constraints, the length of the document, its intended use, and the level of quality. In these traditional methods, ink is transferred from printing plates to paper in one of three ways: offset, letterpress, and gravure. All of these methods use physical ink-transfer mechanisms that create mass production reproductions at high speeds.

Traditional methods all require that you go through a similar process: prepress, bluelines and perhaps color proofs, press run, bindery, and finish work. Bluelines and color proofs will typically be the major proofs you will examine. And all three traditional printing methods eventually lead to a press run (the actual printing) and bindery (folds, staples, binding, etc.).

Offset printing is the most common high-quality printing method. Its name comes from the fact that the printing plate never touches the piece of paper. Instead, the inked image is transferred to a rubber blanket or platen, which in turn transfers the image to the paper. This offers two advantages: First, the plates themselves last much longer and are less easily damaged during the course of a print run. Second, the blanket is more pliable than a plate and can fit more easily into different textured paper surfaces. Offset printing is well suited for intermediate length runs and is less expensive than gravure (described below) while still offering high-quality printing, especially for reproduced text.

Some print shops with large presses use metal plates instead of paper or plastic plates. A metal plate is far more accurate and stable over a press run and offers a much sharper, higher-quality image.

Offset presses are sheetfed, meaning that they use large pieces of paper called parent sheets. A wide variety of textured, coated, uncoated, and recycled papers are available in sheets for your projects. Web presses, which print long magazine, newspaper, or catalog runs, use huge rolls of paper. Specialty printers are often contracted for these types of jobs.

Letterpress or relief printing works by applying ink to the raised surfaces of a plate and then by pressing the plate against paper. The ink is then transferred by pressure to paper or some other material. Like a rubber stamp, the ink moves from the high surfaces of the plate to the paper. Letterpress printing is messy and time-consuming, and it is rarely used except by small printing houses and for short copy runs, usually between 500 and 1,000 copies.

Gravure is a high-volume printing process employing an ink-transfer mechanism that is fundamentally different from that of letterpress or relief printing. In gravure printing, the printing surface is a polished metal cylinder covered with an array of tiny recesses, or cells, that constitute the images to be printed. The cylinder is partially immersed in a reservoir of solvent-based fluid ink. As the cylinder rotates, a steel blade running the entire length of the cylinder wipes the ink from the polished surface, leaving ink only in the cells. The ink is then transferred immediately to a moving roll of paper that is forced against the cylinder under great pressure. Color printing is accomplished by using separate printing cylinders to combine colors. The expense of manufacturing a set of gravure cylinders has restricted its use to long-run jobs (hundreds of thousands or millions of copies).

SPECIAL PROCESSES

Sometimes your project will require special processes instead of or in addition to the usual printing methods.

Die cutting is a process that is used to create unusual shapes, or cutouts. The process can range from something as simple as cutting out rotary card file refills to something as complex as altering the shape of an entire finished piece. Line art in the shape of the desired cut is used to produce a cutting die. The die is used much like a cookie cutter to perform the die cut desired. Die cutting allows a great amount of creativity in producing items with irregular edge shapes that could not be produced with conventional cutting.

Embossing is a graphic process that uses a magnesium, copper, or brass die to reshape the paper fibers into the desired image. The result of raising (embossing) or lowering (debossing) an image in the surface of a paper creates an appealing sense of depth and dimension. Blind embossing simply bumps up the paper fiber without any printed or foil-stamped color. When the paper has a textured or patterned finish, this process also creates the element of contrast.

Foil stamping is a graphic process that uses pressure and controlled heat to bond foil to the material. The foil-stamping process can create a wide range of contrasting effects with paper through color and surface changes. Many foil colors are available, from vivid bright hues to subtle tints and pastels, plus black and white.

Engraving is a process for adding a special look to stationery, for example. The engraving die is created by cutting a reverse image into the metal die. Ink is then applied to the surface of the die, then wiped clean, leaving only the recessed image filled with ink. The paper is then pressed against the die with considerable pressure, leaving the impression elevated from the surface of the paper.

ELECTRONIC METHODS

For short print runs, electronic processes are more economical than traditional print methods. Electronic processes do not use printing plates, and they produce good reproductions without wasting paper. Modern office copiers and many kinds of computer printers can quickly create high-quality copies. Even an inkjet printer can be used to print many kinds of documents.

High-speed, high-resolution digital printers (such as the Xerox Docutech systems) offer the possibility of relatively economical short print runs. This technology offers "print on demand" flexibility, which is particularly useful for material that will change or will need frequent updates, material that has to be customized for several locations, or material that you may want to reorder from time to time. As an added advantage, you can submit your print jobs electronically, either on magnetic media such as diskettes or over the Internet.

Significantly, the emerging relationship between traditional and electronic printing is more complementary than competitive. Even for print runs using traditional printing methods, digital processes are increasingly used to inspect and enhance the appearance of images before they are eventually processed by whichever method is most appropriate.

If you are supplying an electronic file for your publication, check these tips to help ensure that your project is in final form and ready to enter the print production process as quickly as possible.

TIPS: Submitting Electronic Copy

- Make sure the final copy has been proofread carefully. Unless printers are contracted to do so, they aren't going to take the time to proofread the copy you provide. They will print the copy as it is when they receive it. Any typographical errors that are your errors are your responsibility.

- Consider sending your document to the print shop as a Portable Document Format (PDF) file or a postscript or encapsulated postscript file. These formats retain the fonts and formats you have designed into your document, ensuring that you won't encounter unanticipated page reflows or other problems when your document is printed. You will need a program called Adobe Acrobat Distiller to create a PDF file.

- Send graphics in a separate file as well as in the document file.

- Make sure all photographs are handled appropriately. They should be identified, cropped, sized, or scanned as necessary.

- Make sure all inks, screens (the shading of a solid image area), and reverses (white lettering against a solid or heavily shaded black or colored background) are clearly marked on the hard copy.

- Mark clearly any graphics that must be placed in certain areas of each page.

- Make sure any bleeds extend at least one-eighth inch beyond the page margins.

- Check to see whether all the common page elements (rules, margins, page numbers, and so on) are consistent from page to page.

- Make sure that trim, fold, and center marks are indicated.

- Work with the print shop to ensure that your fonts and their metrics (accompanying descriptive and calibration information) are available. If necessary, send fonts as a separate file. Send a test file to the print shop to make certain your files can be correctly read by the printer software.

SUBMITTING YOUR JOB TO A PRINT SHOP

You will have many variables to consider when preparing your job for submission to a print shop. It helps to view your job analytically, considering each of these aspects as early in the project as possible, preferably before you are actually in the production stage.

Some documents are brand new; others are revisions of previously published words; still others are simply reprinted versions of older works. If you are submitting your job to a print shop and you are ordering exact reprints of a previous job, you typically enclose a sample and note the previous job number on your request form. If you need extensive revisions, you should generally provide a new electronic file containing the necessary revisions. (See the tips list, "Submitting Electronic Copy," page 221.) If you are ordering reprints, revisions, or new publications, you need to consider all of the issues that follow to ensure the best quality and a reasonable cost.

Quantity

Often it's easy to predict how many copies of a document you will need, but sometimes it's almost impossible to know precisely. Quantity is tied to purpose and audience. You need to know how the document will be used, for whom it is intended, and for what period of time this particular print run is intended. If you are printing documents that are to be used quickly, then you must determine whether the quantity you print will be enough to last a reasonable length of time.

Determine carefully the quantity you need. Ordering too many or too few copies can be an expensive mistake. One recommended quantity guideline would be to order a six-month to one-year supply. Printing additional copies once the job is on the press is cheaper than placing a second order. However, you need to consider storage and waste when making this decision. If you are not certain how many copies you need, ask for a price for the minimum quantity you think you will need and a price for each additional printing of a thousand copies. You will find that the price per copy will go down as the number of copies increases. This decrease in price happens at certain price-break quantities. Ask your printer to quote you a price for the quantity you need and also for the next higher quantity, in case this proves to be more economical for you in the long term.

Dimensions

The size of a document affects how it will be produced and what it will cost. Printers need to know the physical size of your publication. A document requiring 8½" by 11" paper will present fewer problems than one that must be cut to a smaller size or one that is 11" by 14", for example. You should plan your printing jobs for standard sizes whenever possible. Typical standard sizes are 8½" by 11", 11" by 17", and 8" by 14". The most cost-effective sizes are usually 8½" by 11" and 11" by 17".

Of course, you also need to keep in mind that not all papers are available in all sizes. You can save money if you have your printer print several copies of your document on a single sheet of paper and then trim the paper to size. Not every paper color, weight, and coating is necessarily available in all sizes either. See "Paper," page 224, for a discussion of the many issues you must consider when choosing paper.

Determine early in the planning process what dimensions are best suited to the purpose of the document and the audience. Don't forget to consider postal requirements for those documents you may be shipping. The dimensions of the document may also become crucial if you are mailing the document. (See "Packaging and Mailing," page 231.)

Resolution

Resolution concerns the sharpness of the typography, illustrations, and other design elements in a document. It's essentially a matter of the quality of the image detail rather than image size. Resolution is an important factor, regardless of the medium you choose for publishing.

High resolution, both in raster (screen) displays (see Part Three, "Design and Illustration") and in printing, refers to a high-quality screen or image that reproduces text and graphics with relative clarity and fineness of detail. High resolution is related to the number of pixels (picture elements or dots) that are used to create the image: the more pixels, the greater the possible resolution. In printing, high resolution generally lies in the realm of laser printers and typesetting equipment, in which resolution is specified by the number of dots per inch (dpi) that are printed. A dpi of 300 is a higher resolution than 200 dpi, and 300 dpi is standard for most scanned images.

If you want to print a document of high-resolution quality, you will need a computer printer that is capable of 1,200 dpi or higher or you will need to use a traditional printing method. (See "Traditional Printing Methods," page 218.) In desktop publishing, many computer printers provide the option of allowing you to print a draft-quality (lower-resolution) version of a document to save on ink.

Paper

Paper selection is one of the most important decisions in the printing process. When you decide on the paper for a document, you have to consider the costs, the typical kind of paper(s) used for the genre, the audience, the purpose of the document, and other factors. For many genres you need to consider the impression your paper choices will make. The look and feel of the paper you choose will greatly influence the value your reader places on your message. In successful projects, paper is part of the design. It enhances the printing as well as meets the physical and cost requirements.

The trick is to choose the right paper for the job. Many factors must be considered in the selection of appropriate stock for your document. Is the material to have any aesthetic value? Will special finishing, such as folding, be required? What is the intended use of the document, and what degree of longevity is required? Is the desired stock readily available, or must it be ordered? Is a lighter-weight paper desirable to reduce mailing costs? For what type of press is the job best suited, and can the desired stock be used on that type of press?

There are nine basic categories of paper. These categories are defined by their weight, coatings, and uses. Bond is a grade that has a hard surface and is good for typewriters, impact printers, and pen and ink. Text paper is a high grade that is used for announcements, menus, folders, and brochures. Book paper is a general grade with a wide variety of weights. It's good for printing and cheaper than text paper. Offset paper is of a grade similar to book paper but resists shrinking after exposure to moisture in the offset printing process. Bristol paper is a thick grade that is quite durable. It is used for greeting cards, tickets, and many other applications. Tag, cardboard, or index is an extremely durable grade that is heavy and thick. Newsprint is the cheapest grade available. It has little strength and turns yellow after a period of time. However, this grade is easy to recycle. Coated paper is a grade that can have a smooth coating ranging from dull to glossy on one or both sides. Another type of coated paper is NCR or carbonless paper, which is coated with special pods to produce multiple forms without carbon paper. Coated paper is relatively expensive and high-quality. Cover paper is a heavy grade that is perfect for covers of booklets, brochures, and menus.

The three grades of paper best suited for general use in production of a document are bond, book, and text. The most commonly used paper for forms, memos, letters, procedure manuals, reports, and most general types of printing is bond. It is designed to accept ink readily for printing and has good writing properties and acceptable strength. Some commonly used bond papers are 20-pound bond and 60-pound vellum paper. The 20-pound bond is used for most jobs. However, if the purpose of the printing requires durability or if the ink coverage is heavy, 60-pound vellum paper is recommended.

There are two principal types of bonds: sulfite and rag. Sulfite bonds are rated as No. 1 and No. 4. The No. 1 is of higher quality, although the difference in cost is not great. Sulfite bond paper is reliable and inexpensive, and there is no limit to its uses. Rag bond is widely used in business and industry and is judged by the percentage of cotton used in the paper. A designation of 25 percent rag means that 25 percent of the paper is cotton and 75 percent is wood content. The combination of wood and cotton adds extra strength to the paper. The higher the cotton content is, the stronger and more expensive the paper.

Uncoated book paper is used for general printing of all types because it is versatile, comes in a variety of colors and sizes, and is relatively inexpensive. However, uncoated book paper can sometimes produce fuzzy or hazy pictures. When you are printing detailed instructions or pictures, it is best to use coated book paper. Among the many types of uncoated paper are premium text (used for annual reports, books, calendars, posters, self-mailers, brochures), text, No. 1 book (for books, brochures, calendars, catalogs, flyers, manuals, newsletters, programs), No. 3 book (for catalogs, direct mail, manuals, rate books), groundwood book (for catalogs, direct mail, manuals, rate books), and lightweight book (for Bibles, dictionaries, reference books).

Text paper comes in a large variety of colors and grades, but it is also the most expensive. Because of superior manufacturing and quality, text paper is generally used for prestige-type printing. The quality of the paper enables the print to appear clearly on the paper. Text paper is expensive but often well worth the price.

Cover sheets are extra heavy book or text paper. Use cover sheets for covering books and catalogs, making folders, and printing anything that needs a heavy and durable paper.

Ink

You will need to consider several important factors when choosing ink. First, the type of printing determines the type of ink you will use. Offset and letterpress printing require inks with different characteristics from those used in other printing methods. You may also select special quality inks (for example, high-gloss inks) for both letterpress and offset. The quality of the ink that is used, of course, directly affects the quality of the print job. Second, in addition to the type of ink used, you'll need to consider ink coverage. Printing costs often increase if your design involves heavy ink coverage. Areas of heavy ink coverage not only require more ink, but also require more care in making sure that the ink is applied evenly. Third, color printing presents many additional challenges concerning inks and, of course, presents higher costs. A commercial printer will review all of these factors and others with you when you begin working with them. (See "Budgeting," page 214.)

Color

Color is perhaps the most predominant design element on a page. The human eye has evolved to discern subtle differences in color, and the perception of color can be pleasurable to the viewer. Color brings a message all its own to a text, and successful document designers take advantage of color's power to enhance the effectiveness of their designs. Use color well and you strengthen your message; use color poorly and you may detract from your document's impact or even send a negative message.

If you decide to use colored ink, you must consider additional points. Colored ink should be used sparingly. It can add a great deal to the appearance of your document, but often it isn't necessary. For desktop publishing, you'll need access to a color printer, and if you're printing many copies, it can be expensive to replace the printer cartridge for a color printer. For high-quality mass production, each color must undergo an extra run through the press, requiring new plates and a cleanup charge, so additional expenses add up quickly. Costs also go up if printers must mix special prespecified colors.

TIPS: Using Color

- Use color deliberately, just like any other tool, to help the reader find and understand content. Color should not be used merely for decoration without also considering its intrinsic effect. You should be able to justify each use you make of color on the basis of what you plan to accomplish with it.

- Use color with restraint. Never use color to substitute for poor layout or typography. With color, less is more. Often merely adding a second color is enough to create a distinct look. Using many colors is rarely necessary. Too many colors add to design clutter, which distracts from your message.

- Be judicious in combining colors. Mixing bright colors causes them to lose their impact or worse. Color can undermine your message when used in unpleasant combinations.

- Be especially careful about setting text in color or against colored backgrounds. Color makes text harder to read, not easier.

- Develop an eye for design and color by looking at top-quality magazines or catalogs that are directed at the same audience you want to reach. Write down your reactions to the documents. Then analyze the pages or sections that appeal to you most strongly so that you can discover how your reactions were evoked. Be aware that designers make careful and deliberate choices to elicit your reactions.

- Use color to highlight a single important element, such as your firm's logo. If your company calls for color in its style manual, follow those requirements.

- Color is described by its hue, value (or sometimes lightness), and saturation. *Hue* means the color itself. Blue and orange are hues. *Value* is a measure of how much color is in the particular hue. For example, 90 percent blue has a higher value of blue and looks darker than 30 percent. *Saturation* describes the brilliance of a color. Coordinate similar items by highlighting them with identical hues. Show differences between items by using higher and lower values of the same hues to represent variety in a single quality. Vary saturation to emphasize importance or draw the reader's attention toward an item.

- When using color, be aware of common color associations, such as blue for water, green or brown for land, and yellow for caution. Do not expect that these associations will hold for other cultures and nationalities, however.

- Use typographic cues to supplement color so that readers who are color-blind or who are reading monochromatic photocopies will recog-

(continued)

Using Color *(continued)*

nize the distinctions you are making with your use of color. In line graphs, for example, have each line appear slightly different; use dashed lines, dotted lines, and other broken lines to differentiate each variable you are mapping. For colored text, consider using bold type or a different typeface along with colored type.

Photographs and Other Illustrations

Including photographs and other illustrations in your document often increases your printing costs. Of course, the cost depends on how you submit the photographs and other illustrations to the printer. For example, if you scan all of your photographs, adjust them so that they are consistent in appearance, and submit them to your printer on a disk, you will save some money. Providing consistent photos in this manner is an important time-saver for your printer.

Expect to pay more if you are going to provide the printer with black-and-white photographs that require conversion to halftones for printing. Halftones, also referred to as halftone shots or halftone negatives, are a method of reproducing photographs for printing. The image is composed of dots of various sizes, which simulate the range of grays or colors in the original photograph. A halftone is always printed in one color. Each photograph increases the cost of your project if you expect the printer to prepare halftones and resize photographs.

Other illustrations (for example, bar graphs, pie graphs, diagrams, technical drawings) can also increase the cost of printing because they require more ink and more care.

Covers

The cover (if your document has one) is usually the part of your document that makes the first impression on the reader. Many covers for your college assignments or corporate work won't be fancy. The cover will often be one of the usual assortment that you can purchase at any college bookstore or office supply shop. However, covers take on major importance in annual reports or software documentation suites, for example. In these examples and many others, the cover is an important part of the image you want to project or the message you want to send. In addition to being a matter of design, covers are a matter of paper stock, of which there are different thicknesses, textures, sizes, colors, and so on. (See Part Three, "Design and Illustration," for more information on the design and illustration principles you should follow in creating a cover.)

Proofs

There are several types of proofs, ranging from laser proofs to bluelines prepared from the actual printing plates. You need to decide what kind of proofs you require for your particular print job.

Page proofs show you how your text, illustrations, and other page elements will look when they are printed. A blueline is a one-color representation of what your final job will look like. It is the final stage at which you can make corrections, and the corrections you make should be minor. Any content changes at this point usually require changing the electronic file, creating film negatives again, and making a revised blueline. These last-minute changes are costly and usually delay delivery dates.

You should also know a little about editing proofs. Catching errors at this stage will save you time and money later.

TIPS: **Editing Proofs**

- Use standard proofreading notation when marking up page proofs or bluelines to help ensure that your changes are communicated accurately.
- Remember to use a contrasting color pen or pencil to mark changes and to write extensive changes in the margins of the page proofs to make sure they can be read. Any comments that are circled are treated as instructions, not as type to be set.
- Confirm that corrections marked on earlier proofs have been made.
- Check photographs to make sure they are in the right place and that they have been scaled, cropped, and positioned properly.
- Check trim and folds; the blueline should be folded and trimmed as requested for the final product.
- Check copy blocks to make sure there are no missing lines at the end of columns or pages and that there is no broken type.
- Circle any blemish, flaw, or spot with a marker.
- Review page numbering and sequence.

Binding

The binding of your document should be determined based on the purpose of the document, the environment in which the document will be used, and, where possible, the preferences of the user.

Perfect binding, also called glued binding or adhesive binding, is used for documents that need to have a professional look. The pages are held together and fixed to the cover with a flexible glue to give the publication a square back or spine. Perfect binding is commonly used for telephone books and paperbacks, for example. This type of binding is expensive but rather flimsy. Use perfect or glued binding when appearance is important.

Stitched binding is assembled by stapling through the crease of the spine or near one edge of the sheets. Three different types of stitching methods can be used to assemble your document: saddle stitching, loop stitching, and side stitching. With saddle stitching, staples go through the crease of the spine. Saddle stitching is used for most magazines. This method of binding is inexpensive and quick, but documents that are longer than eighty pages should use another type of binding. Loop stitching is similar to saddle stitching. This type of binding is less sturdy but allows for an accumulation of documents. Side stitching is very durable and relatively inexpensive. However, it puts staples through the entire stack of paper near the edge. This approach does not allow the document to lie flat.

Thread-sewn binding is dependable and aesthetically pleasing. It is durable, but it is impossible to add new material, the document will not lie flat, and the cost is high.

The two types of mechanical binding are permanent and loose-leaf. Mechanical binding is generally cheap and easy to do. These simple methods are suitable for many types of documents. In a mechanical or loose-leaf binding, the pages are held together mechanically, as the name suggests, usually with a metal or plastic coil.

Velo-bind binding is a permanent mechanical binding that is inexpensive and will last a relatively long time. This type of binding is very professional looking for the price, but the document will not lie flat, and it is very difficult to add pages.

Plastic comb is an inexpensive permanent binding that is reliable to a certain extent. However, it is plastic and will deteriorate with heavy use and over time. Plastic comb, also known as GBC binding, is suitable for a document that is not used often. The most favorable attribute of plastic comb is the price; it is the cheapest of all bindings.

Spiral binding allows your document to be doubled over and is inexpensive too. However, there is a lot of play between the pages, and you cannot add new pages to your document. Spiral binding can be crushed or damaged, and illustrations do not line up evenly across pages.

Wire-O binding consists of doubled loops of wire, crunched together, which allow the document to be doubled over. It is very durable, and pages can be lined up with each other. This binding cannot accept new material and has a maximum capacity of seven-eighths inch of paper. However, it is relatively inexpensive and is stronger than spiral binding.

Three-hole punched binders are sensible for most documents. This type of binding allows the document to lie flat, text can be inserted at any time, and it is relatively inexpensive. Loose-leaf binders are appropriate for manuals and other documents that may need to be updated from time to time.

In sewn bindings, the final product looks similar to perfect binding, but the pages are sewn together with thread and the cover is glued on. This method is commonly used for textbooks. Sewn books may have a soft paper cover (as on a paperback book) or a hard cover. Hardcover books are also called case-bound books.

Packaging and Mailing

Finally, you will need to consider various packing and mailing issues. For example, will the document be shipped in boxes? If so, what size? How will these boxes be labeled? Which service will you use to handle the boxes and what are the packing requirements for this service? If you are mass mailing individual copies of the document, you have other factors to consider. For example, does your mailing list meet U.S. Postal Service (USPS) address standardization requirements for automation discounts? Does the design conform to requirements for OCR read area, barcode clear zone, and fold placement? Will the piece be designed as a self-mailer? Does the design meet the USPS guidelines for shape, size, paper, and closure? Do you need a return card or envelope? If so, how will it be included?

Make sure that your printer knows exactly how your documents will be used so that your paper will fit its intended purpose. Use your printer's experience and knowledge to shape your production decisions efficiently and appropriately.

WORD PROCESSING AND DESKTOP PUBLISHING

Today it's easier than ever before to publish your own documents and to give them a professional look. Tremendous improvements in desktop publishing programs (from full-featured word-processing programs such as Microsoft Word and Corel WordPerfect to heavyweight publication programs such as Adobe PageMaker and Adobe FrameMaker), Web editors (Netscape Composer, Hot Dog Pro, and Adobe PageMill), and authoring programs for CD-ROM (Macromedia Authorware and Director) have given anyone who is willing to learn the tools the ability to create very appealing documents.

Electronic publishing isn't just for individuals. Many major publishers have switched to sophisticated electronic publishing systems. Increasingly, authors are submitting their manuscripts to publishers on disk or sending the manuscript files via an FTP program (see Part Six, "The Internet"), publishers are contracting third-party copyeditors to edit a printed copy of the manuscript or the electronic file, and authors are asked to review the copyedited hard copy or files. Typographers then take the manuscript disk, make changes according to the copyedited manuscript and the publisher's publication standards, send the author page proofs, receive corrected page proofs from the author, make changes, and send the corrected files electronically to the publisher or printer for printing.

Desktop publishing enables you to print your publication or to publish it as an online file, an HTML file, or in some other electronic form. Desktop publishing programs, ranging from limited-feature programs such as Microsoft Publisher to full-featured programs such as Adobe PageMaker or FrameMaker, enable you to publish electronically. Programs for creating online help files, such as RoboHELP and Doc-2-Help, are yet other ways to publish documents electronically. (See Part Three, "Design and Illustration," for a discussion of various tools for creating online help screens.)

Desktop publishing offers several advantages over traditional publication methods, including lower production costs, ease of updating or revising your document, and convenience. With no more than a decent computer, a few good-quality peripherals, two or three software packages, and a bit of experience, anyone can create in a few hours what may have taken a team of artists and typesetters several days just a decade ago. Pasteup work is but one of the tasks that have been all but eliminated because these desktop publishing programs help even those with minimal design skills to create professional-looking documents. However, even the best desktop publishing equipment can't guarantee a professional final product. You will still need to know the principles of planning, researching, writing and revising, designing and illustrating, and so on to create your best documents. Even the best desktop publishing tools can only help you to implement what you already know.

Word-Processing Tools

Computers have made the task of writing much easier from a mechanical standpoint. With a full-featured word-processing program, such as Microsoft Word or Corel WordPerfect, you can simplify and control innumerable writing tasks and check for spelling, grammar, and style errors while you write. Only you can decide which of these tools works best for you. Regardless of which tools you use, you need to learn them well if you are going to take full advantage of the tool's features. Of course, tools alone don't make you a better writer. They can save you lots of time and definitely make the process of writing much more convenient, but a keen ability to write is more essential than any tool.

Although many products perform similar tasks, you will normally use and be familiar with only a few. Often you will not have a choice as to which tools you will be using. Your employer may have a site license for a particular application, your school may provide you with access to a writing lab, or you may own a computer that is preloaded with software. Familiarize yourself with the features of the particular software you will be using. Increasingly, software documentation is available in both printed and online form. In fact, many software developers are forgoing printed documentation entirely, or they provide documentation in the form of online manuals that you can download and print out.

LIMITED WORD-PROCESSING PROGRAMS

Aside from their obvious usefulness in creating and modifying relatively simple documents, text processors are becoming popular as more people try their hand at creating and maintaining Web pages. For those who want to control their HTML tags directly, nothing beats an ASCII text processor such as Notepad or even WordPad. Notepad is a handy, but limited text-processing program that is preinstalled on computers with a Microsoft operating system. Notepad is a convenient program for cutting and pasting short text files, instead of opening and using a full-featured word processing program.

You can use Notepad to create or edit text files that do not require formatting and are smaller than 64K. Notepad opens and saves text in ASCII (text-only) format only. To create or edit files that require formatting or are larger than 64K, use WordPad.

WordPad handles files that are too large for Notepad. WordPad differs from Notepad in that you can open and save files in several formats besides ASCII text: RTF (RichText Format), Word, and Write. In addition, WordPad has more features than Notepad. You can embed and insert objects into a WordPad document, align and indent paragraphs, create bulleted lists, and format your text in a limited way.

SuperPad <http://www.zdnet.com.au/pcmag/download/utils/spad .htm> and TextPad <http://www.textpad.com/> are two programs that improve on Notepad. SuperPad can handle larger files than Notepad. TextPad provides additional features: a spelling checker, unlimited undo and redo, a macro recorder, and a Warm Start option that allows you to interrupt your work and return to it where you left off.

FULL-FEATURED WORD-PROCESSING PROGRAMS

The full-featured word-processing programs, such as Microsoft Word and Corel WordPerfect, are the heavyweights of word processing. These programs have become very sophisticated over the past few years and will continue to as many new features are added with each new version. Of course, along with their sophistication, they bring hefty storage, memory, and processing-speed requirements. If you are using an older, smaller, slower computer, you probably should not try to run one of these giants. With a full-featured word processor, you can expect to have not only the powerful features that enable you to create attractive, effective documents either on paper or on screen, but also quite a lot of help for using these features. Word processors no longer depend on printed documentation to provide help for users. Context-sensitive help—which provides information about the feature you are currently using—is almost always available. In addition to context-sensitive help, most word processors are loaded with helper applications, called wizards, that step you through the process of preparing some types of documents. You can find wizards to help you create a memo or résumé, send a fax, design a calendar, and organize an agenda.

Styles are one of the most powerful features of word processors. If you learn to use only one tool, it should probably be styles. Styles are at once completely basic and extremely sophisticated. They control the appearance of your text. When you set up particular styles for your document, you control the consistency of your headings, body text, tabs, borders, indentations, alignment, spacing, and many other features. When you modify a particular style, the change is carried throughout the document. Using styles ensures uniform treatment of corresponding elements in your text. You can save a lot of time and energy by learning to use styles and using them regularly.

You will likely not use many of the features of your word processor. Word processors have become so feature heavy that it is almost impossible to learn all there is to know about them. Spend some time experimenting with the various features that come with your word processor. Even if you don't use them often, some day you may find that one of the features you read about would be just the thing you need to create a wonderful effect in your document.

Desktop Publishing Tools

Desktop publishing (DTP) software provides the computer equivalent of traditional layout and design tools that have been used for years in the publishing business to create newspapers, magazines, books, and promotional materials. Noncomputer tools include scissors, paste, and the like; their computer counterparts offer an interface designed to provide the same functions with keyboard and mouse. All these programs have a fairly steep learning curve as a consequence of their many options. A full-featured DTP package offers sophisticated text handling (fonts, paragraph styles, special effects such as drop caps and pull quotes, HTML support, spelling checkers), layout capabilities (story threading, graphics placement, grids, master pages), rudimentary graphics creation possibilities, and powerful master document features (embedded indexing, table of contents creation, and so on). A DTP program can easily create files in formats used by printer service bureaus, prepare color separations, and indicate page trims. DTP programs can import text from most word processors but, unlike word processors, have a more flexible interface and, once learned, are much easier to use for publishing applications.

The most widely used DTP programs are Adobe PageMaker <http://www.adobe.com>, Adobe FrameMaker <http://www.adobe .com>, QuarkXPress <http://www.quark.com/>, Corel Ventura <http: //www.corel.com>, and Microsoft Publisher <http://www.microsoft .com>. These programs differ somewhat in their strengths. Some perform better for book-length works (FrameMaker and Quark-XPress, for example), others are stronger in creating shorter, more stylized works such as brochures (Adobe Pagemaker), and still others are more geared to Web publishing (Microsoft Publisher). You may not have a choice as to which package you use. Your employer will already have chosen for you, or your school will have installed one of these in its lab. Once you have learned to use one of these programs, your skills can be adapted to other software. They all offer many of the same features, and to an extent, once you have learned to use one, learning another is not too difficult.

The novice DTP software user can easily get carried away when faced with the broad array of options these programs offer. Some typical mistakes include using too many fonts (often two are enough), using inappropriate or poorly matched fonts together, ruling off too many frames to emphasize blocks of text, formatting with improper typographical punctuation (incorrect use of apostrophes, double hyphens instead of em dashes), extra space between sentences and after colons, too much clip art, too many text wraps (using lines of type to follow the outline of photographs placed between columns or to extend from adjacent columns), styling headlines and subheads in uppercase type, and using underlining instead of italics for emphasis.

Selecting the Most Suitable Tool

When you have the choice of producing your document with a word processor or a page-layout DTP program, consider several factors before making your decision. The first and most important is yourself. Which tools are available, and which of them do you know well? If you believe that a certain tool will suit the job but you don't know how to use it yet, you will have to decide whether you have enough time to learn it, by taking a course, asking a colleague for assistance, or reading a printed or online manual.

If you are equally comfortable with a word processor and a page-layout program, consider the complexity of the task. In general, word processors are suitable for linear text, with some graphics, columns, and header and footer detail. Word processors *can* do more sophisticated page-layout tricks, but that is not their strength. If you know your word processor extremely well and your document is not too demanding, you may find the word processor to be a good choice. For any job with more demanding layout requirements, a page-layout program is probably the better choice. Using a page-layout program, you can place text and illustrations on the page as easily as dragging and resizing them with your mouse. The text, illustrations, and other page elements can easily be manipulated as grouped objects, or they can be modified element by element.

Consider the strengths of each application. Applications do best when they are asked to do what they are designed to do: word processors compose and edit text; page-layout programs arrange page elements.

Most word processors enable you to specify type size in points, just as you would with a desktop publishing program (a point equals approximately $1/72$ of an inch). However, word processors still retain the interline spacing conventions from the days of the typewriter—single spacing, double spacing, and so on. Desktop publishing software uses the spacing conventions and terminology of typesetting. This advantage allows you greater control over the amount of space between lines, as well as between words or between individual characters.

If you want to treat your text as a design element, a full-featured desktop publishing program is your best choice. Word processors are geared for printing out documents in office-type formats: letter or legal-size paper, envelopes, labels, and so on. If you send a file created in your word processor to your service bureau, you may find it unable or unwilling to deal with your file.

The ability to deal with color is another key features that distinguishes high-end DTP programs from word processors. Handling color usually means preparing a file so that a printer or service bureau can create color separations from which color printing plates are made to print each color. High-end desktop programs such as QuarkXPress or PageMaker include better color handling and are far more capable of dealing with color than a word processor.

ONLINE ELECTRONIC PUBLISHING

Web Publishing

Over the past few years Web publishing has become a commonplace phenomenon. Most major corporations have a corporate page, most organizations have an organization page, and countless individuals have personal home pages. Web publishing has put a printing press in practically every home, or at least every home with a computer and the right software. With just a little effort, you can design and publish a Web page that tells everyone with access to the Web who you are, what your interests are, and so on. Many colleges and universities offer courses for credit over the Web (providing lectures, assignments, and tests), companies and organizations offer information and freebies, and individuals publish just about anything that comes to mind.

As a college student, you may be involved with a Web publishing assignment at some time. As an entry-level employee, you will increasingly access all kinds of company documents through your company's intranet or extranet (see Part Six, "The Internet"), and you may be asked to publish documents of one kind or another on your company's Web site.

Web publishing is a tremendous new technology with all kinds of potential effects on the publishing industry in general.

TIPS: Publishing a Web Site

- Try out your prototype page layouts on some test users. See whether they think the pages look good and are user friendly.

- Look at your Web pages on as many different computer platforms and configurations as possible. What looks good on one machine may look disappointing on another. The two main computer platforms in the public arena are Microsoft Windows–compatible PCs and Apple Macintosh. You should also test different browsers in several versions. The two main browsers to test are Microsoft Internet Explorer and Netscape Navigator or Communicator.

- Try to break up the ownership of all the pages at your Web site. That way, the work can be split into manageable portions that are the responsibility of different people. Running a Web site is similar to running a newspaper or magazine, and you may need to assign similar roles of responsibility (editor, art designer, and so on).

- Check all links periodically to make sure they still lead to where you expect and the linked pages have not been deleted or moved.

- Be sure your Web pages are backed up periodically in case a disk drive goes bad or some pages are accidentally corrupted.

- Remember that your Web site can be viewed by anyone on the Internet. Consider how the site reflects on you and your organization.

- Realize that home pages must reside on a computer permanently attached to the Internet if they are to be available twenty-four hours a day, seven days a week. Internet Service Providers (ISPs) offer that service and typically charge $15–40 a month. An account usually includes at least 5 megabytes of online storage for your home page(s) on the central computer. This amount of space is ample for dozens of pages and graphics. If your Web site is set up to advertise a product or company, it will be considered a commercial site. Rates for commercial sites are usually higher than those for sites for educational or personal use. In addition, if your site attracts a great many "hits" (visitors), you may be asked to move to another server or pay a higher fee to stay where you are.

- Register a custom domain name if you want to have a name such as "MyCompany.com." This now costs $70 for the first two years and $35 each year thereafter, over and above costs for Internet services. Most providers will help with the registration process. Having your own domain name has the additional advantage that if you decide to move to a different ISP, your e-mail and business addresses won't change.

- Investigate packaged business plans that offer more storage and services than consumer accounts and might be more appropriate for larger sites. Many ISPs offer such plans.

- Be aware that support, maintenance, and backup costs add up. Some companies want to install their own server. If your company is already hard-wired to the Internet, adding a Web server may be a matter of simply plugging another machine into the local area network (LAN). Establishing your own server (and staff) on premises is not cheap, but it gives the greatest degree of control and benefits such as e-mail and Internet access for employees. The trick is maintaining reliable operations while keeping hardware, telecommunications, and staffing costs under control. The advantage is that you can create your own business and offer video or sound streams or other custom features to consumers while subletting your space to other clients.

TIPS: Promoting Your Web Site

- Consider putting your Web page in an Internet mall such as OpenMarket, Branch Information Services, WebCom, or MarketNet. These services host from a few dozen to a few hundred businesses, charge $20–$500 a month and can help draw new or casual Web users to your site. Some services provide help with order taking and other aspects of electronic commerce.

- Publicize your site. Most people won't find your Web page by random browsing. It must be listed with one or more search engines. Put your site into as many categories and classifications as each catalog service allows. The big ones, such as Yahoo!, Excite, Lycos, and Webcrawler, are

(continued)

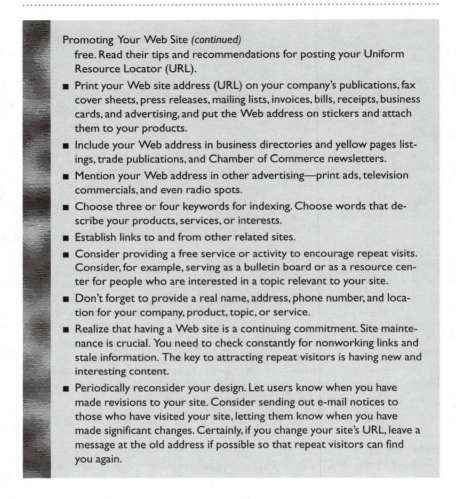

Promoting Your Web Site *(continued)*
free. Read their tips and recommendations for posting your Uniform Resource Locator (URL).

■ Print your Web site address (URL) on your company's publications, fax cover sheets, press releases, mailing lists, invoices, bills, receipts, business cards, and advertising, and put the Web address on stickers and attach them to your products.

■ Include your Web address in business directories and yellow pages listings, trade publications, and Chamber of Commerce newsletters.

■ Mention your Web address in other advertising—print ads, television commercials, and even radio spots.

■ Choose three or four keywords for indexing. Choose words that describe your products, services, or interests.

■ Establish links to and from other related sites.

■ Consider providing a free service or activity to encourage repeat visits. Consider, for example, serving as a bulletin board or as a resource center for people who are interested in a topic relevant to your site.

■ Don't forget to provide a real name, address, phone number, and location for your company, product, topic, or service.

■ Realize that having a Web site is a continuing commitment. Site maintenance is crucial. You need to check constantly for nonworking links and stale information. The key to attracting repeat visitors is having new and interesting content.

■ Periodically reconsider your design. Let users know when you have made revisions to your site. Consider sending out e-mail notices to those who have visited your site, letting them know when you have made significant changes. Certainly, if you change your site's URL, leave a message at the old address if possible so that repeat visitors can find you again.

Many Web editing programs help you not only to create your Web site but also to publish and maintain it when you're ready. Microsoft FrontPage is a good example of a typical program with these features. (See Part Six, "The Internet," for a discussion of various Web editors and publishing programs.)

CD-ROM Publishing

CD-ROMs are the medium of choice when you want to store or handle large amounts of data quickly. The newest CD-ROM drives transfer data into your computer extremely quickly, so animation, sound, and a variety of interactive multimedia have become feasible for a wide range of applications. Multimedia computers can process the huge quantity of information that is contained in these files, providing exceptional responsiveness and interactivity. Because CD-ROMs allow random access to the information they contain, the applications can respond to users' needs in a flexible manner.

Because they can hold huge amounts of information and present that information very rapidly, CD-ROM devices have become the de facto standard for distribution of a wide variety of digital data. Music, video, software, training modules, promotional materials, product catalogs, and databases of news, financial information, artwork, directories, and graphics files can easily be stored and accessed on this sturdy medium. Unlike magnetic media, such as floppy disks, CDs are immune to erasure and the effects of temperature changes and mild environmental abuse (coffee spills and the like). They are easy to store and can be used for a wide variety of content material. Storage limitations and processing speeds previously limited real-time user–computer interactions, but CD-ROMs have brought great flexibility to the human–computer interface.

Specifically, technical communication has benefited greatly from this technology. Documents that were too large to be easily transported because of graphics files, photographs, or other visual elements or because of the length and complexity of their textual material, now present few packaging problems, even for a relative novice.

The random-access possibilities of the medium have made multimedia interactivity flexible and powerful. Also, each time a particular module is accessed, users can be guided in different sequences through the data.

CD-ROM publishing is a complex topic. Multimedia tools make it easier for individuals to create impressive multimedia documents. However, you still must master many different and difficult elements, and you must know how to bring all of these elements together to create and produce an effective multimedia program on CD-ROM. A good place to start is with some of the best books on the subject, for example, Tay Vaughan's *Multimedia: Making It Work*. This book and others (see Resources at the end of this part) will tell you what you need to know about the many choices of basic tools and the challenges presented by combining text, sound, images, animation, and video. You'll also need to know a great deal about planning, estimating costs, designing, producing, and delivering this relatively new medium. A book such as Vaughan's will offer you experienced advice on publishing the CD-ROM once you've created it.

Numerous multimedia authoring programs are available for helping you to create and publish documents on CD-ROM. Macromedia Authorware is one good example. Authorware <http://www.macromedia.com/software/authorware/> is an online authoring tool that is useful for creating an interactive, multimedia presentation for the classroom, the workplace, or a consumer kiosk. Designed around a flowline metaphor, Authorware doesn't require much expertise to produce simple modules, but it has the power to help you create extremely sophisticated works once you have become familiar with its features. Click and drag elements onto the flowline to create a multimedia lesson or presentation or, for more complicated interactivity, use functions and variables to supplement the built-in icons.

With Authorware you can incorporate sound files, video files, graphics, and animation into a presentation. The user can interact with the "lesson" in a variety of ways, including keyboard entries, checkboxes, hotspots, and buttons. Depending on how the module is designed, a high degree of interactivity can be achieved. If the user enters an incorrect response, the module can offer a variety of feedback replies, ranging from a simple notice to a repeat of the relevant material to a visit to a remedial lesson. Visually, Authorware offers a tremendous variety of effects: fade-outs, sound effects, video clip art, animations, and others. For the module designer who wants to customize the appearance of the screen elements, Authorware incorporates a powerful button editor and the ability to import graphics and media files in many different formats. Many logical operators and built-in library functions expand the capabilities of this graphical authoring tool. One of Authorware's greatest strengths is cross-platform capability, which enables the designer to create modules that run equally well on Windows and Macintosh platforms. And with Shockwave, Authorware pieces can be run online over the Internet, using fast-streaming technology to play large media files while they are being downloaded onto the user's workstation. Another powerful authoring tool used for similar purposes is Icon Author.

Macromedia Director is a popular authoring tool used for advertisements and training applications. It is organized around a stage metaphor, with cast members (or screen elements) appearing, interacting, and exiting in a sequence of frames.

Authorware and Director are sophisticated, multifeatured, specialized software applications. To use them, expect to devote time and energy to mastering their use. As with any complex tool, the more you use and experiment with their features, the better you will be able to take advantage of their strengths.

RESOURCES

Beach, Mark, and Eric Kenly. *Getting It Printed*. 3rd ed. Cincinnati: North Light, 1999.

Dinucci, Darcy, Maria Giudici, and Lynn Stiles. *Elements of Web Design: The Designer's Guide to a New Medium*. 2nd ed. Berkeley: Peachpit, 1998.

Garland, Ken. *Graphics, Design, and Printing Terms: An International Dictionary*. New York: Design, 1989.

Parker, Roger C., and Patrick Berry. *Looking Good in Print*. 4th ed. Chapel Hill: Ventana, 1998.

Pickens, Judy. *The Copy-to-Press Handbook: Preparing Words and Art for Print*. New York: Wiley, 1985.

Pirouz, Raymond. *Click Here: Web Communication Design*. Indianapolis: New Riders, 1997.

Rice, Stanley. *Book Design: Text Format Models*. New York: Bowker, 1978.

Saltman, David. *Paper Basics: Forestry, Manufacture, Selection, Purchasing, Mathematics and Metrics, Recycling*. Malabar: Krieger, 1991.

Siegel, David. *Creating Killer Web Sites: The Art of Third-Generation Site Design*. 2nd ed. Indianapolis: Hayden, 1997.

Vaughan, Tay. *Multimedia: Making It Work*. 4th ed. New York: McGraw-Hill, 1998.

Williams, Robin, and John Tollett. *The Non-Designer's Web Book: An Easy Guide to Creating, Designing, and Posting Your Own Web Site*. Berkeley: Peachpit, 1998.

The Internet

Illustrations

Tips

The Internet

*T*he Internet is a vast network of computers consisting of many smaller networks of computers. In fact, the Internet involves so many networks of computers that no one really knows how large it is. We do know that it links computers of many different types, sizes, and operating systems and that it links people in more than 170 countries throughout the world.

The Internet began in 1969 under the name ARPAnet (Advanced Research Projects Agency). ARPAnet was part of the Department of Defense. By 1972, the ARPAnet connected fifty universities and research facilities. Today the Internet consists of hundreds of thousands of networks and, by some estimates, over 40 million daily users.

Because of the rapid growth of the Internet and related technologies, the ability to use the Internet—both to find information and to communicate—has become an essential skill for any college student or professional. College students are increasingly required to have a variety of Internet skills to complete various course requirements (and to find jobs!), and employees must increasingly have considerable Internet skills to communicate within a company's network, to communicate with clients, and to find essential information. Using the Internet, you can

- send and receive e-mail all over the world
- join discussion groups about a common subject using Usenet newsgroups and e-mail mailing lists
- talk by using text-based or audio-based programs (or programs that combine audio and text) with another person who has the same talk program

245

- chat by keyboard in "rooms" provided by commercial Internet providers or with Internet Relay Chat (IRC), which enables small groups of users to meet in conference to "talk" to each other by typing on their keyboards
- access computer bulletin boards where you can read or leave messages, research information, network with others, and download files
- browse the World Wide Web for entertainment and download programs
- research the Web for all kinds of text documents, software, graphics, audio, and video
- connect to thousands of computers using gopher menu systems, which make navigation from one site to another easy
- access, research, and retrieve information from many kinds of libraries all over the world using the Web or gopher (an information-gathering service)
- get or exchange software and files with the File Transfer Protocol (FTP)
- find other people who are on the Internet by using a finger program
- log onto and use many computers around the world using telnet (a terminal emulation program) to use the resources on a remote computer such as games, databases, library catalogs, and many other interesting things

This section provides an overview of the key topics you should understand to use the Internet in your work: basic Internet terms, Internet addresses, asynchronous and synchronous communication, computer bulletin boards, the World Wide Web, other Internet sources, research and the Internet, and Internet tools.

BASIC INTERNET TERMINOLOGY

Many terms are explained in the subsections that follow, but it's helpful to know some other basic terms as well. (See Figure 6.1.)

Numerous dictionaries of computer terms are available both in print and on the Web. Some of the more notable computer dictionaries on the Web are at these sites:

<http://www.delphi.com/navnet/faq/glossary.html#usenet>
<http://www.ipswitch.com/Support/glossary.html>

<http://www.abacom.com/innomagi/online/internet/internt1.htm
   #glossary>
<http://www.netlingo.com/>

Glossary of Internet Terms

ARPAnet (Advanced Research Projects Agency network). The network of linked computers established in 1969 by the Department of Defense that eventually became the Internet.

Asynchronous. Not occurring simultaneously or in real time. Asynchronous communications are initiated at one time and accessed at a later time.

Bandwidth. The information-carrying capacity of a connection.

BBS (Bulletin Board System). An online forum that is used to disseminate files and information and to exchange ideas.

Browser. A program that displays or saves the contents of Web sites. Text browsers display only text and no graphics. Graphical browsers may display text, graphics, animation, video, and sound.

Chat. A method of communicating with others online by typing a conversation. Chats take place in "rooms" or "channels" provided by commercial Internet providers or by dedicated software, such as Internet Relay Chat (IRC). See also **Talk.**

Discussion group. An online forum for exchanging ideas and information on particular topics. Discussion groups can operate through mailing lists, bulletin boards, or newsgroups.

DNS (Domain Name System). The correspondence between an online site name (for example, www.amazon.com) and its Internet Protocol address (for example, 208.216.182.15).

E-mail. Electronic mail.

E-mail address. A code that uniquely identifies a person or organization on the Internet. The e-mail address consists of a username, the @ symbol, and the domain name, which is the Internet provider or host. The e-mail address specifies the "location" to which e-mail is sent.

E-mail program. Software used to write and read e-mail. Some popular e-mail programs are Pine, Pegasus E-Mail, and Eudora.

Emoticon. A typed symbol used in e-mail to represent the writer's feelings or to add visual commentary to a text message. The most widely used emoticon is the smiley, created with a colon, hyphen, and a close parenthesis. :-) When viewed from the side, it resembles a smiling face. Other emoticons depict other facial expressions.

Extranet. A corporate intranet that is open to the public as well as employees and vendors. Extranets combine the advantages of tight security with the ability to reach out to remote employees, business partners, and customers.

FIGURE 6.1
Glossary of Internet Terms

FTP (File Transfer Protocol). An Internet utility program that is used to transfer files from one location to another.

Finger. An Internet utility program that is used to ascertain whether a particular person is currently logged onto the Internet.

FAQ (Frequently Asked Questions). A list of questions and answers maintained by a discussion group to provide basic information on the topic for newcomers to the group.

Gopher. An early form of search engine, using nested menus to lead to more specific information.

HTML (Hypertext Markup Language). A system of codes that is used to describe standard text, graphics, and multimedia files so they can be displayed by a browser; the language used for creating Web pages.

HTTP (Hypertext Transfer Protocol). The communications protocol that identifies an Internet document as being one made for the World Wide Web.

Hyperlink. A text or graphic on a Web page that, when selected by a mouse or by a keyboard action, transfers the display to another location or transfers the action to another program (such as a movie player).

IP (Internet Protocol) **address.** The "street address" for e-mail messages. IP addresses consist of four numbers, each number ranging from 0 to 255. See also **DNS.**

Internet. An international network of linked computers.

Intranet. A private, Internet-like network that is set up behind a firewall (secure gateway) for use by corporate employees. Intranets are the computer networks within a company or a business that use Internet technology to share information, even though they are not connected to the Internet.

LAN (Local Area Network). A network of linked computers in one building or organization.

Mailing List. An Internet discussion group. Mailing lists are similar to newsgroups, but mailing lists send e-mail messages directly to its subscribers, while newsgroups must be read by using a newsreader.

MUD (Multi-User Dungeon or Multi-User Dimension). An online role-playing or simulation game or environment that develops over time.

Netiquette. Internet etiquette.

Newsgroup. An online forum organized around a common interest or theme. Newsgroups are members of the Usenet system, which operates similarly to a BBS. News and tidbits posted (sent) by other people are accessible by using a newsreader or, increasingly, a browser.

FIGURE 6.1
Glossary of Internet Terms *(continued)*

Newsreader. A program that is used to read articles in newsgroups. Many of the major browsers provide built-in newsreaders.

Search engine. A tool used to locate information and resources on the Web. Search engines are a public service offered by a handful of companies and frequently display paid advertisements to cover their costs. They all use a slightly different method to collect, index, search, and display their information. Some are organized by general topics, some are specific to names or titles, and still others are organized by qualities or characteristics.

Signature. Contact information or quotation at the end of a piece of electronic correspondence.

Synchronous. Occurring simultaneously or in "real time." Synchronous communications are immediately interactive. Both parties to the communication are participating at the same time.

Talk. A type of synchronous online communication. Talks take place online between two or more people who type their conversation using a talk program. There are text-based talk programs, audio-based talk programs, and combination text- and audio-based talk programs. See also **Chat.**

Telnet. A terminal emulation program that is used to connect to another computer system without using the remote computer's dialup connections. Telnet programs typically incorporate terminal emulators, various file transfer protocols such as zmodem and Kermit, and login scripting capabilities.

Terminal Emulation. Software that mimics the behavior of a particular type of terminal in order to communicate with another computer.

URL (Uniform Resource Locator). A Web address. URLs usually have a form such as http://www.amazon.com.

Usenet. The main newsgroup network. See **Newsgroup.**

Virus. Computer code that is designed to infect and possibly damage a computer system.

Whois, Ping, DNS Lookup. Utility programs. Whois provides information about registered domains, Ping verifies whether a domain is responding or "down," and DNS Lookup looks up the Internet Protocol address that corresponds to a given domain name.

WAIS (Wide Area Information Searching). An information retrieval system that is used to search online databases.

WWW (World Wide Web). An Internet information system connected by hypertext links. Its connections link systems, computers, and users around the globe into a giant communication complex.

WYSIWYG (What You See Is What You Get). A visual equivalence between the image on an originating computer screen and subsequent images on paper or other computer screens.

FIGURE 6.1
Glossary of Internet Terms *(continued)*

ASYNCHRONOUS COMMUNICATION

Communication on the Internet is essentially one of two kinds: asynchronous or synchronous. Asynchronous communications are accessed after a delay of some kind. They are not read in real time. Telephoning your friend and leaving a message is asynchronous communication. Mailing a letter or sending an e-mail message is asynchronous communication, too. You and your correspondent are not interacting with each other at the same moment in time. Synchronous communications are immediately interactive. You are communicating with others in real time. Speaking to your friend on the telephone is an example of synchronous communication.

E-Mail

Probably the best-known asynchronous Internet communication is electronic mail or e-mail. (See Part Eight, "Correspondence and the Job Search," for other information on this popular medium.)

E-MAIL ADDRESSES

E-mail through the Internet works because each user has a specific address. Just as your home has a street address, which enables your mail to reach you and your friends to visit, so your Internet account has an e-mail address. And just as your letter carrier brings your mail to you by finding your street and then your particular house number on that street, so does your e-mail find you by looking for the "street address" that refers to your Internet provider and then looking for your "house number," which is your particular user identification (also called your user name, your user ID, or your userid).

The "street address," or Internet Protocol (IP) address, is four numbers, each ranging from 0 to 255. For example, a typical address is 132.170.240.15. Because strings of numbers are difficult to remember, you can also access the same numerical address using the Domain Name System (DNS). If you use the domain name when you send e-mail, the corresponding IP address must be found using a DNS lookup. Generally, you will not have to do that yourself. The e-mail software will do the lookup and send the message on its way. For example, the numerical address for the server at the University of Central Florida is 132.170.240.15. The domain name is pegasus.cc.ucf.edu. To send e-mail to Jane Doe at this address, you need to know her user identification and the name of the domain. Jane Doe's username may be Doe, or Jdoe, or Janedoe, or any number of possibilities. For example, Jane's e-mail address at the University of Central Florida might be Janedoe@pegasus.cc.ucf.edu. She may have a second e-mail account, with a username that is totally unconnected to her name, such as wildrose. For example, her e-mail address with America Online might be wildrose@aol.com.

Extensions at the end of Internet addresses indicate the type of connection or institution. For example, .com is for commercial and business connections or for individual users who have a connection via a commercial Internet provider, .edu is for educational institutions such as universities and schools, .gov is for branches of government, and .mil is for military service branches and bases. Extensions can also indicate the country in which the domain is located. For example, .ca means Canada, .au means Australia, .uk means Britain, and so on.

If you send e-mail to an invalid address, the e-mail is usually sent back to you with an explanatory note. When this occurs, the message is said to have "bounced." A message may bounce back to you quickly, or it may take a few days to bounce back to you. Then you'll have to find the correct e-mail address before sending the message again. Some Internet providers do not notify you if a message you have sent is addressed to an invalid username. So if you send an e-mail message to an invalid address on that provider's server, you would have no way of knowing that your message has not been delivered.

MAIL PROGRAMS

Today many kinds of e-mail programs are available, and e-mail programs are also packaged as part of Web browsers such as Netscape Communicator and Microsoft Internet Explorer. Some of the most popular e-mail programs are Pine, Pegasus E-Mail, and Eudora.

Pine is widely available on the Internet, and many service providers make it available as an option. Pine was developed by the University of Washington as an easier alternative to various Unix mail programs. Pine has all the essential features for sending, reading, and managing your e-mail, and it can even be used as a newsreader. Novice users can jump right in, since Pine provides instructions that are always on-screen and is very tolerant of mistakes.

Eudora is an inexpensive commercial e-mail program with many sophisticated features. A less extensive, but still very good, version is available as freeware. Eudora enables you not only to send, receive, and manage your e-mail, but also to do these and other tasks in interesting ways. You can easily attach all kinds of files, make your e-mail secure, create multiple e-mail accounts, filter and sort your e-mail automatically, set up automated responses to common requests, use a built-in spelling checker, customize the program to make it easier to use, and customize your address book to include not only e-mail addresses but also other information such as street addresses and phone and fax numbers.

Pegasus E-Mail is a freeware e-mail program with a variety of good features for sending, receiving, and managing e-mail. Like Eudora, this is a full-featured program with multiple account support and mail filtering, enabling you to manage a large quantity of e-mail with ease.

Some major service providers—America Online and Prodigy, for example—provide their own special e-mail programs. Others—Microsoft Network, for example—use the e-mail program that is available within Microsoft Internet Explorer. The default e-mail program for other providers, such as BellSouth, is the e-mail program available through Netscape Navigator. Some of these providers allow you to use any e-mail program.

Free e-mail accounts are available from various sources. Juno, for example, provides a free dial-up e-mail account to anyone who fills out a questionnaire. HotMail is an example of free Web-based e-mail. In return for the free e-mail account, your outgoing e-mail may carry an advertisement for the company, and you may also have to view advertisements whenever you access your account.

Usenet

One of the most popular areas of the Internet is Usenet. Usenet enables you to communicate with people all over the world who share your interests. It is also a tremendous source for finding all kinds of information from experts who are willing to take time to share their expertise with you. Usenet consists of over 20,000 newsgroups, and more are added daily. Newsgroups are collections of news and tidbits posted (sent) by other people on the Internet. Newsgroups are similar to bulletin boards (see "Computer Bulletin Boards," page 262) in that anyone can post, reply to, or read messages.

SELECTING A NEWSREADER

To access newsgroups, you need a newsreader program. Many major browsers have newsreaders built into them. For example, both Netscape Communicator and Microsoft Internet Explorer have features that enable you to read and subscribe to newsgroups. You may also choose to use a newsreader that is not part of your browser. (See "Usenet Newsreaders," page 277, for a more detailed discussion of various newsreader programs.)

SUBSCRIBING TO NEWSGROUPS

There are newsgroups on almost any topic you can think of. When you use your newsreader for the first time, you may need to sift through the list that your ISP provides and pick out the newsgroups that look interesting to you. Each newsreader has a procedure to indicate the groups you wish to follow. Once you have specified these groups, your newsreader considers you "subscribed" to those groups. Unlike magazine subscriptions, however, newsgroup subscriptions are free. Whenever you use the newsreader after that, you will see only the groups you have previously chosen. You can always add or drop groups, and as new groups become available, your newsreader will give you the opportunity to subscribe to them. Messages in each newsgroup are listed with the username (Internet nickname) of the person who wrote the message, along with a title (the subject of the message). Once you pick a message to read, either you need to save it or your newsreader marks it read (and it disappears forever). Each newsreader has its own way to keep track of which articles you have decided to read, skip, save, or delete. You can also send an article to a friend via e-mail using the forwarding feature available in many newsreaders.

If you have a sincere interest in the group but are not sure of its exact focus, you might want just to read the messages for a while (known as "lurking") to get an idea of the character of the discussions. In fact, experienced newsgroup users suggest lurking for several days on any newsgroup that is new to you.

SEARCHING FOR INFORMATION IN NEWSGROUPS

You can research the postings, or articles, on Usenet using a variety of services or tools:

DejaNews <*http://www.dejanews.com*>. DejaNews searches past and present newsgroup articles. It is useful when searched carefully, because newsgroup articles themselves include information and point to many resources that might not be found through Web searches. For example, it is one of the best ways to find a brand-new Web site that other engines have not yet indexed.

Tile.net <http://www.tile.net>. Tile.net helps you search for individual newsgroups by name or keyword. Some of the major search engines also enable you to search Usenet. For example, AltaVista has this capability.

Usenet FAQ <http://www.cis.ohio-state.edu/hypertext/faq/usenet/ FAQ-List.html>. Usenet frequently asked questions (FAQs) are compiled by the experts who support specific newsgroups. Since the same questions occur again and again as new people start reading newsgroups, these files are developed to provide a ready source of answers. They are generally authoritative, subjected to the review of many participants in a subject field, and frequently updated. However, although the FAQs themselves are updated frequently, the particular archive may not be up-to-date. If the date on an FAQ is not recent, search DejaNews for a later version.

TIPS: Posting to Newsgroups

- Don't post any messages to a newsgroup or mailing list without reading about two weeks of postings first. Each group has its own style and unwritten rules about what is acceptable. Learn these before posting.

- Don't ask questions that have already been asked many times before. Many newsgroups and some mailing lists have frequently asked question (FAQ) files that are posted regularly. Read the FAQ for your group to see whether the answer to your question has already been provided.

- Don't post an article that is off the topic of discussion for the newsgroup or mailing list. No matter how important something seems to you and how sure you are that everyone wants to know it, stick to the topic.

- Don't post chain letters or messages on how to make money fast anywhere.

- Don't repost or forward virus warnings. They are virtually all hoaxes. If in doubt, check <http://www.symantec.com/avcenter/hoax.html> for information before spreading possibly incorrect information and wasting bandwidth.

- Don't post the same article to more than one newsgroup or mailing list unless you have a good reason to do so. Find the most appropriate newsgroup and post it there. Although there will occasionally be exceptions to this, remember that if you post the same message in separate related newsgroups, a lot of people will see your message over and over again. There are ways to cross-post that will handle this problem. When in doubt, post your message just once.

- If you have a product or service that you are sure people want to know about, it may be all right to post a discreet announcement if you follow the rules above. Post it once and find the right place where people will want to know about it. Don't post long, detailed information or use a hard sell. The purpose of some newsgroups is to announce new products and services. Use them, and don't clutter up inappropriate newsgroups. If you want to sell your wares over the Internet, put up your own Web site and entice people to visit it, which is an acceptable approach.

- Keep your articles short. When replying, don't quote any more lines from the previous article than are necessary to convey the essential meaning. Be careful, however, that your quote doesn't change the meaning by too selective a choice of words.

- Don't send "me, too" or "I agree" messages. Send private mail to the author if you want to, but don't clutter up the group with these.

- Don't jump in and answer questions posed by others until you have checked to see whether they have already been answered. Consider sending your answer privately by e-mail to the asker, unless you are sure your answer is unique and of general interest to the newsgroup readers as a whole.

- Don't use large signatures (trailing contact information or quotations) that fill half a screen. In particular, don't use large signatures over and over again. The same people, for the most part, are going to see your signature each time, and the more they see of a large signature, the more you are going to annoy people. Four lines is the suggested maximum for a regular signature.

- Don't publicly respond to others who break the rules. Some people do it just to get the reactions (it's called trolling). The best thing you can do is ignore them.

- Above all, treat people with respect. Consider your words carefully, and use emoticons (symbolic representations of facial expressions) appropriately to indicate your tone. If you feel like flaming someone (sending an abusive or angry message), you should delay sending your message for at least a day to see whether you change your mind the next day. Once your message has been sent, it's too late to change your mind. (Newsgroup postings can be canceled, but many providers do not process cancels, so your angry message will be available to millions of people even if you try to cancel it. E-mail can't be canceled.)

Mailing Lists

Mailing lists are Internet discussion groups. They are similar to newsgroups, but whereas newsgroups disseminate information by making it available to your newsreader, mailing lists e-mail the information directly to your electronic mailbox. When you join a list, you will start getting the messages from the list in your e-mail inbox. When you want to contribute to the discussion, you send out a single e-mail message, which is then distributed to all the other members. Each e-mail list has an owner (one or more people who are responsible for the list), a list address (the place where you send e-mail for the whole group), and an administration address (the place where you send e-mail to have your name added to or removed from the list or to change your list options).

Mailing lists differ from newsgroups in several ways. First, it's harder to participate in a Usenet discussion group because you have to learn how to use a newsreader program (however, these programs have become much easier to use over the past few years). To belong to a mailing list, all you need to know is how to send and receive e-mail. Second, discussions in a mailing list are sent directly to you. You can subscribe to your favorite mailing lists and read their messages at your leisure as they arrive. Usenet newsgroups require more effort. You have to use your newsreader program every so often to see what has arrived. Usenet does have the advantage of requiring you to participate only when you want to. With a mailing list, you have to send a special message to be put on the list and send another message to be taken off the list, and the messages can pile up in your mailbox when you don't have time to read them.

MANUAL MAILING LISTS

Most e-mail programs enable the user to create mailing lists and directories of e-mail addresses. This is usually done by assigning a keyword or nickname to represent a group of addresses. The addresses are stored in this directory, and all messages sent to the keyword address are distributed to all addresses in the group.

For example, let's say you want to send an e-mail message to your three friends who play chess. So you use the keyword "chess" to define the group. Then you would enter the addresses of the individuals in the group (for example, bill@digital.net, ted@aol.com, and wayne@netcom.com). These would be stored under the keyword as a group so that a mail sent to "chess" would be distributed to Bill, Ted, and Wayne.

AUTOMATED MAILING LISTS

Most mailing lists are automated and require users to use certain commands to participate in the list. Listserv is probably the most common listserver (automated list manager). Other systems include majordomo, listproc, and almanac.

A listserver may manage any number of mailing lists. Each mailing list will have a list name of up to thirty-two characters in length. However, most lists have only eight characters in their name and commonly end with '-L' (to readily identify them as listserver mailing lists). A list's e-mail address consists of its list name, which serves as its username, and the name of the server that hosts it, which is its domain name.

The volume and formality of individual lists vary enormously, and you will find a tremendous range in the type of discourse found on mailing lists. The listowner (or moderator in the case of moderated lists) sets the rules pertaining to what topics can and cannot be addressed on the list, what level of formality is appropriate, what tone is acceptable, and what constitutes legitimate list business. Some lists are moderated, in which case each message must be approved by the listowner before it can be sent to the list members. When you subscribe, you will probably be sent a list of the rules of the list, instructions on how to access the list archives if any, and information on how to change your subscription options. You won't be charged for subscribing to any of these mailing lists, and it doesn't take much effort to add your name. If you want to learn more about mailing lists, try some out.

SUBSCRIBING TO A MAILING LIST

You subscribe to a mailing list by sending an e-mail message to the list's administrative address. Once you have subscribed, all messages sent to the list members come to your mailbox. Most lists are open to all, but some are restricted to the people who have been accepted by the listowner. Many lists have a digest option so that you get a compiled set of messages one or more times a day, or less often, instead of receiving each message separately.

Exact procedures for joining may vary because several different systems are used for this purpose. In general, though, you usually leave the subject line of the e-mail message blank and send a message such as the following to the listserver: subscribe [name of list] [your first name] [your last name].

Don't send your requests to the list itself, but to the automated list manager. When you send your request to the list itself, your subscribe command is picked up by the system and distributed to all the members of the list as though it were a contribution to the discussion; this is a good way to annoy hundreds or thousands of people. Of course, if you are not yet subscribed, your message will bounce back to you with an error message. So as a rule of thumb, before you send off a message to or for a mailing list, double-check the address to which it will go, even if you have sent dozens of messages to it before.

Once you send your subscription message, you will almost always get a confirmation via e-mail within a few minutes saying that your name has been added to the list. Very high-volume lists may require you to send a second message confirming your e-mail address. After you send the confirmation message, you will receive a welcome message that provides useful information about various aspects of the mailing list. Keep this welcome message because it also includes instructions for getting off the list when you want to quit.

TIPS: Mailing List Etiquette

There are a few basic rules for using mailing lists:

- Read the list for a while before posting. Each list has its own culture, so "listen" for a while before you "speak up."

- Privacy policies differ among lists. Find out whether it is appropriate to copy parts of postings off the list before you do it. In general, you need the writer's permission before you use the posting elsewhere or send it to someone who is not on the list.

- If you respond to a posting, pay attention to whether you are addressing the whole list or just the author. A common mistake made in mailing list interaction is sending messages meant for one person to the entire list.

- Don't send advertisements to the list unless you have been encouraged to do so. For many lists, sending advertisements is grounds for dropping your membership.

- Respect the opinions of others. Disagreeing with the opinions of others does not entitle you to insult them.

- Keep your postings on topic. Digressions are tolerated to a degree, but in excess they just clutter the list and annoy those who want to get on with the topic at hand.

SYNCHRONOUS COMMUNICATION

Synchronous communications differ from asynchronous communications in that they are immediately interactive. In synchronous communications you are communicating with others in real time. The three major types of synchronous communications on the Internet are talk, Internet Relay Chat, and multiuser dimensions (MUDs).

Talk

TEXT-BASED TALK PROGRAMS

You can use a few talk programs on the Internet to talk to other people via text. All you need to know is whether the other person is logged on at that time, and then you need to use one of these talk programs to "talk." Typically, you type your message at the top of the screen while your correspondent reads what you have to say and responds by typing words at the bottom of the screen. With these talk programs, you can talk to people on the other side of the world. Two popular talk programs are "talk" and "ytalk" for Unix and WinTalk for Windows. To use these programs, you need to know the e-mail address of the person you want to talk to. Some Internet providers have their own talk programs, and using them to talk to subscribers of the same service is as easy as clicking a button and typing a username or "screen name."

AUDIO-BASED TALK PROGRAMS

Many freeware talk programs are available for download from the Internet. Increasingly, talk programs allow you to speak via a microphone and speaker to other people on the Internet who are using the same software, regardless of where in the world they are located. These programs set up an interstate or overseas conversation between two parties via the Internet for the cost of a local call. Depending on your modem speed, the talk program you are using, and your computer processor, the clarity of speech will vary remarkably.

COMBINATION TEXT- AND AUDIO-BASED TALK PROGRAMS

Some programs enable you to use text and audio to talk to others over the Internet. For example, Netscape CoolTalk is a conferencing or collaborative system that works over the Web. With CoolTalk, you can make keyboard or audio calls to other CoolTalk users on the Web. One advantage of this particular program is its widespread availability. Unlike CyberPhone, InternetPhone, WebPhone, or any other Web-based telephone products, it comes bundled with Netscape Navigator. There isn't much point in being able to talk to someone if the rest of the world can't talk back. CoolTalk may well become common enough to be useful.

CoolTalk is more than just an Internet telephone. It also has a "shared whiteboard." You and the person you're talking to can draw and write on the same document. CoolTalk isn't the only software that enables you to collaborate with other people over the Internet. Netscape has another suite of products called Collabra aimed at the business market. Microsoft Netmeeting is another collaborative system.

Internet Relay Chat

Internet Relay Chat (IRC) is a worldwide network of servers that enables people from all over the world to talk to each other through the Internet. IRC is the Internetwide version of, for example, America Online's chat rooms. In theory, IRC is a way for individuals around the world to have stimulating online discussions. Unfortunately, IRC is too often just a more convenient way for people to waste time. As with other Internet services, IRC is what you make it. If you can find interesting people to have interesting discussions with, IRC is worthwhile.

Like every other Internet service, IRC has client programs and server programs. The client is the program, such as mIRC, that you run on your local machine (or perhaps on your provider's system). In concept, an IRC server resembles a large switchboard, receiving everything you type and sending your messages to other users and vice versa. What's more, all the different servers are in constant contact with each other. As a result, words that you type at one server are relayed to the other servers, so the entire IRC world is one big, chatty family.

The difference between IRC and "talk" is that while "talk" allows communication between only two people and is text based, IRC allows what are called chat rooms or channels, each of which can have a large number of people who are communicating with each other. In addition, IRC programs also have the capability of transferring files between users and even playing sound files on some Windows-based systems. IRC is also a very handy tool for getting information and data on almost all subjects in real time from many knowledge-able users.

IRC consists of two main components: the user (identified by a nickname) and the conference rooms. When you initially log on to IRC, you will need to choose a unique nickname so that you can be more easily identified on IRC at that time. It has been estimated that over 10,000 people are on IRC at any one time, so try to choose a name you think would not be commonly used. Once you are logged on to IRC, to start chatting, you simply join a channel (enter a chat room). Once you have joined a room with other people, you will see anything that the other people in that room type, and anything you type will be seen by the others in the same room. You can also com-municate privately with one or more of them.

Just as you need a telephone to connect to your phone line, you will need to download some software before you can use the IRC on the Internet. Netscape Chat is one of the easiest IRC programs to understand and use, and mIRC <http://www.mirc.co.uk/> is another very popular IRC program.

TIPS: IRC Netiquette

Of course, you should observe here the same netiquette you observe for your e-mail and other Internet communications.

- When you join a channel, follow the conversation for a few minutes until you pick up the thread of the discussion (or discussions). Don't im-mediately start saying "Hello!" to everyone.

- Don't use all capital letters; it's similar to shouting.

- Speak (type) the language that's appropriate for the channel and on the topic the channel is about (if any).

- Don't send inane messages, such as "Does anyone want to talk?" If you want people to talk to you, say something interesting. It doesn't have to be anything too sophisticated. "What about those Mets!" or "Cold enough for you?" should do. "I'd like to talk about movies" or "Anybody like cooking?" would be good, too.

- If you have something to say to someone that isn't of general interest to the group, send it privately.

- Don't flood (send lots of messages really fast so that no one else can get a word in edgewise).

MUDs

MUD stands for "multi-user dungeon" (from the popular Dungeons & Dragons role-playing game) or "multi-user dimension." MUDs are sites on the Internet where people can visit to talk and interact. There are many types of MUDs. There are social MUDs, where you go to talk or socialize, similar to the chatting you do in IRC, and there are role-playing MUDs, where you select a character to play and you (on the MUD) act out that character. Many MUDs have a specific theme.

When a MUD has a specific theme, the people who go on there assume certain roles, like characters in a play. If you went to a Star-Wars MUD, you would assume a role from the Star Wars universe and interact with the other characters there as if you were that person. It's like a game or a play: You interact with the environment, something happens, and the game goes on.

MUDs consist only of text that you or the other users write. MUDs do not have any graphics. If you've played with MUDs a little or you plan to, you might encounter the terms MUCK, MUSH, MOO, and perhaps others. Even though there are many of these, they're all basically still MUDs, and they essentially work the same.

You can access MUDs through telnet or via the Web. Lynx, a standard text browser, can access them directly, and Netscape can access them if a viewer is configured. Telnet accesses them directly also.

COMPUTER BULLETIN BOARDS

A computer bulletin board (or bulletin board system or BBS) is a computer that uses a special program to store e-mail messages. People use a computer bulletin board to post messages and send and receive programs and files. Bulletin boards enable users to converse by typing messages back and forth to each other, so a computer bulletin board may be synchronous at times and asynchronous at other times. Computer bulletin boards do not have to be connected to the Internet, although many of them are. Some companies maintain private bulletin boards for their employees or customers, and these may have to be reached by direct-dial telephone from your computer and modem into the company's computer.

Computer bulletin boards are very much like regular bulletin boards. You post information on both or find information on both. However, a computer bulletin board helps you handle much more information much more effectively. Users can create their own topical areas (sometimes referred to as threads), where they can post messages. Other users can review the bulletin board by topic and see questions, notes, advice, or anything else posted by others. Then they can respond if they wish. Many bulletin boards are maintained by companies as a service to their customers, who can write in to receive technical support, documentation, or upgrades to software. Topical bulletin boards are useful for providing members with an open discussion forum on a variety of issues of importance to them, including potential new legislation, products, regulations, or committee issues. They are easy to search, since the user can key in on the single topic of interest. They also do not require staff maintenance. Computer bulletin boards have been around since the mid-1970s. No matter what becomes of the Internet, computer bulletin boards are not going to disappear from the electronic horizon any time soon.

WORLD WIDE WEB

The World Wide Web was created in 1989 by a physicist named Tim Berners-Lee, who worked for an organization called CERN (European Nuclear Research Center) in Switzerland. The World Wide Web, often referred to as WWW or the Web, is essentially an Internet information system connected by hypertext links. The Web consists of Web sites, each comprising one or more Web pages. The term *World Wide Web* is an apt one, since its connections link systems, computers, and users around the globe into a giant communication complex.

HTML, URL, and HTTP

Web pages are created in what is called HyperText Markup Language (HTML). Each Web page has its own Uniform Resource Locator (URL) address. This address is how a browser, such as Netscape Navigator or Internet Explorer, locates a Web site and displays or saves its contents.

The URL format for browsing the Web is in the following form: <http://www.ford.com.> The "www.ford.com" is the site's domain name. Most site domain names start with "www" and end with "edu," "com," "gov," "net," or "org," depending on whether the site belongs to an educational institution, a commercial company, a governmental agency, a network, or a nonprofit organization. In other words, many of the major companies on the Web can be found by simply entering <http://www.companyname.com>, where "companyname" can be Ford, IBM, or America Online, for example. There is also a two-letter notation on some domain names representing the country where the site is located. Alternatively, many browsers have a feature that enables you to type an instruction such as "go" where the URL normally would be, followed by as much of the site name as you know. A search engine will then attempt to locate the correct URL for you.

HTTP stands for HyperText Transfer Protocol, the communications protocol that identifies an Internet document as being one made for the World Wide Web. In other words, the "http" prefix on an Internet address tells you and your browser that you are accessing a World Wide Web document.

The general movement of the Internet is to phase out the use of HTTP on a user level. Browsers such as Netscape Navigator, Mosaic, and Microsoft Internet Explorer will enable you to simply type in an address without the "http" prefix. For example, <http://www.in.on.ca> could be typed in simply as www.in.on.ca in the address field of your browser.

Hyperlinks connect sites or entities to one another on the Web. If you click on a picture or text hyperlink, you will be connected to another page, section, or file.

Browsers are programs that display or save the contents of Web sites. If the site includes text coded in HyperText Markup Language (HTML), the browser may display text, graphics, animation, video, and sound. Browsers are either text or graphical browsers. A text browser shows you only text and no graphics when you look at a Web page. A popular text browser is Lynx. A graphical browser may display text, graphics, animation, video, and sound. An early graphical browser was Mosaic. Netscape Navigator was developed not long after Mosaic, and Microsoft Internet Explorer soon followed. Now at least twenty browsers are available on the market, many having a variety of sophisticated features.

TEXT BROWSERS

Text-only browsers have existed for some time and, for people without the right hardware and high-speed connections to access graphics, are still quite usable for browsing the Web. The pictures are missing, but the text content is still accessible in most cases. Lynx is the most common text browser. To access a Lynx browser or for more information on Lynx, see <http://lynx.browser.org/>.

If you use a text browser to access a site that is high in graphical content or multimedia effects, you will be excluded from many of the effects and may not be able to access all of the features the site offers.

MULTIMEDIA BROWSERS

The two most popular graphical Web browsers are Netscape Navigator and Microsoft Internet Explorer. Just a few years ago, both of these products were relatively simple Web browsers. However, they have evolved dramatically just over the past few years, and both now come with a full array of features, including newsreaders and e-mail programs.

Many people who go online using one of the many major commercial providers, such as America Online, Prodigy, BellSouth, AT&T, and Microsoft Network, think they are limited to the browser they are given by the provider. This assumption is not entirely correct. If you want to see and use the channels offered by America Online, for example, you *will* have to use the special software provided by America Online. However, you can also connect to the Internet using America Online, bypass the main menu of America Online, and use another browser, such as Netscape Navigator, to explore the Web.

Only a few years ago, college students were limited to using a text browser, such as Lynx, for accessing the Web. This text browser is still widely used today. However, because Netscape and Microsoft's graphical browsers are free to everyone, they are increasingly becoming the norm on many college campuses as well as on company computer networks. Selecting one browser over another is not an easy choice to make, and many computer users have both Netscape Navigator or Communicator and Microsoft Internet Explorer on their computers. However, usually one browser or the other is the default browser on a given system. Sometimes you finally have to decide which browser you will use most often. Your decision will be based on which one has the features you use most often and which is arranged the most conveniently. Sometimes your decision will be based on which one was installed on your system when you bought it. As Microsoft makes its browser the Macintosh default, inevitably more Mac users will tend to stay with Microsoft's product.

Search Engines

A search engine is a tool that is used to locate information and resources on the Web. You enter keywords, which the search engine software uses to hunt through its database to locate Internet references. A summary of the files containing the keywords is presented in a list of URLs that link to Internet sites. Search engines are a public service offered by a handful of companies and frequently display paid advertisements to cover their costs.

With the rapidly growing number of sites on the Internet, especially Web sites, it can be very difficult to locate what you are looking for. Fortunately, many powerful and flexible search engines are available to help you locate a particular information resource. Search engines are computer programs that enable you to type in a search word or phrase, which the program uses to search the Internet for documents (and, depending on the search engine, graphics and audio files) that contain that term.

MAJOR SEARCH ENGINES

Each search engine has some important differences from the others. They use slightly different methods to collect, index, search, and display their information. Some are organized by general topics, some are specific to names or titles, and still others are organized by qualities or characteristics. It is valuable to read the help sections that accompany the different search engines. The trick for you is to find a search engine that works best for the kinds of searches you do and learn how to compose efficient searches. If you don't find what you're looking for on one search engine, try the same search on a different search engine. Some programs, such as WebFerret <http://www.ferretsoft.com/>, enable you to have several search engines work for you at the same time.

The following are brief general descriptions of some of the major search engines. For further information, you should go to the URLs provided. The capabilities of these search engines are changing so rapidly that almost any information you read in print will not be up-to-date.

AltaVista <http://www.altavista.com> by Digital uses its software to browse the Web automatically and to catalog what it finds, maintaining a huge database that is constantly updated. You can use it to search either Web sites or Usenet articles.

Excite <http://www.excite.com> uses technology similar to that used by AltaVista. What makes Excite unique is the way it sorts information by concepts. It is a very good starting place to find information on a specific topic. Excite searches Web pages, Usenet newsgroups, Web site reviews, and newspaper and magazine news sources.

Infoseek <http://www.infoseek.com> is an excellent, extensive, multipurpose search engine. It uses spider software (programs that explore the Web and retrieve documents) to constantly update its database. Infoseek searches well-chosen Web pages plus gopher and FTP sites. It also searches newswires, major newspapers, Usenet newsgroups, and Usenet FAQs (frequently asked questions files) as separate choices. Don't overlook the FAQ search as an excellent source on its own. Ultra Seek <http://ultra.infoseek.com/> is a larger, more extensive version of Infoseek. Its huge database is likely to find useful links on many topics. Be prepared to deal with a large number of references even if your search is directed and specific.

Lycos <http://www.lycos.com> provides search results that resemble a cross between Yahoo! (See Yahoo!, page 268) and AltaVista. In addition to searching Web pages, it also searches gopher and FTP sites. It searches for graphics or sound files as separate choices. Lycos Pro has more advanced features, which are available through a Java application (a small computer program that your browser downloads and runs).

Netscape Net Search <http://home.netscape.com/home/internet search.html> is an all-in-one search engine page. It includes links to several major search engines and has its own directories and search engines.

W3 Search Engines <http://cuiwww.unige.ch/meta-index.html> interfaces with many excellent search engines, including several of those just described, all in one location. It is a wonderful directory of search engines.

Northern Lights <http://www.nlsearch.com> is a newer search engine that offers full-text search of "every page out there" as well as licensed sources from entire magazines, reference works, and newswires not found on the Web. Northern Lights also categorizes all documents that it indexes. When an initial search is performed, Custom Search Folders listed on the left side of the screen can be used to see the results by a particular subject, type, or source.

HotBot <http://www.hotbot.com/> is one of the largest search engines in terms of number of pages indexed. It features simple and advanced searches using a variety of parameters. It is a joint venture of HotWired, producers of *Wired* magazine, and Inktomi, a computer company. HotBot is noted for its speed and ease of use.

SEARCH INDEXES OR CATALOGS

One major difference between a search engine and a general index (also called a directory or catalog) is that an index will not list your URL if you do not register it with that particular service. Another major difference is that these indexes do not use the software programs (for example, spiders, robots, and scooters) that search engines use to find what is available on the Web. Indexes are usually subdivided into categories, and you have to submit your URL under the most appropriate heading.

Most indexes organize their listings into subject search trees. At the top of the search tree, you select a general topic. This takes you to a list of subtopics. You then select the appropriate subtopic. Repeat until you reach a list of sites for the topic you are interested in. Subject trees are easy to use and give good results if the topic you are interested in happens to appear in the subject tree.

Yahoo! <http://www.yahoo.com> is the biggest of the subject-matter organized directories and is probably the best. It is very useful for finding good collections of resources for a topic. It has an advanced search mode, too. Yahoo is like an Internet yellow pages. The Web sites are provided by people who have found an interesting or useful site and who have completed a form suggesting where the link should be placed in the Yahoo! directory. You can search by browsing this hierarchical directory or by entering names or keywords. The directory and classification system are both very good, and Yahoo! is a good place to start for almost any kind of Internet or Web search. There is a children's version called Yahooligans <http://www.yahooligans.com/>.

META SEARCH ENGINES

The meta search engines, which combine searches of a number of the individual engines, have distinct limitations. They generally use the most basic searches of each of the major engines and do not handle complex search constructs well. If you are looking for a one-of-a-kind item that can be uniquely defined with a few choice words, they may serve well. If you are looking for the best of many items, they are not likely to be appropriate. However, they are gaining the ability to use the more sophisticated search modes of each platform.

A meta search engine doesn't actually do any searching; it takes a query, submits it to various search engines, and presents you with the results. These search engines have several disadvantages. First, submitting to multiple search engines is likely to be a waste of resources. There are enough pages out there that a single search engine will usually produce a large number of results. Second, if you really do need to search every possible page, you probably have to use queries that contain features (for example, phrases) that many search engines don't support.

Some people consider MetaCrawler <http://www.metacrawler .com/> to be the best of the meta search engines. It has the option of verifying pages before providing you with the results so that you actually can search for phrases, even though some of the underlying search engines don't support phrases.

WebFerret <http://www.ferretsoft.com/netferret/> is one of a suite of products that offer useful Web functions on your desktop. WebFerret harnesses the power of the major Internet search engines by helping you present your searches simultaneously to many of them, regardless of their individual search syntax rules. When WebFerret finds the same link returned by several search engines, it eliminates the duplicates. Perhaps the most convenient feature of this handy program is that it shows you a pop-up summary of each link as your mouse hovers over the name of each site.

Prime Search <http://www. primecomputing.com/search.htm>, SavvySearch <http://savvy.cs.colostate.edu:2000/>, and Use It <http: //www.enet.it/menu/useit/ useen.htm> are some other meta search engines. Freeware programs are also available that enable you to use four or more search engines at once. Look for more meta searchers at <http://searchenginewatch.com/facts/metacrawlers.html>.

EVALUATING AND USING SEARCH ENGINES

Although many people have a favorite search engine, there is no "best" search engine. Some are simply more useful than others for helping you find the kind of information you need. Spend the necessary time carefully reading the help screens for the various search engines and doing test searches with them. The following criteria will help you decide which search engine to use.

Know what you are looking for. It is important first to have a good idea of what you want to find. Is the information you need on an academic or a more commercial topic? Is it current or historical? Is it personal information such as an e-mail address or telephone number? For all of these factors there will be a search engine that will be more helpful than others, although there is quite a bit of overlap. It may even be that a subject list such as Yahoo! (see "Search Indexes or Catalogs," pages 267 and 268), rather than a search engine, is what you really need to use.

Each search engine collects and indexes information from only a portion of the Internet. No search engine searches the entire Web. Some do not include gophers, Usenet groups, or mailing lists. Many index only the titles or first 500 words of a page, not the full text of the Web pages themselves. Most exclude text files larger than 4 megabytes, sites with lots of multimedia, and sites with various restrictions. Find out what part of the Internet the search engine accesses and how it performs its searches. You can usually learn about these abilities by reading the help file for the search engine. Sometimes you will have to consult reviews of various search engines to see what each search engine can do.

Evaluate the quality of the search engine by determining how you can search it. Can you perform only simple keyword searches, or can you search in more advanced modes for whole phrases, use Boolean operators (connecting words such as *and, or,* and *not*), and truncate terms? Can you revise and edit a completed search using this search engine, or must you retype the search each time? Many search sites that index different portions of the Internet (Web, gopher, Usenet) let you choose any or all of the separate sections to be searched at a given time.

Most search engines rank the results of your search. Find out how the search results are ranked and displayed. In most cases the ranking is based on the number of times your search words appear in the document, how close your search terms are to each other within the document, how close your terms are to the top of the document, or a combination of all three. Many search engines also enable you to choose how your results are displayed: URLs, titles of the documents only, quotations from the actual Web pages, or a combination of the three.

Determine whether the search finds truly relevant information. Probably the most important criterion in choosing a search engine is the relevancy of the sites that the search finds. Does the search engine find not only the most sites, but the best ones as well? The best way to judge which search engine meets your needs is to try some "typical" searches at a variety of search engines. Develop a couple of sample searches around your interests, and test them with a few search engines. See which search engines seem to find the information that you are looking for, and use them often. As you become increasingly familiar with the particular features of one search engine, your search technique and your search results will improve.

Finally, always be prepared to repeat your search with other search engines, refine your search with different, related terms, or both. Don't settle for the first links your search engine turns up. There are undoubtedly other good sites that it won't find. Be persistent.

TIPS: Planning a Search

- Planning a search takes time at the beginning, but it will save you time later. The first thing you need to do is come up with a search keyword or phrase, just as you do when you are searching the online catalog of a library, for example. Some sources of search terms are words or phrases describing the topic, words or phrases that are likely to be used in discussing the topic, and names of people who are likely to be mentioned when discussing the topic.

- Refine your search. If your search doesn't produce the results you hoped for, your next step is to refine the terms. Expect your search to produce many pages that aren't of interest to you because the search term has meanings other than the one you intended or can be used in contexts other than the ones you anticipated. You can handle this problem in several ways, as explained in the next few tips.

- Consider replacing the search term with a synonym. Unfortunately, the synonym may bring up its own set of irrelevant pages, so you may not want to use this approach often.

- Add another search term (using the AND operator if the search engine supports it or putting the terms in quotation marks if they form a meaningful phrase). The additional word can be another term selected to get articles related to the topic, or it can be a search term for a broader topic that includes the topic you are searching for.

- Exclude the unwanted articles directly by devising a search to match the articles you don't want and using the NOT operator to exclude them.

- Determine why some of your searches fail to turn up any helpful links. In addition to selecting Web pages you are not interested in, searches can fail by overlooking articles that you are interested in. When your search selects pages you don't want, you can examine them to see why the search didn't work as you intended. Unfortunately, when it misses pages, you have no way of knowing why you missed these pages or even that you missed them at all.

- Follow other relevant links. If you find one relevant page, you can use it to find other useful pages. The normal way to do this is to go to the relevant page and follow links to other pages. Or you can use the "find related pages" feature provided by some search engines and newer versions of the major browsers.

- If your search engine offers advanced features, use them to control your search in a variety of ways. AltaVista's advanced search capabilities, for example, enable you to control the search by dates, by your own rankings, and by variations of complicated search strings. (For example, in AltaVista, *and, or, not,* and *near* have different meanings in an advanced search than they do in a simple search.)

Weaknesses of the Internet for Research

Although information on the Internet is increasing at a phenomenal rate, be aware that some kinds of information are not yet available. Such information includes many kinds of historical information before 1990, information about many small companies (although many companies are on the Web, many are not), many kinds of dictionaries or business handbooks for specific industries, and so on.

Also, the Internet is nothing like a library when it comes to authenticating information. (See Part One, "Planning and Research," for ways to evaluate information you find on the Internet.) Information is not stable on the Internet; it can appear and disappear from one moment to the next. Even if the URL doesn't change—and it often does—documents can be pulled or changed without leaving a trace of what was formerly there. Although browser companies will tell you otherwise, a determined hacker can break into an Internet location and change the information, so even supposedly reliable sites should be reviewed carefully.

It's also important to remember that most Internet "publishing" is not peer reviewed. Anyone can publish information on the Internet, and the problem is that "anyone" *does*. Anyone can post information and claim it to be authoritative. In most cases, no colleagues or independent judges evaluate what is placed on a Web site. However, increasingly, various online journals, for example, are requiring submissions to be peer reviewed. Unless you know that a particular online article has been peer reviewed, it is safest to view its authority critically.

INTRANETS

An intranet is a private, Internet-like network that is set up behind a firewall (secure gateway) for use by corporate employees. Intranets are the computer networks within a company or a business that use Internet technology to share information but are not connected to the Internet.

Some types of media and functions that can be used on intranets include text, file downloads, static graphics, video, animated graphics, e-mail responses, sound, electronic forms, presentations, search forms, computer-generated documents, workgroup discussions, and interactive computer programs. Organizations with intranets store and update information electronically on a Web server configured on a local area network (LAN). As information changes, the server content can be easily updated with new or revised data. This approach enables organizations to deliver timely, consistent, and accurate information to their employees worldwide, without expensive typesetting, printing, distribution, or mailing charges.

An intranet that is accessible to every company employee may contain key information, such as employee handbooks, internal newsletters, and stock-plan descriptions. Multinational companies may choose to set up a World Wide Web (WWW) site at each geographical location. Depending on their size and specific needs, organizations may implement a combination of the above, with a corporate WWW server that is accessible to all employees plus dedicated Web servers for individual departments.

An intranet is inexpensive and easy to configure, use, and manage. Organizations can set up a home page for each department or functional area. Because the intranet is configured on a LAN, it is well suited to multimedia applications, such as video and audio, which require higher bandwidth (information-carrying) performance and capacity.

EXTRANETS

An extranet is a corporate intranet that is open to the public as well as to employees and vendors. Extranets are increasingly replacing intranets in the corporate world. They provide the advantages of tight security and the ability to reach out to remote employees, business partners, and customers. Netscape's corporate Web site is a good example of an extranet. Netscape puts its software on the extranet to allow potential customers to download, test, provide feedback for, and buy its products and to provide customer support. Unlike a regular Web site, an extranet offers regulated access to its content that varies according to the authorization level of the user. An extranet gives a business a chance to provide data to the outside world without requiring employee assistance.

INTERNET TOOLS

FTP

File Transfer Protocol (FTP) describes a class of tools that are widely used to transfer files from one location to another across the Internet. In practice, this usually means uploading or downloading programs or data, and it used to be as frustrating a process as can be imagined. Early FTP programs required the user to enter a series of arcane *get* and *put* commands followed by long strings of directory and subdirectory names. With the advent of graphical versions of these programs, transferring files became intuitive and simple.

Two of the most highly regarded of the FTP programs are WS_FTP and CuteFTP. WS_FTP <http://www.ipswitch.com>, available in freeware and commercial versions, enables you to log in at remote sites, browse remote directories, transfer files back and forth, rename or delete files, and create a log of your session. Using WS_FTP, you can see your local directory structure and files next to a remote directory and files and easily move back and forth between them. The program comes with a list of popular anonymous FTP sites, so downloading software is simple. These so-called anonymous FTP sites allow anyone to log in as a user without prearranging an account. Visitors to these sites log in with an account name of "anonymous" and use their e-mail address instead of a password.

CuteFTP <http://www.cuteftp.com/> incorporates many of the features of WS_FTP and adds a few others. Its *stop* command, similar to the stop button found on many Web browsers, enables you to stop a transfer in progress without breaking the connection. CuteFTP also lists file descriptions obtained from the index files found at many anonymous FTP sites. In addition, it stores the directories of sites you have visited recently in a cache, or directory on your computer, making it simpler to reaccess those directories later.

Telnet

Telnet programs are useful when you want to connect to another computer system without using the remote computer's dialup connection. Some operating systems, such as Windows, are equipped with their own telnet programs. For example, HyperTerminal is a telnet program that comes with Windows 95, Windows 98, and Windows NT 4.0. HyperTerminal Private Edition <http://www.hilgraeve. com/htpe.html> is a more powerful version of this program. Many stand-alone versions are also available as freeware or shareware. Telnet programs typically incorporate terminal emulators, various file transfer protocols such a zmodem and Kermit, and login scripting capabilities.

Among the commonly used shareware telnet programs are CRT <http://www.vandyke.com>, NetTerm <http://starbase. neosoft.com/

~zkrr01/>, and ZOC (Zap-O-Comm) <http://www.emtec. com/zoc/index.html>.

Finger, Whois, Ping, DNS Lookup

When you're trying to find out whether a friend, acquaintance, or colleague is currently logged on to the Internet, or if you know someone's e-mail address but want to know to whom it belongs, you will use a finger program. In addition, you may have a dynamically assigned Internet address, that is, one that changes each time you log on to your account. If you want your friends to be able to use an Internet chat or talk program to converse with you online, they will have to have some way of knowing what your particular address is each time they want to initiate a chat session.

Many utility programs provide finger capability, often along with whois (which provides information about registered domains), ping (which verifies whether a domain is responding or "down"), and DNS lookup (which looks up the internet address that corresponds to a given domain name). WsFinger <http://www.empire.net/~jobrien/>, and Here <http://www.cris.com/~Beers/here/> are some of the many finger clients in general use.

WAIS

Wide Area Information Searching (WAIS) is an information retrieval system that is available on the Internet. It's almost like a giant index to the Internet. WAIS makes it possible to access information distributed on servers around the world from clients that may be running on a number of different platforms. Local sites set up servers and make their own information accessible to WAIS users. Every word and document subject is indexed for searching purposes. This information is then passed to the WAIS Directory of Servers, which is a database of information pertaining to all registered WAIS servers.

WAIS is an excellent means of determining sources of information when you have no idea where to begin looking. If you know exactly what you are looking for, WAIS databases can probably provide what you need. Also, many Web sites now offer WAIS search engines. You can use these search engines to locate the specific topic or computer file you're interested in. You can access WAIS on the Web by using the following address: <http:// www.w3.org/Gateways/WAISGate.html>.

Editors

Internet editors, also called HTML editors, take the mystery out of designing and creating Web pages. Even beginners can create professional-looking Web sites that incorporate multimedia and interactivity by using programs that are becoming more and more intuitive to learn and use. Newer editors take advantage of interfaces that resemble the word-processing programs that users are familiar with. Point-and-click and drag-and-drop capabilities make creating even complex Web pages less a matter of learning commands than of imagining an outcome. It's easy to create clickable image maps, animations, tables, frames, and a variety of hyperlinks by using the menus and buttons provided in the Internet editors.

Many editors let you see what your Web page will look like as you work on it. Because simple keystrokes such as the Enter key and the space bar produce the same results on your Web page as they do on a word processor, you can easily visualize the results of your work. Today, the most useful editors are WYSIWYG (What You See Is What You Get) tools; you don't need to worry about typing arcane command tags and then being surprised at the effects they produce. And when you have finished entering your text, the editor's spelling checker will verify your spelling. Then simply post your Web pages to your server, using the editor's publishing (for example, uploading) capability.

Netscape Composer <http://www.netscape.com> belongs to the Netscape Communicator suite, so it seamlessly interfaces with the other programs of the group. PageMill, Adobe's Web authoring tool <http://www.adobe.com>, was a Macintosh application that eventually was released in a Windows version, enabling users on both platforms to benefit from its easy drag-and-drop interface. FrontPage <http://www.microsoft.com> is Microsoft's HTML editor, which was designed to facilitate transfer of images, texts, and tables from other applications onto your Web page. Unlike many of the other editors, Allaire HomeSite <http://www.allaire.com/> is not a WYSIWYG editor, but it makes up for that difference by providing wizards (helper applications) to guide you through the creation of complex HTML elements, such as frames and tables. With HomeSite, you can see exactly what your page looks like in real browsers by previewing your page in Internet Explorer. HotDog Pro <http://www.sausage.com> is another advanced HTML editor without WYSIWYG functionality. However, it provides many HTML features that were formerly seen only on professional pages. HotDog Pro provides its own preview browser, so you can see the effects of the various additions and changes you make to your Web page. Luckman WebEdit <http://www .ditr.com/software/webedit/> is another highly regarded, Windows-based shareware editor.

Usenet Newsreaders

Newsreaders are applications that manage information contained in Usenet newsgroups. When you use a newsreader, you can select which newsgroups you want to subscribe to, and your reader will download the latest postings in those groups. You can see at a glance which articles you have already read and which ones you might like to read, based on date, subject, or author. Newsreaders are also helpful when you want to post your own follow-up to an existing thread or start a thread yourself. Among the features you can expect from a full-function newsreader are spelling check, sorting, filtering, and organizing of postings. You should be able to save and forward messages that are posted on the newsgroup using your newsreader.

Forte Agent and Free Agent <http://www.forteinc.com> are highly respected commercial and freeware newsreaders, respectively. The free version has some limitations, such as not incorporating a spelling checker, searching, or sorting capabilities, but it is still a nice, widely used program.

NewsFerret <http://www.ferretsoft.com/netferret/> serves a slightly different function from the other newsreaders. Other newsreaders enable you to subscribe to newsgroups of interest and manage the messages they produce. NewsFerret operates more as a newsgroup search engine, finding news articles on subjects you specify, regardless of which newsgroup they are in and regardless of whether you have subscribed to that newsgroup at all. In addition, NewsFerret functions as an offline newsreader, downloading articles and saving them for later reading.

RESOURCES

Basch, Reva. *Secrets of the Super Net Searchers: The Reflections, Revelations, and Hard-Won Wisdom of 35 of the World's Top Internet Researchers.* Wilton: Pemberton, 1996.

Calishain, Tara. *Official Netscape Guide to Internet Research.* Research Triangle Park: Ventana, 1997.

Campbell, Dave, and Mary Campbell. *The Student's Guide to Doing Research on the Internet.* Reading: Addison-Wesley, 1995.

Gilster, Paul. *Finding It on the Internet: The Internet Navigator's Guide to Search Tools and Techniques.* 2nd ed. New York: Wiley, 1996.

Glossbrenner, Alfred, and Emily Glossbrenner. *Visual QuickStart Guide: Search Engines for the World Wide Web.* Berkeley: Peachpit, 1998.

Hahn, Harley, and Rick Stout. *The Internet Complete Reference.* 2nd ed. New York: McGraw-Hill, 1996.

Harris, Cheryl. *An Internet Education: A Guide to Doing Research on the Internet.* Belmont: Integrated Media, 1996.

Hedtke, John. *Using Computer Bulletin Boards.* 2nd ed. New York: MIS, 1992.

James-Catalano, Cynthia. *Researching on the World Wide Web.* Rocklin: Prima, 1996.

Krol, Ed. *The Whole Internet: User's Guide and Catalog.* 2nd ed. Sebastopol: O'Reilly, 1994.

Maloy, Timothy K. *The Internet Research Guide: A Concise, Friendly, and Practical Handbook for Anyone Researching in the Wide World of Cyberspace.* New York: Allworth, 1996.

McGuire, Mary, Linda Stillborne, Melinda McAdams, and Laurel Hyatt. *The Internet Handbook for Writers, Researchers, and Journalists.* New York: Guilford, 1997.

McKim, Geoffrey W. *Internet Research Basics.* Indianapolis: Macmillan, 1997.

———. *Internet Research Companion.* Indianapolis: Que, 1996.

Morris, Evan. *The Book Lover's Guide to the Internet.* New York: Fawcett, 1996.

Seltzer, Richard, Eric J. Ray, and Deborah S. Ray. *The AltaVista Search Revolution: How to Find Anything on the Internet.* Berkeley: McGraw-Hill, 1997.

Interpersonal Skills

Tips

Interpersonal Skills

*A*s a college student or someone who is just beginning a career in the workplace, you may think you have all of the essential skills to do well in your studies or to succeed at a job. As a college student, you are intelligent, disciplined, and motivated. What more could you possibly need to succeed? As an employee in the workplace, you know the essential points of your discipline. What more could you possibly need to know? Of course, succeeding in college requires far more than intelligence, discipline, and motivation, just as succeeding on the job requires knowing far more than your discipline. For both, you must also master a wide variety of interpersonal skills. In many situations these interpersonal skills are as important as your writing and technical skills.

This section provides an overview of some of the skills you will need to find a job and to succeed at it; among them are interviews, presentations, and professional and personal development.

DEVELOPING INTERVIEWING SKILLS

Interviewing skills are essential for researching information. (See Part One, "Planning and Research.") The interviewing skills covered here require many of the same approaches you take in researching information through other people. After all, you're still dealing with people, and whenever you're dealing with other people through interviewing or any other kind of communication, you have to use all the skills at your disposal to communicate successfully.

279

The job interview is just one type of interview, but it is probably the one that almost everyone encounters at some time during a working life. Typically, most people are nervous about job interviews. They expect all of the questions to be tough, and they are worried that they will provide the wrong answers. It's perfectly normal to be nervous about a job interview. After all, you've worked hard in college, and now you're looking for a good job at a good salary in a good location. You want everything to go well. The key to a smooth interview is relaxation and preparation. Of course, these two are related: thorough preparation helps you to relax.

The Informational Job Interview

One way to practice your interviewing skills and increase your knowledge about a career field or a specific company (including the people who work there) is to set up a series of informational interviews with knowledgeable corporate contacts.

RESEARCHING FOR THE INFORMATIONAL JOB INTERVIEW

Use the informational interview as a tool for acquiring information that you would not find through other means. Before requesting informational interviews, you should do some self-assessment. Identify your skills, values, and interests. Also, obtain background facts and details on the company so that you can use the informational interview productively.

To prepare for a series of informational interviews, write a list of potential contacts. These individuals may be personal friends, relatives, and associates as well as alumni and others to whom you have been referred. Most people find that it is easier to approach those they know first. Once you have developed a comfortable style for requesting and handling the informational interview, you can begin to seek referrals from other sources.

OBTAINING AN INFORMATIONAL INTERVIEW

Typically, there are two ways to obtain an informational interview. The first is a written letter to unfamiliar contacts asking for assistance and mentioning how you received their name. This letter is followed up with a phone call to establish a date and time for a meeting. The second approach is more direct and is used with people you already know: either call or approach the individual in person to make your request.

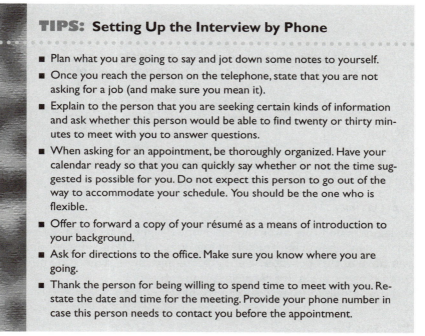

TIPS: Setting Up the Interview by Phone

- Plan what you are going to say and jot down some notes to yourself.
- Once you reach the person on the telephone, state that you are not asking for a job (and make sure you mean it).
- Explain to the person that you are seeking certain kinds of information and ask whether this person would be able to find twenty or thirty minutes to meet with you to answer questions.
- When asking for an appointment, be thoroughly organized. Have your calendar ready so that you can quickly say whether or not the time suggested is possible for you. Do not expect this person to go out of the way to accommodate your schedule. You should be the one who is flexible.
- Offer to forward a copy of your résumé as a means of introduction to your background.
- Ask for directions to the office. Make sure you know where you are going.
- Thank the person for being willing to spend time to meet with you. Restate the date and time for the meeting. Provide your phone number in case this person needs to contact you before the appointment.

PREPARING INFORMATIONAL INTERVIEW QUESTIONS

After you have set up an informational interview, organize an agenda for this meeting. It is usually a good idea to ask general questions when you are seeking information about a company, function, division, or department. Specific questions are useful when speaking with a person about a job or responsibility that is of interest to you.

Regarding a particular job or career field, consider asking about major job responsibilities and skills required, what the individual likes and dislikes about the job, how many hours the person works each week, what kind of schedule is typical, the degree of flexibility in the schedule, the amount of travel required, the degree of interaction with coworkers and clients, fringe benefits, new projects that are under way or planned, types of decisions that are made, other positions that are possible in the field, and advice for beginning a job search.

Regarding a particular company, consider asking how the company differs from competitors, the type of management style, trends within the company, opportunities for professional development, general company policies and expectations, and problems the company is facing.

Regarding a particular department, consider asking about the relationship of the department to the company, its current problems, current priorities, new programs, current staff and their backgrounds, and department goals.

Regarding your background, consider asking about where the individual would see you fitting into the organization, who in the company would be most interested in your skills, what the individual thinks of your résumé, and whether you have the qualities and skills for positions in the department or organization.

EVALUATING THE INTERVIEW

As soon as possible after the interview is complete, evaluate the interview by answering the following questions: What impressions do you have of this position? company? department? How did your contact help you to clarify your career objectives?

Finally, don't forget to send a thank-you note.

The Job Interview

PREPARING FOR THE INTERVIEW

Preparing for the interview involves knowing your own strengths and weaknesses, knowing about the company and the industry it is part of, anticipating the kinds of questions you are likely to be asked, having questions to ask, and allowing plenty of time for your arrival at the interview.

To help yourself relax at least a little, you need to assess seriously your strengths and weaknesses. What are you best at? What qualities do other people seem to admire in you? Why do people find you likable? Now is not the time to be modest. You need to perform an honest assessment of what makes you a good person before you can begin to realize that the qualities you possess also make you a good prospective employee.

Never go to an interview knowing little or nothing about the company you are interviewing with. Ideally, you've already done your homework here and reflected your knowledge of the company in your application letter. Numerous resources can help you find out what you need to know about a company's progress, its future growth, its products, its employees, its benefits packages, and so on.

Having a good knowledge of a company will help you think of other questions as the interview proceeds.

Make a list of questions you can expect to be asked and rehearse your answers. Many interviewers begin an interview, after the typical pleasantries, with a comment such as "So tell me about yourself." Be prepared to tell the interviewer about yourself. Discuss why you went to college, why you chose your discipline, what areas of your discipline interest you the most, and so on. Many interviewers will ask you why you are applying for this particular opening. Have a few good answers ready. You are also likely to be asked about your strengths and weaknesses. Be prepared to accentuate the positive.

Be prepared for the interviewer to ask you whether you have any questions, and make sure you do have some. For example, ask if a job description is available for your desired position, ask who would be your immediate supervisor, ask questions about this supervisor, ask about product development and current products, and ask about benefits and work conditions. In brief, try to obtain answers to all of the questions you will have if the interviewer at the company calls you and tells you that the job is yours. You would no doubt have lots of questions before you would feel comfortable accepting the position.

Should you ask about salary, and, if so, when? There is no easy answer to this question, but one effective strategy is to assume that the salary will be competitive, and, if you are truly interested in the position, do all that you can do to have the company offer the job to you. Once you have reached this point, discussion about salary can be meaningful.

Finally, consider traveling to the interview as part of planning for it. One of the worst mistakes you can make is to arrive late for your interview. Plan to arrive at the receptionist's desk at least fifteen minutes before your scheduled interview. An early arrival allows time for you to relax and collect your thoughts and shows the receptionist as well as the interviewer that you are well organized and professional.

PARTICIPATING IN THE INTERVIEW

To participate in an interview successfully, view it as a three-part process. Whether you or your interviewer is aware of it, most interviews have a beginning, a middle, and an end.

In many respects, your interview starts before it officially begins. The clothing you have chosen to wear, your arrival at the company, your introduction of yourself to the receptionist, your demeanor as you wait, and your first handshake with the interviewer are all the beginning of your interview. Most of this part is nonverbal. Another part of the beginning occurs when you join the interviewer in his or her office or a conference room. Body language is important here, too. Your posture, gait, eye contact, and composure tell the interviewer a great deal about you. How you sit in your chair is another factor. All of these nonverbal signals are part of the beginning, and you need to be aware of them throughout the interview.

When the interviewer begins talking about the weather, the traffic, or anything else, the official beginning of the interview is under way. Now is the time to exchange pleasantries and to talk casually about anything the interviewer cares to talk about. Let the interviewer take the lead and follow his or her cues.

The middle and end parts of the interview are those that you have already prepared for and can anticipate. Remember the list of questions you expect to be asked as well as the questions you want to ask. Usually, when the interviewer asks you whether you have any questions, you are beginning the closing of the interview. Comments on how the job search will be handled at this point and when you can expect to hear from the company are part of the closing.

As you wrap up, don't forget all of the important body language—the eye contact, the firm handshake, the posture, the walk, and all the rest. Your body language tells the interviewer a lot about how interested you are in the job.

FOLLOWING UP ON THE INTERVIEW

Now that you've survived the interview, your "ordeal" still isn't over. You need to follow up. You need to thank your interviewer, remain proactive, continue searching, and be persistent.

At the very least, you need to send a prompt thank-you letter to the person or people who took the time to interview you. The strategies for writing this letter are provided in Part Eight, "Correspondence and the Job Search." The thank-you letter shows your interviewer that you are professional and appreciative. And, of course, this letter helps to keep you visible during the decision-making process.

Many interviewees mistakenly take a passive role after the interview is over. The interviewer may have said something like "I'll call you in two weeks to let you know what we've decided." Many job candidates then wait two or three weeks hoping to hear from the interviewer. You can't be passive and expect a company to hire you. You need to be proactive. Send follow-up correspondence (or e-mail), and don't wait for the interviewer to call you. After a week or two, call the interviewer and say that you are just checking on the status of the job search. With most interviewers at most companies you won't jeopardize your chances of being hired. If you handle yourself professionally during the phone conversation, the interviewer will view you as acting professionally and as someone who is definitely interested in the job.

Conducting the Job Interview

Someday it will be your turn to hire others, and for this process you will need to know how to conduct a job interview. Conducting an effective job interview requires many additional skills. Unfortunately, many candidates are interviewed by interviewers who are not well prepared to interview. Conducting an effective job interview requires that you be familiar with the job correspondence (application letter, résumé, job application form, and other materials) of the candidates you are interviewing.

You need to know what type of person you are looking for. You need to identify clearly the technical skills, the interpersonal skills, and the many intangibles that make a person an ideal candidate, and you need to measure each candidate according to these criteria.

You also need to give each interview a structure—a definite beginning, middle, and end. For the beginning of the interview you need to develop a scale of importance for how you value and rate everything from appearance and cleanliness to handshake, voice, nervousness, and composure. You need to know general questions you typically like to use at the beginning of an interview. Typical beginning-of-the-interview questions might be, "Did you have any problems getting here?" or "Where were you coming from?" or "How long did it take to drive here?" or "How did you hear about this position?" You need to save many of the tougher questions for the middle of the interview. These questions will concern the candidate's preparation and suitability for the job. Typical questions might be, "Have you ever done work on projects similar to ours?" or "How would you describe your usual working pattern?" or "Why did you leave your previous job?" You also need to bring closure to the interview and let the candidate know what to expect after the interview is over.

Be realistic. Most of the candidates will not be ideal. Everyone has different strengths and weaknesses. You need to develop your intuitive sense or gut feeling about which strengths are the most important to enable this person to succeed.

You have a responsibility to hire the best person for the job, and that means hiring the best person for the department, division, or company. You may like some candidates more than others, but maybe one of the candidates whom you find the least personable is actually the best candidate for the job. Also, be careful about other biases during the interview. For example, if you have biases concerning gender or ethnicity, you may fail to hire the best person for the job. You need to look past your own biases or preferences.

DEVELOPING PRESENTATION SKILLS

Few interpersonal skills are more important than the ability to give an effective oral presentation. Many technical people have the skills and knowledge they need to succeed on their jobs, but sometimes they lack the ability for or interest in giving an effective oral presentation. Some people do seem to have a talent for talking before groups of people, but most of us have to work at acquiring this important skill. That's the important point to remember here: You can learn about oral presentation skills the way you learn about anything else, and once you know what these skills are, you can practice and practice until they seem almost second nature to you.

Kinds of Presentations

Most presentations fall into one of three categories: impromptu, outlined, or scripted. Many are informal. You may be called on unexpectedly in a staff meeting just to summarize quickly a few points concerning your progress on a project. You'll have to respond with an ad hoc or impromptu presentation. You'll have to think quickly of what you want to say, and then you'll have to say it.

Other times you may be asked to prepare a formal presentation for upper management on the progress of a project you're working on. You'll have the time to prepare visuals and other aids for this more formal presentation. You'll probably organize your thoughts according to an outline, keeping in mind that the outline is intended only to help you stay on track with the main points.

Depending on your profession, you may also have the opportunity to present your ideas at a professional meeting or conference. Presentations at conferences also vary from the informal to the formal.

One common kind of formal presentation is the scripted presentation. In a scripted presentation you write down word for word every sentence you will read to your audience or that you will memorize and recite for your audience.

Whether your presentation is impromptu, outlined, or scripted, you need to know certain strategies that will help you to give the most effective presentation you can give.

Preparation

The best way to cope with nervousness is to prepare for and practice delivering your presentation.

TIPS: Preparing for Your Presentation

- Plan your presentation. Make sure the content is what it needs to be. Always keep in mind what the audience needs to know. What you want to talk about and what your audience needs to know are not always the same thing.

- Research your presentation. You may be asked to speak about a topic that you know about entirely from personal experience, but more often than not you will be asked to speak on a topic that you will have to investigate further or that you have been investigating or working on for some time.

- Plan to use language that is appropriate to your audience. Make sure you will explain your points clearly to your audience. Use clear, concise, accurate, and descriptive language.

- Organize your presentation. Remember, your presentation should have a distinct beginning, middle, and end. Decide what order is the best for your subject and audience: chronological, topical, causal, problem–solution, and so on.

- Prepare a good beginning. A good opening can boost your confidence in a way that will help you through the remainder of the presentation. If necessary or appropriate, introduce yourself first. Establish your credibility. Prepare at least one specific strategy for getting your audience's attention: humor, an anecdote, a statistic, a quotation, a dramatization. Tell your audience the topic and purpose of your presentation. Give your audience a preview by telling them what you are going to cover in your talk.

(continued)

Preparing for Your Presentation *(continued)*

■ Prepare a good middle, or body. Remember that for most class presentations you barely have time to cover five main points, and most often you have time to cover only two or three. Let the audience know when you have moved to the middle part of your presentation. Say something like "For my first point...." Use transitions to help your audience follow what you are saying. Cover each main point in enough detail to suit your purpose. Support each main point adequately with facts, examples, expert testimony, and so on. Use appropriate link words to establish the relationship of each point to the previous one: *nevertheless*, *additionally*, *and yet*, *in contrast*, and the like will act as navigation tools to point out the sequence of your remarks. While these transitions are important in written works, they are especially important in oral presentations when your audience cannot go back and review the earlier sentences.

■ Prepare a good ending. A conclusion is usually less than 10 percent of your presentation, but you should prepare it as carefully as you prepared both your introduction and the body of your presentation. Let your audience know that you have reached your conclusion. If appropriate, review your most important point or points. Use a quotation, a startling statistic, or a dramatization to help make the conclusion memorable. When you have completed your prepared remarks, tell your audience that you will be happy to answer questions and mention the availability of handouts, if any.

■ Prepare by practicing. Practice your presentation several times before you present it in public. Time your talk so that you can stay within the limits you are given. If possible, practice in front of a mirror and critique your gestures. Tape-record your talk and listen to your voice. Do you frequently use meaningless fillers in your speech? If you hear a lot of "ums" and "y'knows," work on eliminating these verbal tics, which distract from what you are saying.

■ Become familiar with the room where you will speak in advance of your talk. If possible, visit the room beforehand and position yourself in the spot where you'll be standing or sitting later. Look out at your imaginary audience and try to imagine them, their interests, their concerns, and what they want to know.

■ Plan to present a well-groomed and professional appearance. If your professor requests that you dress appropriately for your classroom presentation, then you should wear a suit or other business attire. If your professor doesn't make your appearance an issue, you should still come to class dressed and groomed appropriately for your talk. Cut-off jeans, short skirts, T-shirts, and the like aren't proper attire for presentations. Of course, in a corporate setting, where your professional appearance is always an important factor, you must dress appropriately for staff, department, or division presentations. When you are speaking to the company's clients, your appearance takes on an added importance.

■ Prepare the place for your talk. If you will be standing at a lectern, check that it is placed where you want it and that you have sufficient lighting to read your notes. If there is a microphone, adjust its height to accommodate your own and make sure you know how to turn it on and off. Listen to the room acoustics. You want your voice to carry easily to the last row without projecting into the hallway. If you need a screen to project onto, make sure that one is available and verify that it is ready for you. If you have brought a laptop computer to show a presentation, check that you can attach it appropriately to the display hardware in the room. Turn on your presentation equipment and make sure your images are placed and focused properly. Locate the light switches, and make sure you know who will be operating them for you if you need the lights out. If you will be reading notes, verify that they are where you expect them to be and that they are in the correct order. If you plan to use handouts, distribute them to the places where you want to audience to pick them up.

Delivery

Becoming a good speaker requires lots of time and practice. Perhaps it's true that some people have a talent for speaking in public, and perhaps it's true that some people aren't nervous when they give a talk or presentation. However, most of us have to work very hard at becoming good public speakers, and we have to work hard at overcoming our nervousness. Often, what you interpret as someone else's natural public-speaking ability is, in fact, the result of excellent preparation and careful practice. So even if you think you have no talent for speaking in public, you can learn the techniques that will take you through to a successful, effective presentation.

Lots of people have plenty of advice about various ways to deliver a presentation. You should read or listen to what others say and use the information that you think will help you to relax and to give a good presentation. It helps to watch other speakers and notice how they get their messages across to their audiences.

TIPS: Effective Delivery

■ Remember to relax. Now the time for your talk has arrived. You have been introduced, or you must introduce yourself. Before you begin speaking, take a few deep breaths and look at your audience. Remind

(continued)

Effective Delivery *(continued)*

yourself that these people want to hear what you have to say and that they want you to succeed. Control your nervousness. Remember that even the most experienced speakers are nervous before they give a presentation and that nervousness is a natural response to speaking in public. You can learn to control this nervousness in many ways. First, before you begin speaking, find a quiet room where you can calm yourself and take some deep breaths. Imagine that you are standing before your audience, and look at this imaginary audience calmly. Tell yourself that you have much to offer this group and that they are looking forward to benefiting from your knowledge and experience. Second, after you are introduced and before you begin speaking, take a few moments to look at your audience and take a few deep breaths. Third, focus on the fact that you will most likely be the expert about your topic. Try to enjoy this opportunity for sharing what you know with people who want your information. Fourth, keep things in perspective. You are more than likely nervous because you fear that you will make a fool of yourself or that you may make a mistake. Don't focus on yourself. Focus on your message. And don't be afraid of mistakes. If you make them, learn from them. Finally, maintain a sense of humor. If it helps you, imagine your audience sitting in their underwear or with dunce caps on. If you make a mistake and if it's appropriate to do so, laugh at yourself. Make the mistake work in your favor.

- Look at your audience and remember to maintain eye contact during your presentation. Read your audience for signs. With practice you'll be able to tell whether your listeners are bored, interested, motivated, enthusiastic, and so on.

- Project your voice appropriately for the size of the room; your voice will sound different, depending on the room acoustics and the size of the audience. Work on pitch. Pitch is the highness or lowness of your voice, its rise and fall as you speak, its intonation. Be sensitive to the sound you are producing and be prepared to modify it if you find that it is not appropriate for the room.

- Maintain a steady pace or rhythm for your presentation. Don't speak too quickly or too slowly. Cover your points at a pace that your audience can easily follow. Guard against speaking so steadily that your voice descends into a drone. Your speaking pace should settle into a cadence that maintains audience interest with a balance and variety of rhythm.

- Use gestures appropriately. An experienced speaker will use all kinds of gestures for emphasis; many will be so natural that the audience won't even be aware that the speaker is using the gestures. Your hands, shoulders, eyes, mouth, and eyebrows may be used to intensify all kinds of meanings. Of course, your gestures may also be inappropriate to your point. Exaggerated or erratic gestures may help undermine a point instead of intensifying it. You need to learn how to make the gesture that is appropriate to the point and how to time the gesture accordingly.

- Avoid distracting behavior. Examples are shaking your car keys in your pocket, tapping the podium with your pen or pointer, pushing your glasses up too frequently, coughing or clearing your throat too often, drinking from your glass of water too often, shaking or tilting the podium, pacing the room too quickly, or having your back to the audience.

- Tell the audience what you are going to tell them, tell them when you move on to a new point, and tell them what you have told them.

- Give the audience a constant sense of the structure or organization of your talk. Let your audience know when you are moving on to your next important point. Give your audience frequent auditory and visual clues. Use words and phrases such as "And now for my next main point" or "Now that I've summarized this issue, I want to move on to the next issue." Use your visuals to let your audience know where you are in your presentation, too.

- Be prepared for something to go wrong. Most often everything will go smoothly throughout your presentation, but something may surprise you, too. Perhaps the projector bulb will burn out in your overhead or slide projector. Perhaps you will drop your notes unexpectedly. Perhaps you will trip over a wire while you're pacing the room. Perhaps you will discover that your slides are in the wrong order even though you double-checked them just before your presentation. Perhaps you will find yourself unable to suppress a sneeze, cough, or hiccup. These and many other events have occurred to even the most experienced public speakers. Handle such events with grace and a sense of humor. Audiences are typically patient and will wait while a bulb is replaced, notes are reshuffled, and so on. If appropriate, apologize for the oversight and then continue. Don't dwell on what just happened. Get back into the rhythm of your presentation, and deliver the kind of presentation both you and your audience expect.

- Know how to use your presentation tools and keep them from distracting your audience. Professionals use all kinds of tools during their presentations. These tools can become a distraction if the audience pays more attention to them than to you. For example, if you are using overhead transparencies, practice putting up one transparency after the other so that the process appears smooth and natural. If you are using a computer and a program such as Microsoft PowerPoint or Corel Presentations, avoid going from one slide to another too quickly. If you are using a pointer, use it sparingly and appropriately.

(continued)

Effective Delivery *(continued)*

Used well, your tools can enhance your presentation tremendously. Used poorly, your tools can make your audience lose its focus on what you are saying.

■ Don't hide your audience's view of your visuals, keep them simple, and make sure the information on them is clearly readable by someone in the back of the room. Avoid using too few or too many visuals.

■ Let the audience know when you are going to begin your wrap-up. Look at your audience, take a deep breath, and say something like "And now to wrap up" or "Let me conclude by saying" or "In closing, I would like to say. . . ." You'll find that your audience appreciates hearing these words and will be even more attentive. This is a good time to remind your audience of your most important point or points. Remember, your closing is important and can be used effectively to inform, persuade, or motivate your audience.

■ Don't speak longer than you are supposed to. Remember to manage your time effectively. You want the presentation to be just the appropriate length, rather than too short or too long. Many presentations are constrained by time limits. Presentations at conferences may be limited to fifteen minutes per panel speaker, for example. Or perhaps your supervisor simply asked you to speak to other employees in the division for thirty minutes. If your audience knows how long your presentation is supposed to be, make sure you stay within that time limit. Have a wristwatch or stopwatch in front of you where you can refer to it unobtrusively, or locate a wall clock you can glance at while you speak. Audiences aren't very forgiving of speakers who speak longer than they are supposed to.

■ If appropriate, thank your audience for listening. If you have done well, your audience will let you know through applause or some other acknowledgment. They will thank you for your comments, and you should thank them for their participation. It's the professional thing for you to do.

Presentation Tools

A wide variety of presentation tools are available to speakers. You need to decide which tool or tools are best suited to your particular topic and your audience. On one basic level you may choose to use models, a chalkboard, posters, flip charts, an overhead projector, a video, a slide projector, or a computer. There are some important guidelines to keep in mind for each of these media.

MODELS

Models are physical representations of ideas or other kinds of objects. An engineering student, for example, may use a model of a bridge to illustrate various principles of design. A chemistry student may use models to demonstrate the structure of a molecule. Models can be effective for catching the attention of your audience, for dramatizing a particular point or concept, and for quickly making a complex point clear.

CHALKBOARDS AND DRY-ERASE BOARDS

Despite all the other technology available to you, a simple chalkboard, chalk, and an eraser may often prove most useful. Most classrooms and many meeting rooms contain a chalkboard or a dry-erase white board with erasable markers. Either type of board can be used to demonstrate a variety of points quickly. However, if you spend too much time writing on a display board with your back to your audience, your audience's attention may wander. Unless you are a skilled presenter, you should take very little time during your presentation to put information on a board for your audience to read.

POSTERS

To save time during your presentation, you can use a poster to provide the information you want your audience to see or to know. Of course, posters are more suitable to smaller audiences than larger ones. Make sure that whatever you are presenting on your poster is large enough and clear enough to be viewed easily even from the back of the room.

In many of the sciences, poster sessions are common for showing others the results of your research on a particular project. Your aim should be to present visually the key concepts of your research. Use diagrams, arrows, photographs, and other strategies to show what you did rather than try to explain your research using text alone. Your poster should address one central question. State the question in the poster, and instead of elaborating on various issues related to the question with text on the poster, use discussion with the people who stop to view the poster to clarify issues. An effective poster in a poster session provides one major idea for viewers to take home and consider. Any conclusions should be summarized briefly and in a straightforward style. If you must acknowledge any contributors or funding organizations, you may do so using smaller type than elsewhere on the poster. Ideally, any text on the poster should be large enough to be read easily at least six feet away. Group any text you use into sections and use headings. Your headings should be in a large type size—perhaps 36 point or larger—and the font for these headings should be different from the font for supporting text. Supporting text should be in 24 point type or larger. Keep in mind that posters in poster sessions are not publications. If any points need elaboration, bring some handouts to the poster session.

FLIP CHARTS

Like chalkboards, flip charts enable you to sketch your thoughts as you speak, although some speakers prepare a series of sheets on a flip chart and then walk the audience through each sheet. Advantages of flip charts include low cost, low noise, and the possibility of creating visuals on the spot. Disadvantages include their difficulty to reproduce and their "I just thought of this" connotation.

TRANSPARENCIES

Most college classrooms and company meeting rooms have overhead projectors for using transparencies. The projector and transparencies (also called viewgraphs or foils) are some of the most widely used communication tools. Like other media, this technology has both advantages and disadvantages.

Among the advantages of transparencies are that flexible ordering is made easy, visuals can be created on the spot, rehearsals are possible without projection equipment, hard-copy prints can be made easily, the technology is easy to use and take notes from, you control the pace and sequencing, transparencies catch the audience's attention, and they allow clear visual explanation and reinforcement. Disadvantages are that they can't be sequenced rapidly, their large format makes color expensive, projectors are noisy, too many overheads are overkill, acetate material will deteriorate if kept for a long time, and bright and true colors are difficult to achieve.

Some common misuses are having too many transparencies (about ten per forty-minute presentation is adequate), presenting too many lines of print or too small print, covering too much information (use a handout instead), producing unreadable graphs, showing cartoons with unreadable dialogue or captions, and using unprofessionally reproduced photos and other visuals.

When using transparencies, make sure the projector is at the front of the room, with plenty of space on a nearby table for orderly placement of transparencies. This allows you to face your audience and encourages interaction. Dimmed lighting helps. Do not read the lettering on transparencies to your audience. Instead, mention key points and elaborate with statistics or examples. When appropriate, allow ample time for your audience to take notes. Be concise. Use key points in words or phrases, with a maximum of six to eight words per line and seven lines per transparency. Type should be at least 24 point. Check your transparencies from the back of the room to make sure they are legible.

35-MM SLIDES

Someone once said that the three most feared words in the English language are "Next slide, please." Yet slides are a common tool for oral presentations. With slides you can show many kinds of visual material—charts, graphs, bulleted lists, photographs, and a host of others. Although computers have taken over the role formerly held by the slide projector to some degree, there are times when only a slide show will do. Of course, you'll have to plan a slide presentation carefully. Not only do you have to ensure that your slides are set up in the correct order and ready to go, you also have to verify that the projector is working properly and that you know how to use it.

Some advantages of showing slides include the ease of rearranging their sequence, the possibility of remote operation and long- or short-distance projection, the relatively low cost of producing color images in a photographic format, the visual realism that is possible with photographs, the inexpensive production of color visuals in quantity, easy control over the pace of information, flexibility of positioning at the front of a room or in a rear-screen projection, ease of storage and transportation, visibility to a whole room, the ability to point out critical items on a large screen, the availability of special visual effects such as cutaway enlarged views or image distortion to enhance the impact, the possibility of reviewing specific points easily by reversing slides, the ease of quick rearrangement for revision, and the capability of updating and adapting to different audiences or varying the emphasis.

Disadvantages include the long lead time for preparation, diminished effectiveness in high ambient light, the distraction of projector noise, the difficulty of obtaining hard-copy prints, the dimness of lighting required, the inability to show continuous motion or processes, and the susceptibility of slides to getting out of order and causing confusion to your audience.

For the most effective use, make sure visuals do not clutter the slide (avoid overuse of images); carefully integrate slide content with important handouts; make sure images can be seen by all audience members; plan ahead (when requesting a 35-mm projector, also ask for a remote controller and an extension cord if you'll need them). Place the projector a sufficient distance from the screen so that images are relatively large.

VIDEOS

Many people own or have access to camcorders. Families have hours of tapes of family get-togethers and other events. Video can also be just the right tool to make a variety of points during an oral presentation.

Videos are relatively cheap to produce, edit, and purchase, and videotape has a shelf life of five to nineteen years, depending on the frequency of use. It is available in a variety of formats, and it can easily be used to show motion, process, or interpersonal skills.

Among video's advantages are that it allows points to be shown at various speeds and stopped or reversed, as necessary. Video magnifies processes, fine details, and operations, which can be helpful in demonstrations (through "zooming" by video camera). Video allows the same information to be shown simultaneously to audiences in different locations. It provides the opportunity for immediate critique, analysis, or evaluation by replaying segments of the tape. Video allows individualized instruction by sequencing the content interactively with workbooks, guides, texts, or computers. Visual effects are an excellent way to enhance either the information or entertainment value of the presentation. Video allows material to be kept current through the editing or correcting of material. You can be sure of consistent quality in the presentation of the material. Video provides resource information on remote cultures and other countries and offers performances or interviews with noted individuals. Also, video is not novel to an audience that is familiar with television.

Among the disadvantages are that illustrations and lettering are limited to a 3 by 4 ratio (television screen proportions). Audience size is limited unless multiple monitors, closed circuit or cable broadcasts, or video projection systems for large rooms are used. Video can be quite expensive if elaborate productions or a lot of equipment, materials, or personnel are involved.

To use video effectively, preface the material to be viewed by the audience with reference to presentation content and objectives. Prepare follow-up activities to reinforce learning. Make sure video and audio quality are good. Make sure the videotape is not too long. (To allow adequate follow-up, it should not exceed thirty minutes for a fifty-minute time slot.) Make sure the entertainment value doesn't exceed the information value.

COMPUTERS

Increasingly, computers, connected to multimedia projectors to project the computer image onto a large screen for the audience, are used for all kinds of oral presentations. For example, many presenters use portable projectors to project high-quality images from their laptop computers. And many presenters use full-featured presentation software programs to deliver their presentations. Microsoft PowerPoint and Corel Presentations have become favorites of speakers because of the many features these programs offer. They are equipped with preformatted presentation designs or templates and a variety of special effects, such as animations, sweeps, and fades. The slide templates make it easy for even novices to create a succession of text, illustrations, bulleted lists, and other slide formats, complete with graphics, clip art, and photographs. Music and other sound effects make the computer-presented slide show into a complete multimedia experience. The slide shows can proceed either automatically and unattended or at a pace controlled by the presenter.

DEVELOPING PROFESSIONAL SKILLS

Whether you are a student or already employed, you will naturally want to do all you can to be competitive and to gain an advantage for job placement or professional development. You should remember that several hundred other people may be competing for the same job opening or promotion, and you need to do as much as you can to enhance your chances of being selected. Even when you have secured a position, you cannot afford to become complacent. You will have many opportunities to develop a variety of professional skills on the job, and you should take advantage of as many of them as possible.

Professional organizations, subscriptions, conferences, in-house seminars, participating in meetings, networking, mentoring, collaborating, and negotiating are important memberships, activities, or skills that will help you get and keep the job you want and help you stay current in your profession.

Joining Professional Organizations

Almost every discipline or profession has an organization that represents its interests, and many have at least one journal or trade magazine. Also, increasingly many disciplines have at least one mailing list or one discussion group on the Internet. Membership in a professional organization is so important that many companies will pay its employees' membership fees, which can be several hundred dollars a year.

Membership in an organization often includes a subscription to its major publications. In the field of technical communication, for example, membership in the Society for Technical Communication <http://www.stc-va.org/> provides you with subscriptions to the quarterly journal, *Technical Communication;* the monthly magazine, *Intercom;* and the annual membership directory, among other publications, in addition to the other benefits of being a member. People who teach technical communication and those who work in industry find these sources valuable for keeping current with some of the research in the discipline.

In the field of information technology, the International Federation for Information Processing <http://www.ifip.or.at/> acts as an umbrella group for a variety of national organizations, many of which have their own journals and newsletters. Various engineering organizations are collected under the rubric of the Institute of Electrical and Electronics Engineers. Most of these organizations publish newsletters or journals as well. Coordinating more than 180 university departments of computer science and computer engineering, the Computing Research Association <http://cra.org/> supports basic computing research and affiliated professional societies. The Association for Computing Machinery provides services for members of the general organization and its thirty-six special interest groups.

Subscribing to Journals, Mailing Lists, and Newsgroups

Other journals or trade magazines may be essential to your discipline but unrelated to a particular professional society. You need to know what these journals and magazines are, and you should read them regularly.

The Internet has increasingly become one of the best ways to stay current in a discipline. Many disciplines have their own mailing lists, in effect newsletters that are mailed to you as e-mail to keep you informed about issues in your field. And many disciplines have at least one mailing list, discussion group, or newsgroup. Here people post e-mail or Usenet messages to the group about all kinds of issues related to the discipline. (See Part Six, "The Internet," for a detailed discussion of mailing lists and newsgroups.) Some of the mailing lists in technical communication—TECHWR-L, ATTW-L, and CPTSC-L—are good ways to keep track of the concerns of others in this rapidly changing field. Many disciplines have other special interest groups within them. In technical communication there are a variety of special interest groups or SIGS. For example, for those who have a special interest in contract work, there is the Contracting and Consulting mailing list. People who want to share their problems with and concerns about contract work regularly post valuable information to this list. This special interest group also publishes a printed newsletter for its members.

Attending Conferences

Attending conferences is not something just college professors and graduate students do. Attending conferences is one of many things professionals do to help stay current in their disciplines. Many professions have annual meetings, either as part of a larger conference or, if the organization is large enough, as a stand-alone annual conference. These conferences typically have many concurrent sessions in which experts discuss the latest trends and issues.

In technical communication the annual conference for the Society for Technical Communication (STC) is typically held every May in some major U.S. or Canadian city. Almost 3,000 technical communicators, professors, and students attend every year. Each conference's proceedings are published, representing all of the papers for the many panels, presentations, workshops, and so on. STC also has many regional conferences, which may be dedicated to a specific area of concern to technical communicators.

Most professions have societies or other organizations that provide those who work in the discipline with opportunities for networking with their peers. In engineering, the largest professional organization is the Institute of Electrical and Electronics Engineers (IEEE). Engineers in several dozen areas are represented by its many technical groups. Each year, hundreds of meetings sponsored by one or another group are provided for members and interested nonmembers. Some typical annual conferences are the IEEE International Symposium on Information Theory, Visual Communications and Image Processing Conference, Southern Biomedical Engineering Conference, National Radio Science Conference, Conference on Wireless Communications, Southeastern Symposium on System Theory, IEEE Conference on Telecommunications, IEEE International Professional Communication Conference, IEEE International Conference on Acoustics, Speech and Signal Processing, and many, many more. The IEEE Web site <http://www.ieee.com> is a good place to start if you're looking for a conference on a particular topic or in a particular location for this organization.

The corresponding society for computer professionals is the Association for Computing Machinery (ACM), which also sponsors many conferences on several continents. Its Web site <http://www.acm .org> is a good starting place to find conferences and learn about other member benefits in the computing field. The ACM's great strength is its thirty-six special interest groups (SIGs), which sponsor meetings and publish newsletters on a variety of computer-related topics, such as Algorithms and Computational Theory, Applied Computing, Computer Architecture, Artificial Intelligence, Biomedical Computing, Computers and Society, Computer–Human Interaction, Computer Uses in Education, Systems Documentation, Computer Graphics, Groupware, Hypertext/Hypermedia, Multimedia, Programming Languages, Simulation and Modeling, Software Engineering, and many more. In computer science some conferences are the ACM Conference on Human Factors and Computing Systems, Annual Symposium on User Interface Software and Technology, ACM Symposium on Virtual Reality Software and Technology, Computer Supported Cooperative Work, ACM International Multimedia Conference, Technical Symposium on Computer Science Education, Symposium on Applied Computing, and others.

Many of the sciences are represented by one kind of professional meeting or another. To find out what is available in your particular area of interest, check your organization's Web site if it has one, or go to the library and review the professional journals in your field.

Participating in In-House Seminars and Classes

Many people cannot afford to take the time away from work to attend a conference out of town. Another common resource for professional development is in-house seminars. These seminars may be as short as an hour or as long as a few months. Many large companies have their own staff members who provide seminars in professional development. Other companies hire outside experts to lead corporate seminars. At these seminars, employees can learn everything from using a computer tool to giving oral presentations to catching up with some of the latest findings in their discipline.

Participating in Meetings

Employees typically spend countless hours attending meetings in the workplace. Some are mandatory daily or weekly progress meetings. Some are informal; some are formal. Some are on pedestrian matters, and some are on matters of utmost importance or something between these two. Most busy people dread meetings because they are too busy to attend them. Meetings can be a waste of time because they are often poorly run, poorly organized, and too long. Only a little planning is required to run or to participate in much more efficient and productive meetings.

Meetings can be effective ways of reaching decisions; however, they can also be huge wastes of time. When you invest time in a meeting, you should expect a sufficiently large payback to justify that investment.

One of the most effective ways to make a meeting more productive is to use and follow an agenda. The agenda of the meeting shows the aim of the meeting and points of discussion in priority order. In effect, it is a "to do" list for the meeting. Using an agenda helps to focus the meeting and stops it from drifting off topic. Agendas should be composed and sent out in advance so that the participants know what is expected of them. If you circulate the agenda sufficiently far in advance, people have time to prepare fully for the meeting, and the meeting will not stall for lack of information. A well-written agenda minimizes the chance that a key member will show up unprepared for a crucial discussion.

TIPS: Writing and Using an Agenda

- Keep the agenda short. An agenda states the purpose, time, meeting location, and goals of a meeting. It outlines each topic to be covered and often designates an amount of time allotted to each subject. Listing who is responsible for each topic and the type of action needed helps clarify the goals and direction of the meeting for the participants.
- List each item sequentially according to its order on the program. Include who is responsible for the presentation, what kind of action is needed, and the time allotted for the topic.

(continued)

Writing and Using an Agenda *(continued)*

- Include an ending time for the meeting.
- Some items to consider for the agenda are approval of the minutes from the last meeting, committee reports, old business, new business, and guest speakers.
- Set the time of the meeting carefully. You can usefully change the timing of the meeting depending on the habits of the attendees. If people tend to waffle excessively, you can schedule the meeting just before lunch or going home. This gives people an incentive to be brief. If people tend to arrive late, start a meeting at an unusual time, such as nineteen minutes past the hour. This seems to improve punctuality. If possible, ensure that the meeting starts on time. If it starts late, the time of all the attendees is being wasted waiting for the start. If latecomers are not critically needed, start without them.
- Invite only the minimum required number of attendees to a meeting. The more people present, the more who will want to air their views. Similarly, bringing people who are not needed to a meeting wastes their time.
- Before the meeting, ensure that decisions made at previous meetings have been acted on. This guarantees that the meeting will not be seen just as a time to discuss old issues once again.
- At the opening of the meeting, ask for additions to the agenda; get group approval for the agenda before you start the meeting.
- At the end of the meeting, summarize the points discussed and create an action plan from the decisions made. This ensures that everyone understands what has been decided and who will do what.

In addition to following an agenda, a few other guidelines will make meetings more productive:

TIPS: Attending Meetings Productively

- Be on time. Be present only if you are needed.
- Be well prepared.
- Be attentive to the discussion so that your contribution does not repeat what was covered earlier by someone else.
- Be brief, relevant, focused, and courteous in your comments.

Minutes are the notes concerning major issues that were discussed and decisions that were made during a meeting. They are the official record of a meeting, and many companies, organizations, clubs, and so on require that minutes be taken at their meetings. Since minutes are the official record of the proceedings and are often incorporated into the permanent records of the company, accuracy and completeness are essential. In addition, minutes can be helpful for informing all members about what occurred at a meeting. You never know when you may be asked to take the minutes at a meeting, so it is important to know how to take complete and accurate notes.

TIPS: Writing Minutes

- Try to make sure that the meeting has a printed agenda. (See the tips list on writing and using an agenda, page 301, for ideas on how to create an effective agenda for a meeting.)
- Organize the minutes according to the topics and subtopics provided on the agenda.
- Record the date and the starting time of the meeting.
- If appropriate or if requested, provide the names of all of those attending the meeting.
- Listen carefully to what is said, and make sure you understand what is being discussed and what decisions are being made.
- Take careful and thorough notes. (See the tips list on taking notes, page 313, for tips on how to take good notes.)
- If necessary, interrupt the discussion to make sure you are recording various motions and decisions correctly.
- Become familiar with *Robert's Rules of Order*. As the person recording the minutes, you won't usually be the chair of the meeting, but it helps to be familiar with these rules of procedure so that you can more easily understand what is being discussed and why it is being discussed that way.
- Be consistent and specific when referring to people. For example, include the name of any person who makes a motion. Give the exact name of the person and use people's names consistently throughout the minutes.
- Be sure to record all motions word for word. Other matters and discussions may be summarized or paraphrased. Record the disposition for each motion. If the motion is voted on, record the result of the voting (passed, defeated). If the motion is tabled for later discussion, record that too.

(continued)

Writing Minutes *(continued)*
- Record the ending time for the meeting.
- Make sure your notes are complete and accurate. If possible, have someone else who attended the meeting review your minutes while both you and that person have a fresh memory of the meeting.
- Sign your name to the minutes document.
- Provide copies to all members of the group (including those who were not able to attend) as soon as it is convenient to do so after the meeting. Keep a copy on file for future reference and as a record of the meeting.

Networking

Networking is an essential professional development skill. (See Part Eight, "Correspondence and the Job Search," for another discussion of this topic.) Networking involves forming ties with other people for friendships or professional relationships. Networking is reaching out and making others aware of who you are and what you would like to do or have. You network when you introduce yourself at meetings or conferences and exchange business cards with those you meet. You network when you reach out both to those you know and those you don't know through face-to-face conversations, phone calls, conference calls, video conferencing, e-mail, and so on. Some people have a talent for such networking, and others find it difficult to do. If you are one of the latter, you need to relax and accept that networking is part of what you are expected to do in your job. Networking is making professional ties so that you will have stronger professional relationships in your current job or so that you will have people to call on to help you if you should lose your current position. In sum, networking is a survival skill.

It's a good idea to compile a database of contacts (a database is simply a collection of information that is organized in some way). A common way to keep track of those with whom you network is to keep a list of names, addresses, phone numbers, fax numbers, and e-mail addresses. Some people prefer to use a rotary card file, some an address book, some a personal information manager or a database program on their computer, and some an electronic handheld organizer. Computer programs such as Sidekick or Microsoft Outlook help you to manage your networking file. Database programs such as Microsoft Excel can do the same thing. The important point is to avoid having the names and contact information of your contacts scribbled on all kinds of pieces of paper and the backs of business cards. You will too often find it difficult to track down the one name you need. Before you begin your professional career, begin to collect contacts for your database, and plan to add to this database regularly. You will find it an invaluable source for growing professionally within a company and for moving from job to job.

Mentoring

Mentoring—the providing of guidance by an experienced person to a less experienced person—has proved to be invaluable in a number of professions and activities. In academe, full or associate professors often act as mentors for assistant professors. These mentors help evaluate the assistant professor's teaching, research, committee service, and so on. By working closely with the assistant professor, the mentor is able to help the newcomer improve in teaching, become more productive in research, serve on the appropriate committees, and, in general, navigate the politics of a particular department or administration.

In the corporate world, mentoring is also essential for the success of any new employee. Often such mentoring is very unstructured. The new employee essentially finds out who holds the real power or who can be depended on for information or action. However, this common approach is not truly mentoring, and it requires far too much of the new employee's time. A true mentorship implies some kind of structured relationship. Certain mentors are assigned to specific people, and guidelines are followed. The two key requirements for establishing such a relationship are: Pair a skilled (senior) professional with a novice, and make sure the relationship is outside formal channels of supervision and evaluation. Mentoring can also occur between colleagues who may not be working in the same location or even at the same company. A novice who is lucky enough to find someone who is willing to act as a mentor will be able to bridge many of the difficulties newcomers face when entering a field.

TIPS: Mentoring

- Realize that high-quality mentoring requires a commitment of your time, energy, and expertise. Good mentoring is hard work.
- Get to know the person or people you are mentoring.
- Establish mentoring guidelines and goals and make sure the person being mentored understands both.

(continued)

Mentoring *(continued)*

■ Promote the growth of the protégé through counseling and empathy, but also remain detached, avoid judgment, steer clear of personal problems, and set limits.

■ Be a good communicator and a good teacher—well-organized, flexible, committed, and patient.

■ Be honest, particularly in providing candid feedback.

■ Always practice what you preach, since the person being mentored will often emulate your technique and style.

■ Establish specific time allocations for assistance. Flexibility is essential to accommodate individual schedules; on the other hand, there should be enough structure to make sure that meetings occur regularly (at least once a month). Possible venues are business lunches, coffee after work or after a professional association meeting, or a tour of the workplace. Telephone calls and e-mail are useful supplements to—but not replacements for—face-to-face sessions.

■ At the end of the mentoring period, back away. Although this separation can be somewhat painful for the person being mentored, a successful disengagement is important because it leads to a redefinition of roles and a new relationship as professional colleagues. Failure to disengage can lead to dependency and erode the success of the mentoring.

TIPS: Being Mentored

■ Work to make the relationship a professional one.

■ Remember that your mentor is a guide, a colleague, a friend. Your mentor is not your parent or your banker. The value of the relationship is in guiding you through the early stages of your employment and not in solving your personal, social, or financial problems.

■ Remember that your mentor is more experienced than you are.

■ Keep in mind that the goal is not to make you a clone of your mentor but to help you realize your own strengths.

■ Schedule frequent meetings with your mentor.

■ Be honest, open, and receptive to benefit fully from the relationship.

■ Make certain the mentor is aware of your concerns, fears, or worries.

■ Keep in mind that suggestions and criticisms are meant to be helpful and to benefit you, so take them seriously and work toward improving your performance.

■ When you disagree with your mentor, settle any disagreement amicably or at least amicably agree to disagree.

Despite the best intentions of both parties, sometimes a mentorship may fail. The mentor and protégé may have been mismatched from the beginning. They may feel uneasy with one another or have unrealistic expectations of each other. They may not have the strong trust a mentorship relationship must have. In these situations it's best to be candid about the problems and terminate the mentorship promptly. After all, the person being mentored is certainly not realizing a benefit from the relationship, and the time spent in the failed mentorship could be better spent paired with another mentor.

Collaborating

Collaborating means working with others, and it is an essential job and social skill. Unfortunately, some people don't work effectively with other people. They don't want to share the workload because they don't like to delegate, they fear that the quality of the work will suffer if they allow others to share the workload, they don't like having to deal with the egos of others, or they don't like conflict. There are many other reasons too.

Most of the time, collaboration is not a choice—it is a requirement. A professor may assign a group project, or an employer may form a team to work on a particular undertaking. Therefore, like it or not, teamwork plays a big role both in college and in the workplace. Students in a variety of disciplines may be required to work together on teams to produce a project. Employees in the workplace have to work in teams because of the complexity of the products they are creating or modifying. Although the drawbacks of collaboration are what you would expect when any group of individuals have to pull together as a team, the rewards can be more than commensurate with the aggravations. Since collaboration is a fact of working life, learn to appreciate its advantages and deal with its disadvantages.

The most obvious advantage of collaboration is the sharing of the workload. The resulting work may be of much higher quality as the talents of the group members combine to improve the end result. In a true, equitable collaborative effort, team members contribute equal value and effort to the project. Although the division of labor is not always as equal as it should be, the effect of shared work—lightening the burden—should not be ignored. As a corollary to sharing the workload, the time required to finish a job is often shortened. Unfortunately, this advantage can sometimes fail to materialize if the coordination of deliverables takes too much time, but in general, a collaborative effort should take less time than an individual one.

Also, teamwork can be superior to individual labor because team members will bring a range of skills to the table and can view critically and helpfully the work of the other participants. There are disadvantages in having to monitor team members to ensure that they do their fair share and work together in harmony, but awareness of the potential gains and a determination to make the process work well can carry the project through to its desired conclusion. Cultivating the skills necessary for successful collaboration is a worthwhile effort that will pay off tangibly in your work.

TYPES OF COLLABORATION

There are several types of collaboration. In a horizontal model, team members divide a project into tasks so that each member is responsible for one part of the project. Students collaborating in this way would divide a project into three or four parts according to the number of group members. The results of such an approach are mixed because students will have their own styles, methods, attention to detail, and so on.

Another common model is often referred to as the sequential model. In this model one person would typically start a project by planning, designing, writing, editing, revising, and so on. Then another person would take this work and review and evaluate it. A third person would further refine it and pass it on to a fourth. Overall, the results are more uniform than those of the horizontal model because everyone is looking at the same document and making contributions.

A third common model is the vertical division model or stratification model. In this model, team members are responsible for the tasks they do best. One member would draft the document, one would edit and revise, another would contribute the design and illustrations, a fourth would oversee production, and a fifth would serve as a project manager, coordinating the activities of everyone involved. Usually, at least in technical communication, at most five people are on the team.

The stratification model has the chief advantage of specialization. Each person is doing what that person does best. The entire technical document benefits from the special strengths of each team member. Another advantage of this method is that it follows the project team approach. Each person reports directly to the project manager, and team members feel that they are contributing equally instead of competing as in the sequential model.

An important point to remember about collaborative approaches is that they are often much more effective than individual efforts. If it is well planned and managed, a collaborative approach can be a much more fun way to work on a project or a number of projects.

HANDLING CONFLICT

Even the best teams using the stratified model may encounter conflict. However, conflict can be both positive and negative. People can learn from genuine disagreements with each other, and considering the views of others can often make every team member more creative. Conflict can also motivate people to work harder on a project to make sure that the job is done well and that they are contributing their best effort. Conflict can also pressure team members to meet deadlines that they might not otherwise meet.

Unfortunately, if a group is poorly managed—and at times even despite good management—members working on a project are sometimes just not able to get along. Some people may not be willing or able to do their fair share of the work. Some may fail their team by not delivering an essential part of a project when they promised to deliver it. Some team members may simply lack the interpersonal skills to communicate their views and their feelings effectively to other members.

There are no easy solutions for resolving conflict, but one step is to distinguish the positive kinds of conflict from the negative. Another step is to make sure there are a variety of ways to communicate concerns, opposing views, and so on.

TIPS: Handling Conflict

- Accept the fact that differences of opinion will occur and often lead to good solutions.
- Give everyone on the team the opportunity to express ideas so that all members know they are contributing to the project.
- If you have to resolve conflict among members of a team, collect enough information from all sides of the disagreement. Focus on the current disagreement, and don't let team members digress from the subject at hand to the history of current conflicts.
- Don't jump to conclusions about the views of any one side. Appearances can be deceptive.

(continued)

Handling Conflict *(continued)*

■ Don't assume that you know how one side feels about the matter.

■ Assess the emotional state of those who are involved in the conflict. You may need to calm people who are highly agitated.

■ Show sympathy toward all. Try to understand the situation from all points of view.

■ Avoid blaming one person or one side in the conflict. Your goal should not be to alienate anyone but to resolve the conflict in the best interests of all involved.

■ Be sincere and honest in your attempt to resolve the conflict. Treat both sides equitably.

■ Make sure the conflict is resolved to everyone's satisfaction, or at least that the resolution is something everyone can accept.

Negotiating

Negotiating involves communicating with others in an attempt to resolve a conflict or some other matter. Successful negotiating is a difficult skill to master, but whether in life or in your work negotiating is not something you have a choice about. You will have to negotiate often about all kinds of decisions and situations. You continually have to negotiate with your classmates, friends, dates, significant other, spouse, kids, colleagues, and many other people. So the issue is not whether you will negotiate. The issue is whether you will negotiate well or poorly.

TIPS: Negotiating

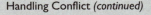

■ Learn how to listen. To be a good listener, you have to know how to understand what the other person is trying to tell you instead of just trying to hear what you want to hear. You have to know how to read both verbal and nonverbal cues. Knowing how to read nonverbal cues can give you valuable information about the attitudes and emotions behind the words you are hearing. You have to know how to recognize qualities in the other party's voice. Listen to the person's words, but pay attention to the tone of voice, how the person's voice is pitched, the facial expressions, and the person's stance. These are all clues to how the person actually feels about the situation you are both involved in. Careful listening is not merely a matter of giving the other party a chance to speak. Careful listening is the best strategy for you to acquire all of the information you need to know to negotiate with the other person.

- Focus your attention on the other party. Maintain eye contact. Learn how to read what the eyes may be telling you. Your focus should not be on persuading but on perceiving. The better you understand what the other person is trying to say, the better position you will be in to respond appropriately. Don't focus on what your next statement will be in response to what the other party is saying. Instead, focus on what the other party is saying and how it is said.

- Put yourself in the other person's shoes. As you listen, ask yourself why this person sees things this way, and ask yourself whether you would see things this way if you were this person.

- Let the other person explain his or her position completely. Don't interrupt. Now is the time for you to gather information that you can use later in the negotiating process.

- Be empathetic.

- Explain your own position logically and clearly. Facts and a well-organized presentation will help your cause. To have the facts you need, you will have to prepare in advance, but this preparation will pay off when it's your turn to speak. Make sure you cover every point that needs covering, and make sure that your facts are honest and accurate. If necessary, modify your presentation to take into account points raised by the other person.

- Avoid arguing or shouting or other hostility or personal attacks. Keep the focus on the message. Nothing will be accomplished by other strategies. For example, if you become angry, you give the other party an advantage. Anger can be used to your advantage, but you need to know when to use it and how.

- Think win–win rather than win–lose. Many people think that one person wins a negotiation and the other loses. However, most often, if the negotiations are successful, both parties win. During your discussions focus not only on what you want to gain, but also on what the other party has to gain to feel good about the negotiating.

- Be willing to compromise. You won't get anywhere in your negotiations if you're not willing to be flexible on any points. There should always be a little room for flexibility. Even when big issues are at stake, compromise can usually be found by agreeing, for example, to phase in slowly what you agree on rather than doing it all at once. Negotiation often requires that you give something to get something. Think of what you can give in return for what you want.

(continued)

Negotiating *(continued)*

■ Expect compromise. Just as you should be willing to compromise, you have a right to expect the other party to compromise. Both sides must be flexible.

■ Be willing to follow through. Don't offer anything that you are not sincere about giving after the deal is finalized. You have to keep your word.

■ Be patient. After negotiations have been proceeding for a while, both sides may lose patience with each other. This is the time to be more patient than ever. Think of patience as a strategy to help you obtain what you want.

■ Recognize negotiation as an ongoing process. If negotiations are particularly long and complex, it's a good idea to write everything down. Otherwise, you and the other party may later disagree about what you agreed on earlier.

DEVELOPING PERSONAL SKILLS

Enhancing Listening Skills

Just as most writers focus on what they want to say, rather than on what the audience needs to know, most people are preoccupied with their own thoughts and not with what others have to say. Most people are poor listeners. They may listen intently for a few minutes, but they soon let their thoughts drift to something else. Of course, many people are difficult to listen to because of their voice or possibly because it's taking them a long time to get to the point. You should be able to enhance your listening skills nevertheless. (See the tips list for negotiating, page 310.)

Taking Notes

Note taking is a skill that must be learned. Many people are poor note takers. They try to write down everything they hear and end up missing many important points, or they write down too little and miss important points, or they write illegible notes and have difficulty later reading and understanding what they wrote. You'll find that you have to take notes in all kinds of situations—during research (see Part One, "Planning and Research," for a discussion of various methods for taking research notes), during lectures, during staff meetings, during conference presentations, during interviews, and so on. Sharpening your note-taking skills can make a big difference in how well you perform in a college course or on the job.

Notes are the keywords, phrases, paraphrases, facts, statistics, and quotations that you listen to in a staff meeting or in a college lecture. You have to make sure you are recording what you hear accurately, selectively, legibly, and completely. If your notes are not accurate, then they are useless. If you haven't been selective in your note taking, then you are most likely trying to write down too much information. If you haven't written legible notes, then your notes will not be helpful to you or to anyone else. Finally, if you have written an incomplete set of notes, then you may be missing information that could prove vital later.

TIPS: Taking Notes

Methods for taking notes vary considerably from one person to the next, but some techniques can make your notes more complete and useful.

- Use the materials or equipment that is best suited to your method of taking notes. Some people take notes on anything that's handy and later have trouble making sense of the order and the topics. Decide on one approach, such as using a notepad and pen, and stick with the method that works best for you.

- Don't become dependent on audio recording for lectures, meetings, interviews, and so on. Recording the occasional lecture or interview is fine (as long as you have the permission of the people whose voices you are recording), but it's not a good idea to use this approach too often. Most people don't like to have their voices recorded, and relying on recording voices won't do anything to sharpen your note-taking skills.

- If an agenda or lecture outline is provided, try to organize your notes according to the main points provided in the agenda or outline.

- Try not to fill every page with notes. In fact, if you can use just the middle four or five inches of each page for your notes, you may find it easier to focus on what is important and easier to read your notes later.

- Try to write legibly. Consider practicing your printing skills if your habitual handwriting is difficult even for you to understand.

- Use abbreviations frequently, but don't invent cryptic abbreviations that you won't understand later. Teach yourself a system for abbreviating common words and phrases, particularly those typical in your field.

- Don't try to write down everything you hear. Be selective. Write down just the most important points. Underline the points that seem the

(continued)

Taking Notes *(continued)*

most useful and important. Place marks in the margin reacting to those points. For example, if something may prove particularly helpful to you later, place a dash or an asterisk beside it. Information that you consider surprising or unexpected may merit an exclamation mark in the margin.

- If you're not sure of the accuracy of a note, write a question mark by it so you remember to check it later.

- When appropriate, try to compare your notes with the notes others have taken. Learn from any tricks or techniques that others may use.

- After you finish taking your notes, don't wait long before reviewing them and filling in any blanks while the lecture, meeting, or interview is fresh in your memory. If your handwriting is a problem, transcribe the notes into a more legible form.

- Organize your notes so that they will be easier for you to study later and, if necessary, memorize or recall.

Acting Ethically

Ethics, simply defined, is the study of right and wrong conduct or behavior. Developing or adhering to your sense of ethics and learning how to deal with people who are unethical are essential interpersonal skills. Working in today's highly technological society often challenges personal values and convictions, and you need to have a good sense of what your values are if you are to keep your integrity and character intact.

There are many kinds of ethical systems, ranging from those rooted in a religion to those that are completely secular. You have probably already decided which ethical system is best suited for you. Whatever you choose to adhere to, you are essentially choosing a system or hierarchy of values to live and act by. It's important to choose these values wisely and to adhere to them conscientiously. If you have thought consciously about the value system you will follow, you will face ethical challenges armed with your convictions. Deciding what your values are on an ad hoc or situational basis can leave you unsure of where you stand and how you want to behave when ethical challenges present themselves.

CAMPUS HANDBOOKS

Many college campuses have handbooks concerning the behavior of students, faculty, and staff on campus. These handbooks typically provide guidelines about everything from attendance to disruptive behavior. Handbooks are helpful because they establish minimal ethical standards and give all concerned at least some understanding of what is expected. They cannot ensure that everyone will follow the rules, but they create a climate that helps to foster ethical behavior.

PROFESSIONAL CODES OF ETHICS

Many professions have a code of ethics that they expect their members to adhere to in their day-to-day activities. Providing a code of ethics creates a climate of expectations for professional standards and behavior. Some members may choose to ignore the code of ethics of their profession; you cannot make people who are inclined to act unethically behave ethically. However, a code of ethics can lend guidance to people, especially newcomers to a profession.

The code of ethics for a profession or organization typically appears in membership directories for the profession's major society or societies, in promotional literature (for example, brochures, annual reports), and increasingly on Web sites. The Society for Technical Communication, for example, publishes its code of ethics in its annual fall membership directory.

UNDERSTANDING ETHICAL SITUATIONS

College students and people working in the technical professions face ethical challenges just as anyone else does. As a college student you may see other students who are cheating on an exam, plagiarizing all or part of a research paper, stealing from their roommates, or lying to their friends. (See "Research Ethics," page 38.) As a technical professional, you may be asked to misrepresent a product to some clients, you may know of others who falsely take credit for another's achievement, you may have colleagues who do not contribute their fair share to a team project, or you may see others steal supplies from the office and take them home. These and innumerable other possible ethical situations occur daily on college campuses and in offices throughout the world. How do you do what is right and avoid doing what is wrong? In a complex situation in which you may be faced with two equally tempting actions, how do you determine which action is best for you and others?

There are no easy answers to these questions. In addition to the codes of ethics mentioned above, many ethical systems offer ways to resolve these conflicts. You will have to choose the ethical system that makes the most sense to you, one you think you can adhere to no matter how difficult the choices you must continually face. Living an ethical life as a college student, as an employee, and as a person isn't easy. However, you can have an ethical life if you are determined to have one.

RESOURCES

Badaracco, Joseph L., Jr. *Defining Moments: When Managers Must Choose Between Right and Right*. Boston: Harvard Business School P, 1997.

Barker, Thomas, and Karen Steele. *Getting Started in Consulting and Independent Contracting*. Society for Technical Communication. <http://english.ttu.edu/GScic/DEfault.htm>. Last updated May 1, 1999.

Bergin, Francis. *Successful Presentations*. New York: Prentice Hall, 1996.

Biederman, Patricia Ward. *Organizing Genius: The Secrets of Creative Collaboration*. Boston: Perseus, 1998.

Bishop, Sue. *The Complete Guide to People Skills*. Oxford: Gower, 1997.

Brown, William. *Interpersonal Skills for Leadership*. New York: Prentice Hall, 1998.

Chang, Richard Y. *Success through Teamwork: A Practical Guide to Interpersonal Team Dynamics*. Irvine: Chang, 1994.

Dombrowski, Paul. *Ethics in Technical Communication*. Boston: Allyn & Bacon, 2000.

Hager, Peter J., and H. J. Scheiber. *Designing and Delivering Scientific, Technical, and Managerial Presentations*. New York: Wiley, 1997.

Hargrove, Robert. *Mastering the Art of Creative Collaboration*. New York: McGraw-Hill, 1998.

Humphrey, Watts S. *Managing Technical People: Innovation, Teamwork, and the Software Process*. Boston: Addison-Wesley, 1996.

Kent, Peter. *Making Money in Technical Writing: Turn Your Writing Skills into $100,000 a Year*. New York: Macmillan, 1998.

Kramer, Marc. *Power Networking: Using the Contacts You Don't Even Know You Have to Succeed in the Job You Want*. Lincolnwood: VGM Career Horizons, 1998.

Kratz, Dennis M. *Effective Listening Skills*. Toronto: Irwin, 1995.

Moore, David J. *Job Search for the Technical Professional*. New York: Wiley, 1991.

Morrisey, George L. *Loud and Clear: How to Prepare and Deliver Effective Business and Technical Presentations*. Boston: Perseus, 1997.

Rees, Fran. *Teamwork from Start to Finish: 10 Steps to Results*. San Diego: Pfeiffer, 1997.

Reimold, Cheryl. "How to Deliver Winning Presentations." *IEEE PCS Newsletter*. Mar/Apr.–Nov./Dec. 1997. <http://www.ieee.com/society/pcs/creimol2.html>.

———. "Preparing Outstanding Presentations." *IEEE PCS Newsletter*. Jan./Feb. 1996–Jan/Feb. 1997. <http://www.ieee.com/society/pcs/creimold.html>.

Rothstein, Michael F. *Ace the Technical Interview*. 3rd ed. New York: McGraw-Hill, 1998.

Stark, Peter B. *It's Negotiable*. San Diego: Pfeiffer, 1994.

Syer, John. *How Teamwork Works: The Dynamics of Effective Team Development*. New York: McGraw-Hill, 1996.

Voss, Dan, and Lori Allen. *Ethics in Technical Communication: Shades of Gray*. New York: Wiley, 1997.

Wellins, Richard S. *Inside Teams: How 20 World-Class Organizations Are Winning through Teamwork*. San Diego: Jossey-Bass, 1994.

Yager, Dexter. *Dynamic People Skills*. Wheaton: Tyndale House, 1997.

Correspondence and the Job Search

Illustrations

Tips

Correspondence and the Job Search

*T*he majority of your daily writing will be letters, memos, or e-mail in which you communicate directly with other people and try to provide what they want or need to know. These forms are often no more than one page long, but don't let this brevity fool you. Writing an effective letter, memo, or e-mail requires many of the same steps that are involved in longer communications such as reports or proposals. In many cases you will still have to brainstorm or use some other technique to help develop your ideas (see Part One, "Planning and Research," for prewriting techniques), you will still want to establish your purpose and narrow the scope of your message, and you may still outline or organize to state the points in the order that will be the most effective for your reader. You will also find that careful control of tone and style makes the difference between an effective letter that achieves your purpose and one that fails.

LETTERS

Letter writing is a vitally important skill in the business world. Letters are a basic person-to-person communication that takes place in a wide variety of settings and situations. Whether or not you ever have to write reports, proposals, or manuals on the job, you will certainly have to write letters. Since letters tend to be much briefer than many other business communications, you have a limited opportunity to make your point and achieve your purpose in a letter. Therefore, more than for most other business communication, every word counts. Moreover, how you format your letter becomes part of the message.

You will have to write many kinds of letters for a variety of professional, social, and personal occasions. The requirements for letters differ depending on their purpose and audience. A letter of application has different requirements than a letter of complaint, and a personal letter to a close friend has different requirements than an invitation letter to acquaintances. Most of your letters will be professional letters, and you must learn to use the correct formats, the appropriate strategies, and the widely accepted conventions for such letters. Your professional letters say a great deal about you and, if you are an employee, about the company you work for. Letters that ignore accepted formats, use poor strategies for the kind of letter you are writing, or misuse conventions will fail to communicate the intended message. Therefore, understanding the basics of effective letter writing is essential to any college student or employee.

Direct and Indirect Patterns

In correspondence—letters, memos, and e-mail—the terms *direct* and *indirect* refer to the order in which you present good or bad news or delicate information. These patterns are also sometimes referred to as the "yes/no" order. If you are offering good news, then present it in the opening of the letter (the direct pattern). If you are providing bad news, then place it toward the end of the middle or in the closing of the letter (the indirect pattern). The indirect pattern is a challenging way to organize the information of a letter. In addition to stating the purpose of your letter in the opening (as you should try to do in all your correspondence), you are preparing the reader for the bad news that you will state late in the letter. This type of delay is often necessary and will make your letter more effective than if you simply state the bad or blunt or delicate news without preparing the reader for it.

An indirect pattern is demanding because it constantly challenges you to see your subject and purpose from the reader's point of view. You have to consider carefully a number of questions, such as these: "Have I provided an adequate buffer in the opening of my letter?" "Have I considered possible concerns or objections the reader may have or make concerning this bad or delicate news?" "Have I stated the bad or delicate news as tactfully as possible in the best place in the letter?" "Have I been convincing?" Knowing how and when to use an indirect pattern will help you in many kinds of correspondence situations. (See Figure 8.1.)

Davies Consulting

1404 Gardenview Lane; Suite 2000; Roanoke, VA 24011
540-555-4147 (phone); 540-555-5498 (fax)

June 23, 1999

Mr. Bill Sharp
Vice President
Mills Photography
8004 Gilcrest Lane
New York, NY 10019

Dear Mr. Sharp:

As you know, we have done business together for many years. Your company has continually provided me with high-quality professional photographs for numerous technical manuals. I appreciate the way you promptly handle all orders. I can truly say I have been very satisfied with your work. However, I am writing to explain a decision I have made concerning future dealings with your company.

Recently, I had to reassess all of the businesses I use for my consulting work, and I have decided to patronize only those companies that can provide me the best work for the lowest price. After soliciting numerous other bids, I decided that a photographer here in Roanoke will provide the photographs I need for the technical manuals I produce. I am writing to let you know that although I have been pleased and impressed with your service, I, regrettably, will no longer use your service. Enclosed is my payment for the final invoice you sent to me. You may now close my account.

Thank you for your fine work over the years.

Sincerely,

Phil Davies

Mr. Phil Davies, Owner

FIGURE 8.1
Indirect Pattern Letter

Parts of a Letter

Letters have eight commonly used parts: a heading, a date line, an inside address, a salutation, a body, a complimentary close, a signature, and, when necessary, an end notation or end notations. However, other parts (including an attention line, a subject line, and an identification line) may sometimes be necessary. In professional correspondence you must use and provide these parts appropriately and correctly. (See Figure 8.2.)

HEADING

Headings (also called return addresses or outside addresses) differ, depending on whether they occur on company or organization stationery or on personal stationery. Official correspondence between

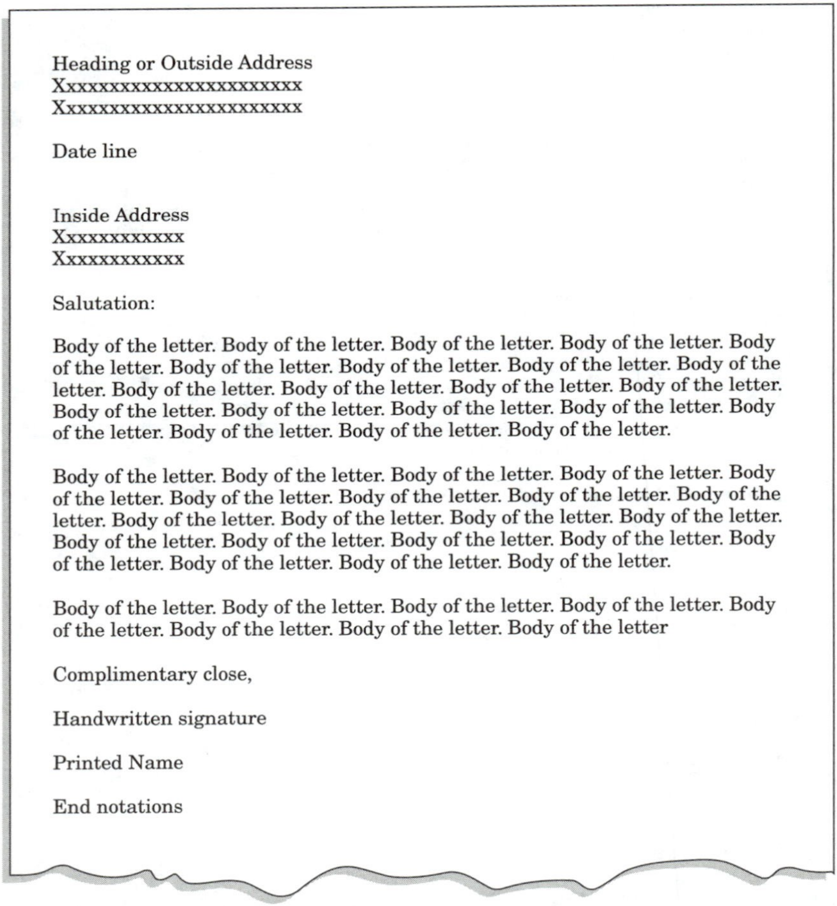

Heading or Outside Address
Xxxxxxxxxxxxxxxxxxxxxxxx
Xxxxxxxxxxxxxxxxxxxxxxxx

Date line

Inside Address
Xxxxxxxxxxxx
Xxxxxxxxxxxx

Salutation:

Body of the letter. Body of the letter.

Body of the letter. Body of the letter.

Body of the letter. Body of the letter. Body of the letter. Body of the letter. Body of the letter. Body of the letter. Body of the letter. Body of the letter

Complimentary close,

Handwritten signature

Printed Name

End notations

FIGURE 8.2
Parts of a Letter

an employee of the company and someone outside the company requires the use of letterhead stationery. Personal correspondence may be more appropriately carried on through electronic mail or on personal notepaper. Typically, a heading on business stationery consists of a company or organization name, street address or post office box, city, state, zip code, and sometimes phone number and fax number. Increasingly, companies and organizations are also providing a company e-mail address and Web-site address. Many companies and organizations combine all of these elements with a logo and other design elements at the top of the page, either flush left or centered. Some provide much of the same information at the bottom of the page and provide just the company or organization logo at the top of the page. Because you will normally not be able to modify a company letterhead, you need only recognize when its use is appropriate.

Now that desktop publishing is available to virtually anyone with a computer and a printer, people can easily create their own personal stationery and add all kinds of individual touches. For example, it's not uncommon for some people to print their name, address, city, state, zip code, e-mail address, and Web-site address, perhaps with a logo, at the top of the page.

DATE LINE

If you are using company organization letterhead, you usually skip at least two lines below the address and provide the date for the letter either flush left at the left margin or flush left at the center line.

On personal stationery, people usually provide the date as part of the heading, but since the advent of desktop publishing, people are increasingly treating their personal letterhead as preprinted stationery and are providing the date two or more lines below their address.

The typical order for the date is month, day, and year: December 1, 1999. Spell out the month to avoid confusion. In the United States, 12/1/1999 means December 1, 1999, but in many countries it means January 12, 1999. In the military the order is day, month, and year with no commas: 1 December 1999.

INSIDE ADDRESS

The inside address is the address of the person, company, or organization to which you are sending your letter. Whenever possible, you should address your letter to a specific person rather than just to the company or organization as a whole. Typically, an inside address includes the name of the person you are writing to (including a courtesy title for the person and the person's job title), the company or organization name, the street address or post office box, the city, the state (either spelled out in full or abbreviated with the official postal service two-letter state code), and the zip code:

Mr. John Doe, Manager

Office Mart

555 Wimberly Road

Orlando, FL 32816

A courtesy title is the "Mr.," "Ms.," "Mrs.," "Miss," "Professor," "Reverend," or whatever is appropriate for the person you are writing to. If the person has a one-word job title, place this title on the same line; if the title is more than one word, place the job title on the line just below the person's name.

If you have questions about forms of address in the inside address and correct salutations for clerical and religious orders, college and university faculty and officials, consular officers, diplomats, foreign heads of state, government officials, military ranks, various professional titles, multiple addresses, and an assortment of special titles (for example, Doctor, Esquire, Honorable, Professor), consult a reference work such as *Webster's Guide to Business Correspondence,* listed in the Resources at the end of this chapter.

ATTENTION LINE

Attention lines are sometimes necessary in business correspondence if you want to address an organization in general and also bring your letter to the attention of a particular individual within the organization. If you address a letter to a company with an attention line for a particular person and that person is unavailable to read your letter, then anyone representing the company may read it. If you address a letter to an individual and also provide an attention line to another person at the company, the letter will be forwarded to the person in the attention line if the addressee is unavailable.

Because a letter using an attention line is actually written to the company, the salutation following the attention line should be

"Ladies and Gentlemen," "Gentlemen," or "Ladies" depending on who owns the company. See "Salutation" below.

Place the attention line against the left margin in all letter formats, two lines below the inside address. The word *Attention* should be spelled out, and the first letter should be capitalized. Placing a colon after the word is optional.

SALUTATION

The salutation is your greeting to the person with whom you are corresponding. It should be placed against the left margin regardless of the letter format you are using, and it should be two line spaces below the inside address or the attention line. The salutation is followed by a colon except in personal letters, where a comma is commonly used. Typical salutations include the following:

Dear Mr. Doe: Dear Mrs. Smith:

There are a variety of salutations and various guidelines concerning their use. The courtesy title in the salutation must agree with the courtesy title that you use in the inside address. If you're writing to Mr. John Doe, for example, you should address him as "Dear Mr. Doe." If you know John Doe and have a good relationship with him, you may simply use "Dear John" for your salutation.

If you are writing to a company and do not know the name or gender of the specific person you are addressing, you have several choices. You can begin with "Dear Office Mart Manager" or "Dear Personnel Supervisor," using the person's title as if it were a name. If the owners of a company are both men and women, you may use "Ladies and Gentlemen" or "Dear Sir or Madam" for your salutation. If the letter is addressed to an all-female organization, the salutation "Ladies" is commonly used. If you have the name of an individual but you are uncertain of the gender, you might want to use "Dear Pat Smith."

"To Whom It May Concern" may be used if your letter is not addressed to any particular person or company. If at all possible, find a different salutation because this one is considered very impersonal. Letters of recommendation sometimes use this salutation because they are often sent to unknown addressees and are often distributed throughout a company or organization.

The guidelines concerning whether to use "Miss," "Ms.," or "Mrs." are complex. If you do not know the preferred title of address for the person you are writing to, you should use "Dear Ms. Smith." If this person has indicated a preference for "Miss" or "Mrs.," then you should use what she prefers.

In academic settings it is usually safest to write "Dear Dr. Smith" or "Dear Dr. Doe," even if you are not sure the person has a doctorate. Few people will be offended by being called "Doctor" inappropriately, but some who do hold doctorates will resent an omission of the title in correspondence.

SUBJECT LINE

A subject line is often used in professional correspondence to provide the content of the letter in a few key words. The subject line should be placed two spaces below the salutation either against the left margin or indented, depending on the format of the letter you are using. For letters with indented paragraphs, indent the subject line. The word *Subject* should be followed by a colon. You may also provide a subject line and omit the word *Subject*. In some situations, notably in legal correspondence, *Re:* may be used instead of *Subject*.

BODY

The body of a letter, also referred to as the message, consists of the text of the letter from the first sentence of the opening paragraph to the last sentence of the closing paragraph. If you want to write effective letters, remember that every letter has a definite structure and that even brief letters have an opening, a middle, and a closing.

Openings in letters vary according to purpose and audience. For many kinds of letters you need to provide context for the letter and to state your purpose for writing in the opening. Providing context can mean acknowledging previous correspondence if you are writing in response to a letter or establishing a common ground if, for example, you are writing a letter of complaint. Also, don't keep your reader guessing about why you wrote the letter. In one of the first few sentences, state why you are writing. (See "Direct and Indirect Patterns," page 320, for a description of a situation in which you state your purpose in writing early but delay conveying bad news until later.) One of the most common mistakes people make in writing letters is to bury their purpose for writing somewhere in the middle of the letter. In general, letters are brief, and your reader needs to be told why you are writing. Put yourself in your reader's shoes and ask "Why am I reading this? Why was this letter sent to me? What does this writer want?" Most letters should answer these questions in the opening paragraph or two.

The middle of a letter may be only a sentence or two, or it may be several detailed paragraphs. This is the part of the letter in which you must use facts, statistics, examples, anecdotes, testimony, dates, names, reasons, and so on to support the purpose of your letter. In a brief letter on an ordinary or commonplace topic the middle won't be very involved. In a letter concerning a complicated or delicate issue the middle may be quite extensive. Whether the middle is brief or long, you must make sure it's appropriate to your purpose for writing.

How you end your letter also depends a great deal on your purpose for writing. See the discussion of various letters below for various closing conventions. If you are using the indirect pattern (see "Indirect and Direct Patterns," page 320), then you may delay revealing your news until the end of the middle of your letter or the beginning of the closing. The last few sentences of a letter typically enhance its professional tone. For example, you may write, "Please contact me if you have any questions" or "I am looking forward to hearing from you soon." However, some people view these kinds of phrases and others as trite or commonplace closings and suggest that you avoid them. They believe that these closings have been so widely used that they don't communicate anything anymore. You might want to try to be a little more original. At the very least, use the closing that seems most natural to you and what you want to say.

COMPLIMENTARY CLOSE

Another convention of correspondence is the complimentary close. The most common closes are the one-word closes "Sincerely" and "Cordially." Two-word and three-word complimentary closes such as "Sincerely yours" or "Very truly yours" are much less commonly used in correspondence today. Avoid original closes that are not appropriate in professional correspondence, such as "Later," "Signing off," or "Your humble servant." Now that e-mail has become so commonplace, the complimentary close has taken off into a whole new direction. Abbreviations, which would be totally inappropriate in a conventional business letter, are quite usual at the end of an e-mail message. (See "Emoticons and Abbreviations," page 352.) International correspondence presents other problems where more formal closings, salutations, and so on, are common practice and in fact considered necessary in the well-written business letter. For example, as a rule, French business letters always end with a formula such as "Please accept, my dear sir or madam, the expression of my highest esteem." To omit this type of close in a formal French business letter is considered poor form.

SIGNATURE

Normally, your letter will end with a complimentary close and a signature block. The complimentary close is usually two lines below the last line of the last paragraph in the letter. The signature block, consisting of your signature and your typed name, follows at the bottom of the letter. Skip four lines between the complimentary close and the typed name. Both the complimentary close and the signature block should be aligned in the same way as the date at the top of the letter. If the date is flush left on the left margin of the page, so should your complimentary close and signature block be. If the date is flush left on the center line of the page, your complimentary close and signature block should be too. Your signature may be as flamboyant or illegible as you care to make it (that's why you also provide your typed name), but it's a good idea not to have a signature that's too large or too small. For business letters your signature should be written in blue or black ink, never in other colors and certainly never in pencil.

In much business correspondence it's common practice to provide the company name as part of the signature block. The company name should be placed about two lines below the complimentary close. Then skip two lines and provide the letter writer's signature, printed name, and job title.

IDENTIFICATION LINE

The identification line informs the reader who wrote or dictated, who signed, and who transcribed or typed a business letter. Sometimes assistants draft a letter for their employers to sign, and often a typist types letters written or dictated by employers. The initials of the assistant who drafted the letter for your signature should be provided, aligned against the left margin two line spaces below the last line of the signature block. If John Smith drafted the letter for Jane Doe's signature, the identification line would read JS: JD. If someone else typed the letter, it's common practice for the typist to provide his or her initials after the writer's at the bottom of the letter. For example, if the writer's initials are ABC and the typist's initials are XYZ, the initials ABC:xyz would appear at the end of the letter, aligned with the left margin of the page.

END NOTATIONS

Several major end notations are commonly used in correspondence. If you are providing enclosures, you can write "Enclosures" or "enc." (without the quotation marks) at the bottom of the letter, against the left margin, and even indicate the number of enclosures: "Enclosures: 2" or "enc. 2." The end notation "cc" (which formerly meant "carbon copy" but now can be taken to mean "courtesy copy") indicates who else is receiving a copy of the letter.

Letter Formats

The professional appearance of your letter is essential. Many readers won't bother to read a letter that looks unprofessional. So you need to devote some attention to margins, line spacing, fonts, paragraph length, and so on. There are five common letter formats: block or full-block, modified block, semiblock, simplified, and personal.

BLOCK OR FULL-BLOCK

In a block or full-block style, all of the text is aligned with the left margin. The date line is placed two to six line spaces below the last line of the heading or letterhead. Placement of the inside address varies depending on the length of the letter. A common spacing is four line spaces below the date line. The salutation is placed two lines below the attention line (if an attention line is provided), and the first line of the body is placed two lines below an attention line or two to four lines below the last inside address line. Paragraphs are single spaced, with a double space between paragraphs. (See Figure 8.3 on page 330.)

MODIFIED BLOCK

In a modified block letter the inside address, salutation, body, identification line, and enclosure notation are flush left. The paragraphs are not indented in the body of the letter. The date line, complimentary close, and typed name are aligned flush left at the center line. (See Figure 8.4 on page 331.)

SEMIBLOCK

In a semiblock style, also known as modified semiblock style, the date line may be aligned either slightly to the right of dead center or

Orion Consulting
200 Orange Avenue
Orlando, FL 33816

May 2, 1999

Mr. Mike Kendrall
Vice President of Operations
Loci Corporation
63 Canal Center Plaza
Suite 710
Alexandria, Virginia 22314

Dear Mr. Kendrall:

Thank you for your letter of April 28 providing me a copy of a Confidentiality
Agreement with your company. As you requested, I kept a copy for my files, and I am
enclosing a copy for you. I am also writing concerning some changes in possible
starting dates for the seminars because of a change in my schedule.

First, because I have signed this agreement, I hope that you can now send me sample
correspondence (memos and letters), sample reports, and any other relevant
documents from those who will participate in the writing seminars. After I review
these materials, I will have a better idea of what areas need to be addressed, and
then I will promptly send you a proposal for the writing seminars. I hope this plan is
okay with you.

Second, during our conference call on April 15, we discussed some possible dates in
early June for starting these seminars. Because of recent changes in my schedule,
the earliest I can conduct the first four-hour (9:00 a.m. to 1:00 p.m.) seminar is
Friday, June 19, instead of Friday, June 5 or June 12.

During our conference call, I know there was some concern about the availability of
participants for the latter part of June. I could offer the first seminar on June 19 (if
this date is convenient), and if Friday, June 26, is not convenient for the second
seminar, I will send you a proposal offering seminars for Friday, June 19, and several
Fridays in September or October (for a total of three seminars). As I recall, we
excluded conducting any seminars in July or August. Please let me know what Fridays
are most convenient for you, and I will do my best to adjust my schedule accordingly.

Sincerely,

John Smith

John Smith

FIGURE 8.3
Sample Letter in Full-Block Format

flush right. The inside address and salutation are aligned along the
left margin, and the first line of each paragraph in the body is indented five spaces. The complimentary close and signature block are
aligned under the date. Identification initials and various end notations are aligned with the left margin. (See Figure 8.5 on page 332.)

Orion Consulting
200 Orange Avenue
Orlando, FL 33816

May 2, 1999

Mr. Mike Kendrall
Vice President of Operations
Loci Corporation
63 Canal Center Plaza
Suite 710
Alexandria, Virginia 22314

Dear Mr. Kendrall:

Thank you for your letter of April 28 providing me a copy of a Confidentiality Agreement with your company. As you requested, I kept a copy for my files, and I am enclosing a copy for you. I am also writing concerning some changes in possible starting dates for the seminars because of a change in my schedule.

First, because I have signed this agreement, I hope that you can now send me sample correspondence (memos and letters), sample reports, and any other relevant documents from those who will participate in the writing seminars. After I review these materials, I will have a better idea of what areas need to be addressed, and then I will promptly send you a proposal for the writing seminars. I hope this plan is okay with you.

Second, during our conference call on April 15, we discussed some possible dates in early June for starting these seminars. Because of recent changes in my schedule, the earliest I can conduct the first four-hour (9:00 a.m. to 1:00 p.m.) seminar is Friday, June 19, instead of Friday, June 5 or June 12.

During our conference call, I know there was some concern about the availability of participants for the latter part of June. I could offer the first seminar on June 19 (if this date is convenient), and if Friday, June 26, is not convenient for the second seminar, I will send you a proposal offering seminars for Friday, June 19, and several Fridays in September or October (for a total of three seminars). As I recall, we excluded conducting any seminars in July or August. Please let me know what Fridays are most convenient for you, and I will do my best to adjust my schedule accordingly.

Sincerely,

John Smith

John Smith

FIGURE 8.4
Sample Letter in Modified Block Format

SIMPLIFIED

The simplified letter format is common in professional correspondence. Its features are a block format and fewer internal parts. The traditional salutation is replaced by an all-caps subject line aligned with the left margin. All paragraphs are aligned with the left margin

Orion Consulting
200 Orange Avenue
Orlando, FL 33816

May 2, 1999

Mr. Mike Kendrall
Vice President of Operations
Loci Corporation
63 Canal Center Plaza
Suite 710
Alexandria, Virginia 22314

Dear Mr. Kendrall:

Thank you for your letter of April 28 providing me a copy of a Confidentiality Agreement with your company. As you requested, I kept a copy for my files, and I am enclosing a copy for you. I am also writing concerning some changes in possible starting dates for the seminars because of a change in my schedule.

First, because I have signed this agreement, I hope that you can now send me sample correspondence (memos and letters), sample reports, and any other relevant documents from those who will participate in the writing seminars. After I review these materials, I will have a better idea of what areas need to be addressed, and then I will promptly send you a proposal for the writing seminars. I hope this plan is okay with you.

Second, during our conference call on April 15, we discussed some possible dates in early June for starting these seminars. Because of recent changes in my schedule, the earliest I can conduct the first four-hour (9:00 a.m. to 1:00 p.m.) seminar is Friday, June 19, instead of Friday, June 5 or June 12.

During our conference call, I know there was some concern about the availability of participants for the latter part of June. I could offer the first seminar on June 19 (if this date is convenient), and if Friday, June 26, is not convenient for the second seminar, I will send you a proposal offering seminars for Friday, June 19, and several Fridays in September or October (for a total of three seminars). As I recall, we excluded conducting any seminars in July or August. Please let me know what Fridays are most convenient for you, and I will do my best to adjust my schedule accordingly.

Sincerely,

John Smith

John Smith

FIGURE 8.5
Sample Letter in Semiblock Format

and single spaced, with a double space between paragraphs. There is also no complimentary close. The writer's name and business title, if necessary, are aligned with the left margin and typed in all caps at least five line spaces below the last line of the body or message of the letter. (See Figure 8.6.)

Orion Consulting
200 Orange Avenue
Orlando, FL 33816

May 2, 1999

Mr. Mike Kendrall
Vice President of Operations
Loci Corporation
63 Canal Center Plaza
Suite 710
Alexandria, Virginia 22314

STARTING DATES

Thank you for your letter of April 28 providing me a copy of a Confidentiality Agreement with your company. As you requested, I kept a copy for my files, and I am enclosing a copy for you. I am also writing concerning some changes in possible starting dates for the seminars because of a change in my schedule.

First, because I have signed this agreement, I hope that you can now send me sample correspondence (memos and letters), sample reports, and any other relevant documents from those who will participate in the writing seminars. After I review these materials, I will have a better idea of what areas need to be addressed, and then I will promptly send you a proposal for the writing seminars. I hope this plan is okay with you.

Second, during our conference call on April 15, we discussed some possible dates in early June for starting these seminars. Because of recent changes in my schedule, the earliest I can conduct the first four-hour (9:00 a.m. to 1:00 p.m.) seminar is Friday, June 19, instead of Friday, June 5 or June 12.

During our conference call, I know there was some concern about the availability of participants for the latter part of June. I could offer the first seminar on June 19 (if this date is convenient), and if Friday, June 26, is not convenient for the second seminar, I will send you a proposal offering seminars for Friday, June 19, and several Fridays in September or October (for a total of three seminars). As I recall, we excluded conducting any seminars in July or August. Please let me know what Fridays are most convenient for you, and I will do my best to adjust my schedule accordingly.

John Smith

JOHN SMITH

FIGURE 8.6
Sample Letter in Simplified Format

PERSONAL

Personal correspondence is much more informal in format and tone than professional correspondence. In personal letters a letterhead or outside address is not required. Typically, the following is used instead: a heading providing the street address or post office box on

229 Peachtree Hills Avenue
Atlanta, GA 30355
December 20, 1999

Dear Phil,

Thanks again for your phone call. As always, I enjoyed talking to you and catching up on news.

I'm enclosing a map to our place. If you have any problems with the directions, give me a call.

We look forward to your visit.

Your buddy,

Dave

FIGURE 8.7
Sample Letter in Personal Format

the first line; the city, state, and zip code on the second line; and the date on the third line. Sometimes the date line is provided two line spaces below the city, state, and zip code. In general, other requirements that are typical of professional correspondence are omitted. No inside address is required. Using a first name only in the salutation (followed by a comma) is fine, and providing a personal complimentary close is common, too. (See Figure 8.7.)

TIPS: Formatting Letters

- Letters should have a minimum of a one-inch margin on all sides, and typically, a brief letter should not be perfectly centered from top to bottom on the page; it should be slightly higher on the page.

- Spacing between the different parts of the letter depends on the kind of letter you're writing (the spacing between various parts in professional correspondence is more standardized than it is in personal correspondence). Paragraphs in letters are usually single spaced with a double space between paragraphs.

- Don't use too many different fonts. In fact, unless your letter is very long and you have provided descriptive headings for various sections, the entire letter should use one typeface. If you have provided headings, they may be in a second font that complements the body text. Use conservative-looking fonts that do not distract the reader from your message. (See "Typography," page 113.)

- Styles are italics, bold, and underline and should be used only to provide emphasis for a word, phrase, or short passage. If every sentence contains styled text, it loses its effectiveness, so don't use too many styles in your letter.

- For the text of the letter, don't use a point size smaller than 8 or larger than 14.

- Keep your paragraphs short. Most sources advise you not to exceed ten typed lines in any one paragraph.

- Use good stationery for your correspondence. See Part Five, "Production," for a discussion of kinds of paper.

- Don't use loud colors for the paper. For example, bright pink and yellow are inappropriate in all but personal correspondence. Business paper is usually white, cream, or gray.

- For color printing, be careful about colors for your text and graphics. See Part Five, "Production" for tips on color in print documents and Part Three, "Design and Illustration," for tips on using graphics.

- If you are using letterhead stationery and your letter is longer than one page, your second and subsequent sheets should be on matching, plain (for example, nonletterhead) paper. Provide the proper information at the top of the second and subsequent pages: the name of the person you are writing to with courtesy title, the page number, and the date.

Complaint Letters

Unfortunately, at some time, and probably not just once, you will find yourself having to write a letter complaining about a product, service, or incident. Perhaps the product you purchased was overpriced or is faulty, perhaps you received substandard service, or perhaps a store clerk treated you rudely. Often the only way to get satisfactory action is to write a complaint letter. (See Figure 8.8 on page 336.) You may be angry about the product, service, or incident, but you'll have

1201 Winterwood Boulevard
Oakland, CA 94601
June 25, 1999

Ms. Phyllis A. Dow
Bay Professional Consultants
934 N. Orange Avenue
San Francisco, CA 94103

Dear Ms. Dow:

Over the past few years, you have provided valuable advice to my wife and me concerning our son, Jeff. We appreciate the attention and time you have devoted to our situation. It isn't often that we encounter such professionalism. However, I am writing concerning a billing issue that I hope can be resolved soon.

Recently, we experienced a misunderstanding with your billing office. After our last appointment with you, we stopped by the front desk, where we learned we had a balance due in the amount of $55.00. The staff member did not know what the charge was for and said she would find out on the next day. Because we were in a hurry to pick up our son, I asked my wife to pay the amount, feeling we could sort out the details later.

The next day we found out that the $55.00 is a charge for our supposedly missing an appointment on May 29 at 2:00 p.m. This news was a big surprise to my wife and me. I checked my detailed phone log and verified that I had called your office on the morning of May 27 to cancel the appointment and reschedule it for a later date. In addition to the entry in my log, I specifically recall the telephone conversation.

In sum, my wife and I ask that you and your office staff void the $55.00 charge. It would be unfair to charge us $55.00 when we canceled our May 29 appointment more than 48 hours in advance. I am particularly conscientious about keeping all appointments I make, and if I cannot keep an appointment, I always cancel well before the appointment.

If you would like to discuss this matter with me, please call me at 555-2627.

Considering the amount of time my wife and I have spent with you, this particular incident is relatively minor. We hope it will be resolved amicably, and we look forward to future meetings with you.

Thank you.

Sincerely,

Bob Smith

Bob Smith

FIGURE 8.8
Complaint Letter

to learn how to set some of that anger aside if you want your letter to achieve positive results. You can write an angry letter simply to vent your emotions, but don't expect to receive satisfaction if you mail it. Sometimes you can achieve a satisfactory result by writing the letter and putting it aside. Come back to it in a few days. You may find that you overreacted or that you can express your displeasure more effectively now that you've had some time to cool down.

The purpose of a letter of complaint is to obtain some kind of action concerning the product you purchased, the service you received, or the incident in which you were ill treated. Your purpose may be to obtain a refund, an exchange, or a better product for the same price as the defective product that you purchased. Your purpose may be to have the service provided again for no cost or to have better service provided the next time. Or your purpose may be to obtain an apology for the way you were treated or to have someone reprimanded for the way he or she treated you. These purposes and many others are possible reasons for writing a letter of complaint. You'll need to control carefully the organization, content, style, and tone of your letter to achieve your letter's purpose.

In the opening of the letter, establish a common ground with your correspondent. Identify your relationship with the recipient or company. Tell your reader why you have purchased products or had services provided by this company or organization before. Your goal here is to establish a courteous tone and to build some rapport. In the last sentence or two of the opening, make your purpose for writing clear by simply stating you are writing because of a recent problem or incident that you hope can be resolved.

In the middle of the letter, discuss the reasons for your complaint. Be specific about product names, serial numbers, invoices, dates, and anything else that might help to identify the transaction and persuade your reader to your point of view. Avoid negative phrases such as "your mistake" or "your fault." Avoid insults. Maintain a courteous tone throughout, and you'll more likely achieve the results you seek.

Toward the end of the middle of your letter, state your specific request for action. Request the refund, exchange, apology, better service, or whatever else you wish to receive. If necessary and appropriate, offer a compromise.

In the closing of the letter, look to the future (stating that you would like to continue as a loyal customer, for example), and provide a courteous close.

Adjustment Letters

Adjustment letters are responses to letters of complaint from customers or clients. (See Figure 8.9 on page 338.) In these letters you are providing some kind of adjustment or compromise in response to what was requested in the letter of complaint. Adjustment letters are good opportunities to strengthen relations between the organi-

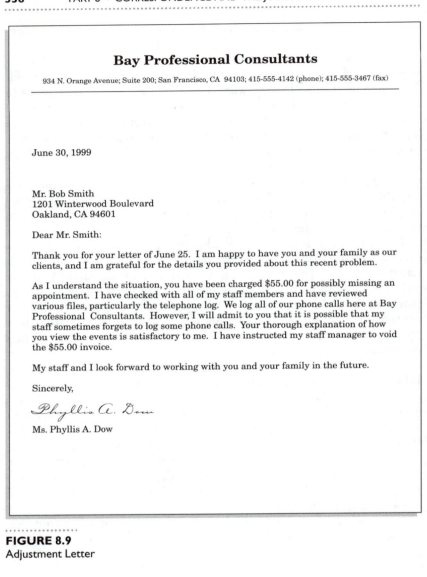

Bay Professional Consultants

934 N. Orange Avenue; Suite 200; San Francisco, CA 94103; 415-555-4142 (phone); 415-555-3467 (fax)

June 30, 1999

Mr. Bob Smith
1201 Winterwood Boulevard
Oakland, CA 94601

Dear Mr. Smith:

Thank you for your letter of June 25. I am happy to have you and your family as our clients, and I am grateful for the details you provided about this recent problem.

As I understand the situation, you have been charged $55.00 for possibly missing an appointment. I have checked with all of my staff members and have reviewed various files, particularly the telephone log. We log all of our phone calls here at Bay Professional Consultants. However, I will admit to you that it is possible that my staff sometimes forgets to log some phone calls. Your thorough explanation of how you view the events is satisfactory to me. I have instructed my staff manager to void the $55.00 invoice.

My staff and I look forward to working with you and your family in the future.

Sincerely,

Phyllis A. Dow

Ms. Phyllis A. Dow

FIGURE 8.9
Adjustment Letter

zation or company and the customer. They can be difficult to write because you may not be able to provide what your reader is requesting, but you will still want to maintain or build goodwill.

In the opening of the letter, acknowledge that you received the reader's letter and state the date of the letter you are responding to.

Thank the reader for taking the time to write to the company, and tell the reader how much his or her business means to the company. In these ways and others you are establishing a courteous and professional tone. If you are granting exactly what the writer of the complaint letter requested, then state in the opening paragraph that you are enclosing a refund, offering a discount for the next visit, or whatever else you are providing an adjustment for. If your response is positive, then use the direct approach or "good news" approach for your reader.

In the middle of the letter, summarize your understanding of the situation as it was described by your reader. This summary shows that you have read the letter closely and that you are paying careful attention to the reader's point of view. After you have finished summarizing the reader's view, offer the company's position on this particular complaint. State whether or not it is company policy to offer an exchange or refund or whatever. If you are free to offer the reader what was requested, then you may state at this point in the letter that you are offering what was requested (if you haven't already done so in the opening). If you must offer a compromise, then state what the compromise is.

Maintain a respectful, professional tone throughout. You aren't representing yourself; you are representing the company, and your goal should be to maintain goodwill as much as possible.

In the closing of the letter, mention that you are looking forward to a continuing relationship with your reader in the future, and provide a courteous ending.

Refusal Letters

In a refusal letter you are declining to offer the writer what was requested in a letter of complaint to your company or organization. (See Figure 8.10 on page 340.) This kind of letter will be more effective if you prepare your reader for your refusal. Open the letter by acknowledging that you received the letter sent to you ("Thank you for your letter of January 23 . . .") and by establishing a common ground. This common ground allows you to provide a buffer for the bad news that you will provide a little later in the letter. It also helps you to establish a professional tone and to maintain a "you attitude" in your letter. Maintaining a "you attitude" (also called a "you" approach) means focusing on the reader and the reader's interests as much as possible. You should use the word "you" frequently instead of "I," "we," and "our." And for the "you attitude" to be successful, your efforts to appeal to the reader must be honest and sincere rather than contrived.

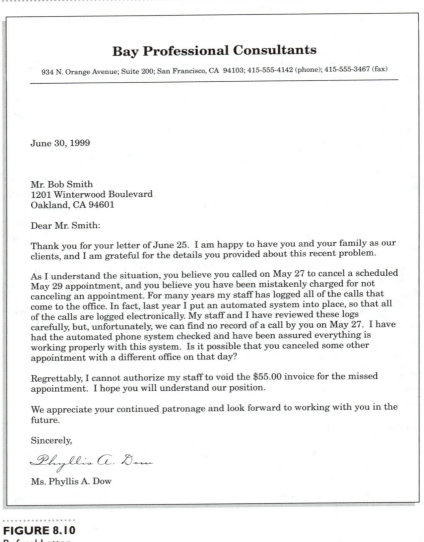

Bay Professional Consultants

934 N. Orange Avenue; Suite 200; San Francisco, CA 94103; 415-555-4142 (phone); 415-555-3467 (fax)

June 30, 1999

Mr. Bob Smith
1201 Winterwood Boulevard
Oakland, CA 94601

Dear Mr. Smith:

Thank you for your letter of June 25. I am happy to have you and your family as our clients, and I am grateful for the details you provided about this recent problem.

As I understand the situation, you believe you called on May 27 to cancel a scheduled May 29 appointment, and you believe you have been mistakenly charged for not canceling an appointment. For many years my staff has logged all of the calls that come to the office. In fact, last year I put an automated system into place, so that all of the calls are logged electronically. My staff and I have reviewed these logs carefully, but, unfortunately, we can find no record of a call by you on May 27. I have had the automated phone system checked and have been assured everything is working properly with this system. Is it possible that you canceled some other appointment with a different office on that day?

Regrettably, I cannot authorize my staff to void the $55.00 invoice for the missed appointment. I hope you will understand our position.

We appreciate your continued patronage and look forward to working with you in the future.

Sincerely,

Phyllis A. Dow

Ms. Phyllis A. Dow

FIGURE 8.10
Refusal Letter

In the middle of your letter, focus on the reasons you must decline the request. Then, after you state these reasons, politely state that you regretfully must decline the request. Offer convincing reasons, and be fair to the reader.

In the closing of the letter, look to the future. Remember that you still want your reader to be a customer. Provide a courteous close.

1404 Victory Garden Lane
New York, NY 10019
October 11, 1999

Mr. Bruce Keyes
Vice President of Operations
Orange Sky Software Company
Suite 3000
1803 Dumont Circle
Anaheim, CA 92801

Dear Mr. Keyes:

I am a graduate student majoring in technical communication. I am working on a
research project concerning different online help programs, and I am writing to ask
you some questions concerning your online help program. If you can take 20 minutes
of your time to answer the attached list of questions and return your answers to me
in the enclosed envelope, I will gladly share the results of my comparative study with
you when it's finished.

To save you time, I have provided a space of several inches below each question. If
you don't mind, just jot down your answers below each question. Also, if it's not too
inconvenient, I will need to receive your response by November 1.

I can assure you that your answers to these questions will be quite helpful for my
study. I look forward to hearing from you.

Sincerely,

Jack Baxter

Jack Baxter

FIGURE 8.11
Inquiry Letter

Inquiry Letters

Often you have to make a formal request to a person, a company, or
an agency to obtain some information. Letters of inquiry are necessary when you need to find information about a person, a product, or
a service, for example, and you haven't been able to find the information through some other avenue. (See Figure 8.11.) Knowing how

to write this kind of letter can be helpful in many kinds of research situations. If no action on the part of the person or company has prompted your letter, then it is considered an unsolicited letter of inquiry. If you are responding to a product ad suggesting that you write for more information, then your letter is a solicited letter of inquiry. The two types of letter require similar and yet slightly different strategies.

In the opening of a solicited or unsolicited letter, state who you are and why you are writing. If you are writing a solicited letter, state how you heard about the opportunity to request information (through a print, television, newspaper, or magazine ad, for example). If you are writing an unsolicited letter, take a more careful approach in the opening, explaining briefly who you are and why you need the information. Use a serious tone and a formal style throughout either letter. Consider offering your reader some kind of compensation for the time it will take to fulfill your request, whether by providing a copy of your research results or by paying for any expenses in addition to postage.

In the middle of either letter, state what kind of information you need. Be direct and clear about the information you need, and provide a deadline so that the person you're requesting information from will know when to respond. If you need to ask a list of questions, enclose the list on a separate sheet and simply use the letter to introduce yourself, state why you need the questions answered, and provide a deadline. Make it easy for the person to respond to your questions, for example, by providing plenty of space below each question so that he or she may simply return your list to you.

Provide a courteous close. In a solicited letter, simply thank the correspondent for sending the information. In an unsolicited letter don't thank the recipient in advance for sending the information, but apologize for any inconvenience your request may cause. Include a self-addressed stamped envelope to make it easier for the person to whom you are writing to respond.

Transmittal Letters

A transmittal letter is a cover letter that is attached to, but is not part of, a proposal, report, manual, or some other document you are sending to someone. (See Figure 8.12.) Transmittal letters are essential for providing a record of how, when, and to whom documents are distributed in an organization or company. Your transmittal letter

June 2, 1999

Dr. Frieda Holley
Interim Associate Vice President
  of Academic Affairs
The Metropolitan State College of Denver
Campus Box 48
P.O. Box 173362
Denver, Colorado 80217-3362

Dear Dr. Holley:

Enclosed is a copy of our *Consultants' Report on the Technical Communications Department at Metropolitan State College of Denver.* Jean Otte and I decided to collaborate on this report as the best way to present our findings to you. We have also sent a copy of the report to Dr. Jane Broida. If you have any questions about the report, please contact me or Jean.

Thank you also for meeting with both of us and answering our questions. As our report shows, you should be proud of the faculty and students of the Technical Communications Department.

Sincerely,

Dan Jones

Dr. Dan Jones
Associate Professor
Technical Communication

Enclosure: Report

FIGURE 8.12
Transmittal Letter

identifies what is being sent, may provide an overview of the contents, identifies to whom the material is being sent, and provides a courteous close. (Also see "Letter of Transmittal," on page 408.)

MEMOS

Even though corporate e-mail has considerably reduced the paper flow at many companies and organizations, many people still write and send memos on all kinds of topics. You will spend a considerable amount of time on the job reading and responding to memos. Unlike letters, which are chiefly intended to be read by people outside of the company, memos are a form of internal correspondence for employees. In addition to the regular contexts in which you will find memos, the memo format is a useful one to use when you need to create a fax cover sheet. The format of memos varies considerably from the format of letters. Formats of memos may also vary from individual to individual within a company or organization.

Heading

The heading for a memo often contains the word "Memo" or "Memorandum" at the top of the memo, either aligned flush left or centered. A fax cover sheet will have "Facsimile" or "Fax Cover Sheet" in place of "Memo" or "Memorandum." Many people omit the word "Memo," feeling that the other heading information makes it clear that the form is a memo. Notice that you don't provide the kind of heading information you provide in a letter (although many companies and organizations have preformatted memos that contain the company name, address, and other information). Many word-processing programs can provide you with various designs for your memos. (See Figure 8.13 for a typical memo.)

Below the word "Memo" or "Memorandum," provide the date line, consisting of the word "Date" followed by a colon. The date for a memo is written in the same way it is written for a letter: December 1, 1999. In place of an inside address, you have an addressee line, labeled "To." You should provide the addressee's full name and, if appropriate or necessary, the person's complete title and address, as you would in the inside address of a letter. If several recipients are listed in the addressee line, each person's copy should have some indication by the name showing to whom that copy is addressed. The indication can be as simple as a check mark placed next to the name or colored highlighting over it. The sender line of a memo takes the place of a signature line used in a letter. After the word "From," provide your full name and your job title if necessary. Traditionally, you handwrite your initials by your name on this line instead of providing a signature block and signature at the bottom of the memo. For some memos you may also use an attention line. An attention line

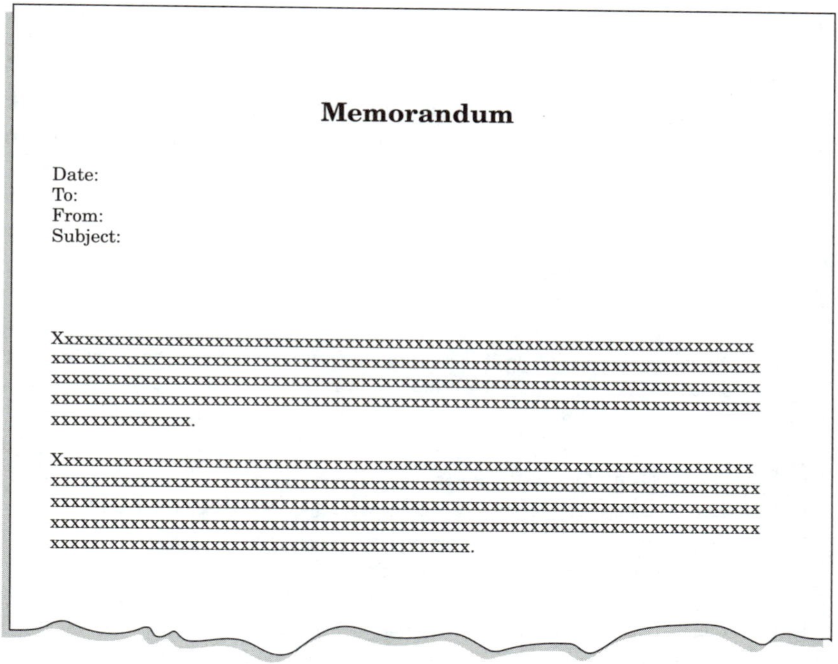
Memorandum

Date:
To:
From:
Subject:

Xxxxxxxxxxxxxxxxxxxxxxxxxxxxxxxxxxxxxxxxxxxxxxxxxxxxxxxxxxxxxxxxx
xxxxxxxxxxxxxxxxxxxxxxxxxxxxxxxxxxxxxxxxxxxxxxxxxxxxxxxxxxxxxxxxxxxx
xxxxxxxxxxxxxxxxxxxxxxxxxxxxxxxxxxxxxxxxxxxxxxxxxxxxxxxxxxxxxxxxxxxx
xxxxxxxxxxxxxxxxxxxxxxxxxxxxxxxxxxxxxxxxxxxxxxxxxxxxxxxxxxxxxxxxxxxx
xxxxxxxxxxxxx.

Xxxxxxxxxxxxxxxxxxxxxxxxxxxxxxxxxxxxxxxxxxxxxxxxxxxxxxxxxxxxxxxxx
xxxxxxxxxxxxxxxxxxxxxxxxxxxxxxxxxxxxxxxxxxxxxxxxxxxxxxxxxxxxxxxxxxxx
xxxxxxxxxxxxxxxxxxxxxxxxxxxxxxxxxxxxxxxxxxxxxxxxxxxxxxxxxxxxxxxxxxxx
xxxxxxxxxxxxxxxxxxxxxxxxxxxxxxxxxxxxxxxxxxxxxxxxxxxxxxxxxxxxxxxxxxxx
xxxxxxxxxxxxxxxxxxxxxxxxxxxxxxxxxxxxxxx.

FIGURE 8.13
Memorandum

may be used in a memo for the same reason it is used in a letter: to ensure that the letter will be read even if the addressee is unavailable. The subject line is one of the most important lines in a memo. You should provide in a few key words the contents of the memo. The subject line is typically the last line to appear before the body of the memo.

Body

The body of a memo is the opening, middle, and closing paragraphs. How you open, develop the middle, and close depends on your purpose for writing. However, even a brief memo has a definite structure.

Some memos also use a signature line, an identification line, enclosure notation, mail notation, and copy notation. Most memos do not use a signature line, but sometimes a company style allows a space for signing memos. Any memo may be initialed, either about two line spaces below the body (a little to the right of the center of the page) or near the printed name after the word "From" near the top of the memo. An identification line in a memo may provide the same information it provides in a letter: identification of who wrote, dictated, signed, sent, transcribed, or typed the memo. Enclosure notations and copy notations in memos provide the same information they provide in letters.

TIPS: Writing Memos

- Keep your memo as brief as possible. Of course, some memos may be many pages long depending on the subject, purpose, scope, audience, and other factors.
- State your purpose in the opening brief paragraph. Don't waste your reader's time. Get to the purpose quickly.
- Keep the paragraphs brief. Paragraphs should not exceed ten typed lines.
- In a longer memo, use headings and subheadings if necessary to help your reader recognize the organization of your memo.
- Use a forecasting statement, if appropriate, to tell your reader what you will cover in the memo. For example, write, "This memo addresses the key issues discussed in our last meeting, provides some new alternatives, and recommends our next course of action."
- If the memo has a second or third page, indicate the recipient at the top of those pages in a continuation page header, as you would do in a letter. If you are sure the pages will remain stapled together, you may omit this header.
- Use a prose style and tone that are appropriate for your subject, audience, and purpose.
- Don't use too many different fonts or styles (italics, underlining, bold).
- Avoid using point sizes for your fonts that are smaller than 8 points or point sizes that are too large.
- Avoid fonts that are too unusual. (See "Typography," page 113.)

E-MAIL

Electronic mail (also known as e-mail, email, or Email) has been available for several decades but has become enormously popular within only the past five years or so. E-mail enables users to send and receive messages electronically. Messages are sent and received

in seconds or minutes—much more quickly than is possible by surface mail ("snail mail"). See Part Six, "The Internet," for more information about using e-mail.

Advantages of E-Mail

E-mail has many advantages over other kinds of correspondence.

- E-mail is easy to learn and easy to use. There are many kinds of
  e-mail programs available (see "Mail Programs," page 251).
- E-mail is convenient. You can just as easily send a message to
  one person or the same message to thousands.
- E-mail is inexpensive. Most people pay for an Internet account of
  one kind or another whether they use America Online, Prodigy,
  CompuServe, Microsoft Network, AT&T, BellSouth, or smaller
  and local commercial providers located in their region or city.
  However, the monthly cost for e-mail and Web access is minimal.
  Most Internet providers offer unlimited access time for a flat fee.
  Many college students benefit from having free accounts
  through their universities and colleges.
- E-mail is efficient. It's possible to handle numerous messages
  in very little time. With just a few keystrokes, you can also manage your correspondence by placing messages in electronic
  mailboxes and by updating your e-mail address book.
- E-mail is quick. Much e-mail is delivered within a few minutes
  or less. Rarely will e-mail require more than a day to arrive.
- E-mail can be useful for recording progress on projects (since
  messages can serve as a log of communications), for getting
  identical information to a large group of people, and for automatically receiving information that may be of interest to you.
- E-mail is good for the environment. In many places, e-mail has
  replaced the massive use of paper to handle the day-to-day
  need to communicate via memo within a company or organization or from one company to another.

Disadvantages of E-Mail

E-mail has some important disadvantages, too, and you should consider these carefully for every e-mail message you send.

- E-mail is so convenient that you may neglect to use print correspondence (a letter, memorandum, or handwritten note) when
  print correspondence may be more appropriate.

■ Tone in e-mail is too difficult to convey. It's easy for a reader to misinterpret a humorous or sarcastic message, for example. To help you convey the appropriate tone, you should become familiar with and use a variety of smileys or emoticons. (See "Emoticons and Abbreviations," page 352.)

■ E-mail can be easily lost. Sometimes a service provider's main computer will go down, not only making it impossible for you to send e-mail until repairs are done, but also sometimes losing e-mail that has been sent to you. Once e-mail is lost, usually it is lost forever. Worse, you will probably not know that your message was not delivered or that you were supposed to receive a message that never arrived.

■ If you don't check your e-mail frequently, your unread messages may create a backlog on your service provider's system. Many e-mail providers set a limit on how much space messages may take up before new messages are refused. Once that limit is reached, new messages will be turned away.

■ Some providers charge a fee for each message you receive. If you are on several high-volume mailing lists, your costs could quickly get out of hand.

■ Service providers with many subscribers may have insufficient resources to let all those who want to use their accounts be online at the same time. If you absolutely must have access to your e-mail account, you may have to change to a provider that can accommodate the traffic.

■ Not everyone has an e-mail address, and it can sometimes be very difficult to find an e-mail address for someone if you don't already have it. Often you can look up e-mail addresses in an online database such as Four11, but if you select the wrong person's address, your e-mail will not reach its intended target. Unfortunately, sometimes your message will go astray. Mistype one letter or number in the recipient's e-mail address, and that person will not receive it. If the mistyping results in a valid address, someone else will receive your message. If your message is not delivered at all, you may or may not receive notification of nondelivery.

■ If your computer is invaded by a computer virus or has some other kind of problem, you risk permanently losing all of your unanswered or saved e-mail and your address book. Ideally, you should keep a backup copy of your e-mail addresses on a backup diskette.

TIPS: Sending E-Mail and Attachments

How you send your e-mail, send attachments, or read the e-mail you receive varies from one e-mail program to another. You need to become familiar with the essential features of the e-mail program you prefer. Regardless of which one you use, there are a few conventions or guidelines you should follow.

- Make certain you have specified the proper e-mail address for the person to whom you are sending your message. Many e-mail programs allow you to insert a person's e-mail address automatically, but you still need to manage your e-mail address book and keep your addresses current.

- Make certain your name and address also appear in your e-mail address. Most e-mail programs insert this information automatically.

- Make certain you provide a clear subject in the subject line of your e-mail message. If you are responding to someone's message, most e-mail programs insert the original subject into the subject line of your e-mail message. Of course, you are free to change this subject line if circumstances warrant the change. If you are composing a new message, be certain to provide a subject in the subject line. E-mail programs will not insert one for you, although many will query you if you forget to provide one.

- Provide a signature for your e-mail when appropriate. A signature may include your name, job title if appropriate, address if you want to provide one, phone numbers, and so on. Or a signature may simply be a favorite quotation. Many e-mail programs automatically insert a signature into the bottom of your e-mail message, and many e-mail programs allow you to have many kinds of signatures. You should use a formal signature for formal e-mail correspondence and a personal signature for informal correspondence to friends and family. You also have the option of providing no signature, and for much e-mail this may be appropriate as well.

- Determine whether you are sending formal or informal e-mail. If you are sending formal e-mail, then you will need to use a salutation ("Dear Professor Jones,"), use a more formal style and tone, and provide a more formal signature. If you know your correspondent, you may simply write "Dear Dan" or "Hello" or provide no salutation at all.

- Make your e-mail easily readable. Try to limit the width of your messages to four inches or less so that your reader can skim your message easily. Keep the prose style relatively simple, and avoid overly long paragraphs. (See Part Two, "Writing.")

(continued)

Sending E-Mail and Attachments *(continued)*

■ Respond to messages appropriately. If you are responding to a long e-mail message, include just the appropriate part or parts instead of the entire message. There is no need to include the entire original message. This point is especially important when you are responding to e-mail messages on Usenet. (See the tips list, "Posting to Newsgroups," page 254.) Traditionally, you should provide the original e-mail message at the top and then your response below the message you are responding to. This approach provides your readers with a context for your message. They will immediately see what original message you are responding to. Most e-mail programs automatically preface this original message with something like "On xx/xx/xx [often including time] you wrote:" or simply "Jane Doe wrote:" This is followed by the original message and then your response. Your response will be followed by your signature.

■ Keep your response short whenever possible. An e-mail message is comparable to a memorandum, and while it's true that a memorandum can be several pages long or longer, most memos are brief and to the point. Follow this practice and keep your e-mail message brief and to the point as well.

■ Be cautious when using colors, fonts, styles, and other design features in some e-mail programs. Increasingly, e-mail programs enable you to use different colors for your text, different fonts and different size fonts, styles such as italics and bold, and even different colors for your background. However, your readers will see none of these features if they do not have an e-mail program that can detect these features. Even if they do, be careful not to mix too many of these elements together. If your e-mail program doesn't have these features, you may put words between asterisks (*) to indicate emphasis or use underscore characters (_) before and after a word or phrase you would like underlined. Some e-mail programs, such as Eudora, handle replies containing styled text differently from text without colors, styles, and the like. It can be quite annoying to reply to a styled e-mail when the program insists on providing an unusual treatment for the cited text. Therefore, use styles in e-mail sparingly.

■ Plan ahead when providing attachments. Make sure you know whether your correspondent can receive attachments. Many e-mail programs make it possible to attach text, graphics, Web pages, audio, and video files. Some do not. For example, if you wanted to attach a recent photograph to an e-mail you are sending to your parents or friends, they must also be using an e-mail program that allows them to view the attachments you send.

■ Be considerate when sending attachments. Consider the size of the attachment. Files smaller than 100KB are generally acceptable. You may want to warn your correspondent with a separate e-mail message if you are sending any attachment that is longer. A 2-MB high-resolution graphic, for example, could take a long time to download and could cause the recipient's Internet mailbox to overflow, preventing further e-mail from being delivered.

■ Be vigilant when sending attachments. While many so-called computer viruses (such as the Good Times Virus) are bogus, others are very real and can be transmitted in e-mail attachments. Before you attach a document to your e-mail message, make sure you have checked it for viruses. The people you're sending it to may not have a virus checker installed on their computers.

■ Proofread your message carefully before sending it. If your e-mail program has a spelling checker, use it.

TIPS: Observing E-Mail Netiquette

In addition to following basic e-mail conventions, you should observe proper e-mail etiquette, often referred to as *netiquette*. These are guidelines for proper behavior concerning you and the e-mail you send.

■ Be aware that what you send in an e-mail message may become public knowledge. Generally, you should assume that your worst enemy is reading everything you post—or at least could if he or she wanted to. Many employers do read their employees' e-mail and have a legal right to do so. So never assume that your e-mail is a private correspondence between you and your reader. All e-mail is saved routinely in system backups by service providers, and these backup copies are not available to you for deletion.

■ Don't send abusive messages. The practice of sending such messages, called flaming, is unfortunately a common practice in many Internet newsgroups and mailing lists. There are even newsgroups devoted to flaming. Basically, you should not write in an e-mail message anything that you would not tell your correspondent in face-to-face conversation. Be considerate of your reader's feelings and how your reader may interpret and react to your e-mail message.

■ Be aware that some people see e-mail as an informal and conversational medium while others see many e-mail communications as formal. Not everyone shares the same opinion of this relatively new medium. Ideas that you may think are to be taken lightly or casually may be taken seriously by others.

(continued)

Observing E-Mail Netiquette *(continued)*

- Use sarcasm cautiously if at all. For example, if you use the expression "Yeah, right," the reader may not know if you are in total agreement or being sarcastic. Sarcasm is a special tone that must be handled carefully in any medium. If you are determined to be sarcastic in your e-mail, make sure you provide enough context and perhaps even use an appropriate emoticon (see below), to help your reader appreciate (or at least recognize) your sarcasm. Look over your e-mail before sending it, in case your wording could be mistakenly interpreted by someone else as being sarcastic.

- Don't shout. Shouting in e-mail is indicated by using ALL CAPS! Some people use all caps because they have a preference for using all caps, and they are totally unaware that others will view their messages as shouting at the reader. All caps are also more difficult to read than mixed case. Avoid shouting in your e-mail messages and use mixed case. The same conventions that are used by writers of printed messages apply to e-mail. Just because you are creating an electronic document, don't imagine that your reader will see it with electronic eyes. Capitalize the beginnings of sentences and the personal pronoun I. Don't use all lowercase; doing so makes it difficult to tell when sentences end or begin.

- Be careful about using an angry tone. If you absolutely must write an angry e-mail message, consider waiting a day before you send it. If you send the message immediately, you may well regret doing so.

- Limit your signature file to a few brief lines. Many newsgroups and mailing lists have rules about the length of signatures. Your readers shouldn't have to wade through fifteen lines of information about you or your favorite joke just to get to the end of your message.

Emoticons and Abbreviations

People on the Internet have become quite adept at inserting emotion and self-expression into computer messages by using typographic characters and other tricks of the keyboard. Figure 8.14 shows a few ways in which you can spice up your messages and let your personality shine through.

Also, many common expressions are abbreviated in e-mail to save time. (See Figure 8.15.)

JOB CORRESPONDENCE

Some people think job correspondence consists of just an application letter and a résumé. However, there are many kinds of letters that should be or may be written during your job search. You'll need to re-

:-)	Smiley	|-)	Sleeping smiley
:)	Smiley without nose	:-o	Surprised smiley
(-:	Left-handed smiley	:-0	Very surprised smiley
;-)	Winking smiley	:-P	Disgusted smiley
8-)	Smiley with glasses	:-D	Very pleased smiley
8-)>	Smiley with glasses and beard	:-X	Keeping lips sealed
		:Q	What?
<:-)	Smiley with dunce's cap	:-}	Ditto
:-|	Serious smiley	:-@	Screaming
:-I	Indifferent smiley	:O	Yelling
:-/	Skeptical	<g>	Grin
:-7	Wry	<l>	Laugh
:->	Sarcastic	<jk>	Just kidding
:-(	Sad smiley	<I>	Irony
:-O	Uh oh!	< >	No comment
;-(	Crying smiley		

FIGURE 8.14
Some Common Smileys or Emoticons

AFAIK	as far as I know	LOL	laughing out loud
BRB	be right back	OBTW	oh, by the way
BTW	by the way	ROTFL	rolling on the floor laughing
CUL	see you later		
FYA	for your amusement	ROTFLMHO	rolling on the floor laughing my head off
FYI	for your information		
HTH	hope this helps	RTFM	read the "fine" manual
IIRC	if I remember correctly		
		TIA	thanks in advance
IMHO	in my humble opinion	TNX	thanks
IMNSHO	in my not-so-humble opinion	YMMV	your mileage may vary

FIGURE 8.15
Common Abbreviations Used in E-Mail

view the "Letters" section (pages 319–343) to understand the basics of writing a good letter. And, of course, it helps to know about direct and indirect orders for letters, another topic that is covered in that section. Understanding tone is also essential for writing effective letters. You'll find a discussion of tone in Part Two, "Writing."

1400 Green Badger Lane
Orlando, FL 32817

December 15, 1999

Ms. Michelle Harris
Director of Personnel
Dynamic Healthcare Services
Orlando, FL 32816

Dear Ms. Harris:

I am applying for the opening you advertised in *The Orlando Sentinel* for an entry-level technical writer. As my enclosed résumé indicates, I am a senior at the University of Central Florida and will graduate in May with a Bachelor of Arts degree in English (concentration: technical writing) and a minor in Biology. As a writer with a strong interest in science, I believe I am strongly suited to work at Dynamic Healthcare Services.

Throughout my college career, I have held a variety of positions that have helped me to fund my college education. Most recently, I have worked for the Office of Sponsored Research and Graduate Studies as an Administrative Assistant. Through my various assignments at OSR, I have gained broad office experience that is essential to a technical writer at your company.

In addition to my part-time work, I have been a full-time student at UCF for the past two years and have achieved a 3.7 GPA in my major. To prepare myself for employment in the technical writing field, I have taken classes that teach advanced writing concepts such as Online Documentation, Techniques of Technical Documentation, Technical Report Writing, Graphics for Technical Writers and Technical Writing Style. I am also pursuing a minor in Biology, which qualifies me further for the position because of my knowledge of biological principles. Since Dynamic Healthcare Services develops clinical information systems for use in laboratory, radiology, anatomic pathology, and anesthesiology applications, I am confident that my technical knowledge would benefit the company.

I have many computer skills and I am able to learn new applications rapidly. I frequently use desktop publishing tools such as Word, Excel, and Adobe PageMaker. I am currently taking a course at UCF named Introduction to Computer Science in order to familiarize myself with general computing principles and applications, operating systems, and networks.

In closing, I want to stress that I have a strong desire to excel. I am very attentive to detail and maintain high standards of quality for my work. I am personable, diplomatic and I enjoy collaborating with other writers. I realize that it is difficult for my résumé and application letter to convey the full breadth of my qualifications and skills. Would it be possible for me to meet with you to discuss what I have to contribute to your company? If so, please call me at (407) 555-7092 or e-mail me at npeters@pegasus.cc.ucf.edu.

Sincerely,

Nichola J. Peters
Nichola J. Peters

FIGURE 8.16
Application Letter

Application Letters

The chief aim of an application or job letter is to help you get an interview. Keeping this aim in mind will help you to focus on what to cover in your job letter. Unlike your résumé, which is a factual listing of your achievements, your job letter is intended to persuade the reader not only that you are a *qualified* applicant for the position, but also that you are the *best* applicant for the position.

Some people have difficulty writing a persuasive cover letter, but that's chiefly due to a lack of extensive work experience. Once you have graduated and you gain more work experience, writing the job letter often becomes easier. (See Figure 8.16 on page 354 for an example of an application letter.)

TIPS: **Writing a Good Job Letter**

- Follow all of the principles of good business letter correspondence. Provide a minimum of one-inch margins on all sides, use professional-quality stationery, avoid unusually bright paper colors (such as green, yellow, pink), use a high-resolution inkjet or laser printer, include all of the major parts of a letter (heading, date line, inside address, salutation, body, complimentary close, and signature block), and don't forget to sign your letter.

- Research the company. Your application letters will be more effective if you can demonstrate in appropriate places in the letter some knowledge of the company, its reputation, its products, and its plans for future growth. You can gather this kind of information in many different ways. Call the company and ask to have an annual report sent to you. Use a search engine to see whether the company has a Web site. Use standard sources in the library, such as the *Moody's* periodicals and *Standard and Poor's* publications, to find essential information on the company, its employees, and its products. Ask your professors and friends whether they know anything about the company that may be useful to you.

- Write to a specific person. Many job ads merely list the title Human Resources Director or Personnel Director. Some provide only the address. Call the company to find out the name of the Human Resources Director or Personnel Director. If you cannot obtain a name, consider obtaining the name of the supervisor or director of the division in the company where you hope to work and addressing your letter to this person. This person can forward your letter to Human Resources or Personnel if what you have written is of interest.

(continued)

Writing a Good Job Letter *(continued)*

■ Follow effective strategies for the opening of the letter. Identify how you learned about the position. For example, state whether someone told you about the job, you saw an ad in the paper (state which paper), you saw the job mentioned on the Web, or you saw it listed in a trade magazine. Name-drop if possible. Perhaps a well-known member of your profession or an employee of the company told you about the opening. Mentioning this person's name will help to make your letter more noticeable than others. State why you are writing by saying something like "I am applying for the opening you advertised in the *Atlanta Constitution* for a senior-level technical writer." Finally, either in the opening paragraph of the letter or in the second paragraph, identify yourself. This strategy may be accomplished simply by writing "I am a senior at the University of _____ and will graduate in May with a degree in _____."

■ Follow effective strategies for the middle of the letter. Mention your enclosed résumé. Often this strategy can be combined with that of identifying yourself. For example, you may write, "As my enclosed résumé indicates, I am a senior at the University of _____ and will graduate in May with a degree in _____." Discuss your relevant work experience. If you have work experience that is pertinent to the position for which you are applying, emphasize the skills and achievements that are most applicable to the current opening. State confidently how you are qualified for the current opening because you have done similar work elsewhere. State how you believe you can make further contributions in these areas for this company. If you have work experience that is less relevant to the particular opening, stress the values of the work experience you have. If you have experience based on unpaid work, such as work for a school course, describe that experience. Remember that all of your full-time and part-time jobs have enhanced your skills in one way or another. If you have little or no work experience, devote more discussion to particular academic strengths.

■ After discussing your work experience, elaborate on your educational strengths, both major and minor, in the middle of the letter. This section may be part of the same paragraph discussing your work experience, or it may be a separate paragraph in the middle of your letter. For example, explain that your major in _____ required _____ semester hours of courses of the _____ semester hours required for a college degree. Discuss any coursework that is particularly germane to the job opening. Discuss any appropriate internships, any significant research work done for any of your professors, or any additional relevant coursework beyond the coursework for your major.

■ While emphasizing both your work and academic experience, mention the reasons for your interest in this particular company. For example, perhaps you have a special interest in one of the company's current or forthcoming products, or perhaps you know that the company has some exceptional groups of employees and you would like to become part of one of these groups. Use the "you attitude." Emphasize how you believe you can make contributions to the company if you are given the opportunity. Have your letter reflect your knowledge of the company and its products wherever possible. Convey an overall enthusiasm for the job opening throughout your letter.

- Follow effective strategies for ending the letter. State that you are available for an interview at the reader's convenience. Don't restrict your availability to certain times or dates or until after final exams are over. If you are invited to an interview, you will make adjustments to your schedule accordingly. Even though all necessary contact information should be on your résumé, provide your phone number or several phone numbers (especially if you will be moving from one address to another during your job search). Sometimes résumés are misplaced, and your cover letter may be all that is available for contacting you. Provide a courteous ending. Emphasize, for example, your interest in working for the company rather than stating something cliché such as "I look forward to hearing from you soon."

- Proofread your letter very carefully. It should represent you in the best way possible. You can't afford to send out an application letter that has typos, spelling errors, punctuation errors, or errors in grammar. There are too many other applicants who will send out professional-quality letters.

- Attach your résumé, and mail the letter and résumé in a properly addressed envelope.

- Follow up with a telephone call if you haven't heard from the company after two weeks. Ask whether your letter arrived and whether the intended recipient has had a chance to look it over. Request an interview if the person you are speaking to sounds receptive to that possibility.

Reference Letters

As a college student, you will often ask others to write a reference letter or a letter of recommendation for you. (See Figure 8.17 on page 358.) Perhaps you need such a letter for a job while you're in college or a full-time job now that you're about to graduate. Or perhaps you need it as a required item for admission into a graduate program. Sometimes you will see these letters and perhaps receive a copy; other times you may never see the letter you ask someone else to write. When you ask others to write reference letters for you, be

May 12, 1999

To Whom It May Concern:

Carolyn B. is applying for admission into the master's program in Women's Studies at Northern State University and asked me to write a letter of recommendation. I am delighted to do so.

I have known Carolyn for ten years. She was a student in three of my technical communication classes while completing a minor in this field and a major in Organizational Communication. The three classes she took with me are ENC 4293 Technical Documentation I, ENC 4295 Technical Documentation III, and LIT 4433 Survey of Technical and Scientific Literature.

I have taught many outstanding students during my 20 years of full-time college teaching, and Carolyn is easily one of the best students I have ever taught. She excelled in all three of my classes. I was so confident in her writing, editing, and leadership abilities that I asked her to serve as class editor for the *Style Guide and Standards* manual created in the Technical Documentation I class. Carolyn's many skills assured that this manual was a quality product and one that was finished on time despite many difficulties.

Carolyn also repeatedly demonstrated her keen intelligence and communication skills in my discussion class, Survey of Technical and Scientific Literature. The course not only examines the varying prose styles of many classics, but also challenges many underlying assumptions concerning science and technology. Carolyn welcomed the class debates and contributed significantly to many lively discussions.

Because I have kept in touch with Carolyn since she graduated in 1990, I have become acquainted with her views on many issues. I can assure you that she definitely has a strong commitment to representing women and women's issues in all kinds of professional circles. And she cerrtainly has the determination, intelligence, discipline, and communication skills necessary to succeed in even the most demanding graduate program. For all of these reasons, I highly recommend that you admit her into your program.

Sincerely,

Dan Jones

Dan Jones
Associate Professor
Technical Communication

FIGURE 8.17
Reference Letter

sure to ask people who you are confident will write strong letters for you. Also, it's helpful to know what kinds of points are typically emphasized in a letter of recommendation in case the person who is writing the letter asks you what should be emphasized.

As you acquire more experience in a company or organization, you may find that others are asking you to provide a letter of reference or recommendation. You'll have to consider, first of all, whether the person requesting the letter deserves a positive recommendation from you. If not, then you'll have to decide whether you should decline to write such a letter, or, if you agree, you'll have to consider the ethics of writing a negative or lukewarm letter for this person.

There is no one formula that will ensure the success of a reference letter, but there are some strategies that will make your letters more effective. Typically, you should state that you have been asked to provide the reference letter, tell the reader how long you have known the person and under what circumstances, discuss the specifics of this person's strengths and how you have observed these strengths, provide an overall assessment of the person, and provide a specific recommendation. Of course, these points need to be covered in specific areas of the letter.

In the opening, state that you have been asked to provide this letter or recommendation and, if it concerns a possible new job, for example, state that the letter relates to this person's qualifications for the job.

In the middle of the letter, discuss your specific observations of and experiences with the person you are recommending. If the person is or was a student of yours, list the classes the student took and his or her performance in these classes. If the person is a current or former employee, state what this person does or did and assess how well he or she performed on the job. In a separate middle paragraph, discuss the overall strengths of this person as they pertain to what he or she is applying for (for example, graduate school, a special program, or another job). Use descriptive adjectives. Finally, provide your overall recommendation stating, for example, something like "For all of these reasons, I highly recommend John for this graduate program."

In the closing, offer to provide further information if requested. Provide any necessary contact information if it is not already available elsewhere on the stationery you are using.

Thank-You Letters

During the job application process you write thank-you letters as a follow-up to an interview. (See Figure 8.18 on page 360.) You thank your interviewer for having taken the time to interview you. The proper time to send a thank-you letter is immediately after the interview.

In the opening, state simply that you are writing to thank the

2705 West Oak Lane
Worcester, MA 01603
August 19, 1999

Mr. Phil Rouse, Manager
Software Documentation Department
Whitcomb Technical Solutions
4703 Haskell Rd.
Boston, MA 02122

Dear Mr. Rouse:

I am writing to thank you, Mr. Steve Jenkins, and Mr. Gray Appleton for the time you took out of your busy schedules to interview me this past Monday.

Once again I must say how impressed I am with your staff, your company products, and your company procedures. I am familiar with most of the companies in the Boston area, and your company is easily the most impressive for what it is doing with its software products and documentation.

During our interview, I forgot to tell you that I was under consideration for election to the President's Leadership Council. I just received a letter today notifying me that I have been selected as a member of this council. Every year this council chooses only a handful of students as members, and I am delighted that I have been fortunate enough to be chosen this year.

I am confident I can contribute a great deal to your company, and I hope I have the opportunity to prove myself.

I look forward to hearing from you.

Sincerely,

Sarah Chase

Sarah Chase

....
FIGURE 8.18
Thank-You Letter

reader for the interview. If possible, try to mention the names of all of the people who were involved in your interview. You may address the letter to the principal interviewer and then, in the opening para- graph, extend your thanks to the others who took time to talk to you and to answer your questions. If you were impressed by the people, various projects or products, and so on, use the middle of this letter to say so. This kind of thank-you letter also provides you with the perfect opportunity to repeat how much you are interested in working for the company and to stress once again your strengths and qualifications for the position. You may want to emphasize these points in the closing of the letter in addition to providing a courteous ending.

Acknowledgment Letters

In job correspondence an acknowledgment letter acknowledges that you have received something. (See Figure 8.19.) You may write such a letter to someone who sent you a copy of a report, a product sample, or a letter confirming interview details. A company's human resources department may send you a letter acknowledging that your letter, résumé, and application for a position have been received.

1501 Northern Way
Lynn, MA 01902
August 21, 1999

Mrs. Stella Warner
Director of Human Resources
Whitcomb Technical Solutions
4703 Haskell Rd.
Boston, MA 02122

Dear Mrs. Warner:

Thank you for sending me a copy of your company's annual report. This information will be quite helpful as I prepare for my upcoming interview with Mr. Phil Rouse.

I appreciate your prompt response to my request.

Sincerely,

Troy Adkins

Troy Adkins

FIGURE 8.19
Acknowledgment Letter

In the opening, simply state that you received the item that was sent to you and the date you received it. In the middle, comment briefly on the helpfulness or usefulness of what was sent to you, and if any action is requested, state how you plan to respond to that request. In the closing, provide a courteous ending.

Update Letters

Like thank-you letters, update letters are helpful for maintaining your visibility to those with whom you interviewed or who are otherwise responsible for hiring someone for the position. (See Figure 8.20.) An update letter may simply inform someone at the company

34 Sala Drive
Boston, MA 02107
August 21, 1999

Mrs. Stella Warner
Director of Human Resources
Whitcomb Technical Solutions
4703 Haskell Rd.
Boston, MA 02122

Dear Mrs. Warner:

As you know, I recently applied for an opening in the Software Documentation Division of your company. I am writing to let you know I have recently changed my address and home phone number.

My new address is 34 Sala Drive, Boston, MA 02107. My new home phone is 617-555-0270. I wanted you to have this information in case you had to get in touch with me before the scheduled interview.

Sincerely,

Troy Adkins

Troy Adkins

FIGURE 8.20
Update Letter

that you have changed your address or phone number. Or it may stress some additional strengths that you forgot to mention during the interview. Or perhaps you were dissatisfied with the way you answered a particular question. The update letter affords you a good opportunity to provide a better answer to the question or to answer a question you didn't know the answer to during the interview. Finally, an update letter may call attention to some achievements that have occurred since the interview. Perhaps you received some award or other distinction at your university, or perhaps you had a recent success with a project at your current job. It's a good idea to write an update letter for any of the reasons provided here.

In the opening, state that you are writing to provide an update on whatever the topic is. In the middle of the letter, elaborate briefly on the topic. In the closing, thank the reader for taking the time to read your correspondence.

Acceptance Letters

In an acceptance letter you accept an interview offer, a job offer, an offer of a promotion, or some other offer for professional advancement. (See Figure 8.21 on page 364.) One of the most common purposes for an acceptance letter is to accept a job offer. Even if you accept the job offer over the phone, you still have to write a letter verifying your start date and starting salary. This letter sometimes precedes and sometimes follows a letter from the company setting out some of the same points.

The acceptance letter is a way of ensuring that you will receive something from the company in writing that verifies the terms of the job offer. It's very important to have this information in writing, and you should make every effort to have a representative of the company provide you this information in tangible form.

In the opening, thank the person you are writing to for the offer. State the date of the offer, and state that you are happy to accept the offer. In the middle of the letter, review any important details such as the start date, meeting place, and, if possible, the negotiated salary. In the closing, state that you are looking forward to beginning your job with the company.

Refusal Letters

The purpose of a refusal letter is to decline a job offer or some other important job opportunity. (See Figure 8.22 on page 365.) As a college student who is about to graduate, you may feel fortunate to be offered any kind of job in your field. However, students in the scientific and techni-

34 Sala Drive
Boston, MA 02107
August 30, 1999

Mrs. Stella Warner
Director of Human Resources
Whitcomb Technical Solutions
4703 Haskell Rd.
Boston, MA 02122

Dear Mrs. Warner:

Thank you for your telephone call earlier today informing me that I have been offered the position as technical writer in the Software Documentation Department of your company. As I told you over the phone, I am delighted to be offered this position, and, again, I gladly accept the offer.

To review the details, you mentioned that my starting date is Tuesday, September 7, and my starting salary is $32,000 a year. Please take a few moments to verify these facts in a letter to me.

I look forward to working at Whitcomb Technical Solutions.

Sincerely,

Troy Adkins

Troy Adkins

FIGURE 8.21
Acceptance Letter

cal professions often face choices between several job offers. Students often ask how they can politely decline one job offer and accept another or how they can decline one job offer even though they have no other offer. Of course, you should not accept a job that isn't right for you (although you should not think that your first job will necessarily be a perfect, high-paying job). There are specific strategies that you can fol-

34 Sala Drive
Boston, MA 02107
August 30, 1999

Mrs. Stella Warner
Director of Human Resources
Whitcomb Technical Solutions
4703 Haskell Rd.
Boston, MA 02122

Dear Mrs. Warner:

Thank you for your letter of August 27 informing me that I have been
offered the position as technical writer in the Software Documentation
Department of your company. I am very pleased that your company has
chosen to extend this offer to me.

Unfortunately, I must regretfully decline this good opportunity. I have a
second interview with Orion Software tomorrow morning. I have been
assured that this second interview is only a formality and that Orion
Software will be extending me an offer tomorrow afternoon.

Both Whitcomb Technical Solutions and Orion have a great deal to offer
any employee, but I feel that Orion Software is the best choice for me.

I want to thank you for taking so much time with me and, again, for
offering me the opportunity to work at such a fine company.

Sincerely,

Troy Adkins

Troy Adkins

·················
FIGURE 8.22
Refusal Letter

low for properly declining a job offer that you feel is not right for you.
Remember, you have an obligation not only to the company you are declining the job offer from, but also to your peers and other graduates
from your university who will follow in your footsteps. If you don't
properly decline a job offer, you could end up making the company reluctant to interview any future candidates from your university.

In this kind of letter you should use an indirect instead of a direct pattern. Don't state that you decline to accept the job offer in your opening paragraph. Instead, open the refusal letter by acknowledging any previous correspondence or other kind of communication (phone conversation, fax, e-mail, and so on) concerning the job offer. Then, in the middle of the letter, discuss any pluses in working for the company. You may then mention, for example, that you have another offer from a different company or that you have chosen to wait for other possible opportunities. In brief, politely and tactfully discuss any good reasons for declining the current opportunity, and then state that because of these reasons, you regretfully decline the job offer. In the closing, look to the future. Stress, for example, that you look forward to seeing the reader and other employees of the company at various trade shows or in other professional circumstances.

Resignation Letters

People typically change jobs often during the course of a career, and many people have more than one career during a lifetime. So it's almost inevitable that you will have to write a letter of resignation at least once. (See Figure 8.23.)

It's generally considered a professional obligation to give your employer at least two weeks' notice if you plan to quit a job, and the letter of resignation is your formal statement about this matter. Also, if the circumstances of your departure allow for it, you should have a face-to-face meeting with your supervisor first to announce your plans before submitting the letter. However, bad feelings on both sides may be involved; if so, a resignation letter suffices to replace any face-to-face meeting.

A resignation letter states that you are resigning, offers reasons for your resignation, states what your last of day of employment will be, and provides a courteous close. Aim for a serious, professional tone throughout. Avoid using this letter as an opportunity to blame your peers or management for anything. Instead, just keep your letter focused on providing the points listed above.

THE JOB SEARCH

Whether you are looking for a job fresh out of school or you have years of experience and want to change jobs, there are several essential credentials you must present to prospective employers. You will certainly need an up-to-date résumé, and you might need a portfolio

2713 Lansdowne Avenue
Alden, PA 19018
August 21, 1999

Mr. John Ray, Supervisor
Software Express
1500 Broad Street
Philadelphia, PA 19101

Dear Mr. Ray:

I am writing to let you know that I must resign from my position as a technical writer for Software Express. My resignation will be effective two weeks from the date of this letter (September 5).

I am resigning this position because I have been offered a full-time job with Telemetric Software in Minneapolis after I graduate next week. This job is an excellent opportunity for me and one that I have hoped for while pursuing my college degree these past few years.

Thank you for helping me to hone my knowledge and skills during the past year. You have contributed a great deal to my education.

Sincerely,

Linda Tobin

Linda Tobin

FIGURE 8.23
Resignation Letter

of your work to show what you can do. Once you have developed these items, you will be ready to look for your new job. Knowing where and how to look for a job is another essential skill in the job-seeking process.

Résumés

Résumés are autobiographical data or fact sheets. They tell readers essential information about your education, work experience, activities, and interests, and they provide contact information. A résumé in an academic setting is usually called a curriculum vitae (c.v.). Résumés and c.v.s are the means by which you present yourself to your prospective employer or a stipend-granting agency. The effective résumé will get you a job, or at least an interview; the ineffective résumé will be discarded, perhaps only partially read. The difficulty of creating an effective résumé is often underestimated. For example, many people represent their work experience poorly on their résumés, many neglect to provide other helpful information, and many don't devote enough attention to making their résumés easily readable. Creating an effective résumé requires numerous drafts before you achieve a balance of content and design.

The two major kinds of résumés are chronological and functional. In a chronological résumé, you provide traditional headings: education, work experience, writing experience, computer skills, personal data, and references—each arranged in reverse chronological order. For example, for work experience you provide your most recent job, then your next most recent job, and so on. (See Figure 8.24.) In this kind of résumé, emphasize the dates and duties of each position. In a functional résumé (also known as a skills and accomplishment résumé) you use headings that draw attention to your key strengths—typically, management skills, technical skills, selected professional achievements, and qualifications. (See Figure 8.25 on pages 370–371.) Sometimes you may use a résumé that is partially chronological and partially functional. Keep in mind that it will often be to your advantage to use more than one kind of résumé and, whenever possible, to tailor a résumé to a particular job opening so that the résumé emphasizes strengths that are specific to that job opening.

PRINT RÉSUMÉS

Be aware of the most common reasons résumés are rejected: They contain distortions or lies, they are too long or too short, they contain typographical errors or misspelled words, they fail to provide enough specific information, they contain irrelevant material, they fail to list job accomplishments, they are poorly designed or otherwise too difficult to read, or they show that the candidate is not qualified for the position.

Avoid being rejected for a job because of a poor résumé. In a chronological résumé, use a traditional order for your headings—typically, name, contact information, job objective (optional), education, work experience, personal data, and references. Use conventional headings, or at least don't stray too far from convention. For example, use

---

Carol Smith

427 Vista Trail; Orlando, FL; (407) 555-2627; csmith@email.com

Objective   Seeking a position as a technical writer or editor.

Education   Bachelor of Arts degree in English. University of Central Florida. May 1986.

**Work
Experience**

1992 to Present   Senior Technical Writer. *Institute for Simulation and Training.* Orlando, FL.
                       Edit a variety of technical documents including grant proposals and reports.

1990–1992   Senior Technical Writer. *Credit Card Software, Inc.* Orlando, FL.
                       Documented credit card software for programmers and computer operators.
                       Included flowcharts, program narratives, files and records, job steps, and
                       control cards. Used Vollie on the IBM mainframe to scan COBOL source
                       listings and copybooks.

1989–1990   Senior Technical Writer/Trainer. *Valencia Community College.* Orlando, FL.
                       Worked as a writer on a federal grant helping develop training materials for
                       in-house employees for a local software company. Involved producing a task-
                       oriented workbook to enhance and explain how the software worked and
                       how customers used it.

1988–1989   Senior Technical Writer. *Software Design Group.* Orlando, FL.
                       Worked as consultant organizing and revising database administrator's
                       guide.

1986–1987   Senior Technical Writer. *Travelers / EBS, Inc.* Maitland, FL.
                       Documented insurance software to produce user guides, documented
                       hardware, and administered new releases.

1984–1986   Technical Writer. *Dynamic Control Corporation.* Longwood, FL.
                       Documented software for hospital systems. Produced user guides,
                       administrative guides, and reports manuals.

1983–1984   Technical Writer. *Assessment Designs, Inc.* Orlando, FL.
                       Developed exercises for assessment centers, included writing scripts. Role
                       played for assessment of candidates for promotion in major companies.

Computer Skills   Word, PowerPoint, RoboHelp, PageMaker, Excel, HTML.

Honors   Phi Beta Kappa and Cum Laude graduate in 1986.

References   Available on request.

FIGURE 8.24
Chronological Résumé

"Work Experience," "Employment History," or "Experience"; don't use headings such as "How I've Paid My Bills."

Use a reverse chronological order (most recent to past) for the information you provide in each section of the chronological résumé

Sarah Adkins

1452 Terrace Drive; Baldwin, NY; (516) 555-6104; Fax 555-2717

Dynamic professional with eight years of progressive experience in computer project planning and management, corporate training program development, documentation, and systems analysis and design. Innovative problem solver who takes initiative, works independently as appropriate, and also enjoys teamwork.

Areas of Expertise

Needs Definition; Training; Communication & Motivation; Program Development & Evaluation; Technical Writing & Editing; Translating Concept to Reality

Accomplishments

Management/Project Implementation

Managed major projects and information systems teams of 3 to 12 members including department managers; managed medical practices.

Planned, installed, and supported mainframe and personal computer systems: analyzed current business, needed outcomes, and costs/benefits; determined policy/procedure changes; designed and tested software; converted and monitored systems and their use. Maintained stable accounts receivable through conversion.

Provided consultant services and 24-hour customer support for 4 hospitals.

Initiated orientation program and manual, enhancing new employee value to organization.

Designed marketing materials which were used in proposals.

Training/Technical Writing/Editing

Developed training and inservice programs: trained classes and individuals; trained 100% staff with 97% success rate.

Wrote training reference manual; wrote organization-wide training proposal.

Coordinated training for 5-state region: designed annual survey of client and internal training needs; published annual training calendar; trained the trainers; designed client newsletter. Enhanced public relations between organization and clients.

Taught computer applications: healthcare-specific; Lotus 1-2-3 2.0; WordPerfect 5.1; Project Workbench; Stress and Time Management.

FIGURE 8.25
Functional Résumé

---

unless there is some good reason to bend this rule. For example, under "Education" list your four-year college degree before your community college degree. Under "Experience," list your most recent job first. However, you may list another job first if it is more relevant to the position for which you are applying.

Avoid using complete sentences in your résumé. Use phrases to make it easier for readers to skim your résumé. Be careful not to use

Adkins/Résumé/2

Community Resource Development

Initiated training, apprenticeship, and placement program for household workers: wrote successful grant proposal; secured generous in-kind contributions from professional, business, and academic sectors, with no refusals; advised curriculum design.

Program objectives were met: skills, pay, and fringe benefits were raised for graduates. Improved skills and increased business knowledge led to three minority-owned businesses. Well-publicized program instrumental in improving the working conditions of household workers throughout the community.

Professional Background

Long Island Institute of Photography, Rockville Centre, NY. 1991 to present.
Registrar & Computer Consultant.

Nassau Community College, Manhasset, NY. 1991 to present.
Adjunct Faculty, Community Education.

Mercy Hospital, Rockville Centre, NY. 1986 to 1990.
Project Leader, Information Services.

Hospital of the University of Pennsylvania, Philadelphia, PA. 1985.
Systems Analyst, Management Information Systems.

Outsourced Medical Systems, Philadelphia, Pa. 1984 to 1985.
Regional Education Coordinator.

Education

Bachelor of Science degree in Public Administration, 1983
Hofstra University
Garden City, NY

References

Available on request.

FIGURE 8.25
Functional Résumé *(continued)*

any potentially unclear or confusing abbreviations—for example, for specialized degrees or professional societies.

Provide accurate, thorough, and honest information including: dates for your degrees, training, certificates, jobs, achievements in college and in the workplace, awards, scholarships, professional memberships, and other distinctions (for example, Dean's List or honor society memberships). Don't lie or distort information. Don't inflate a grade point average if you choose to provide one on your résumé. Don't list degrees you haven't earned, skills you don't possess, memberships you can't claim, or achievements or awards to which you are not entitled.

Don't provide too much information. Try to limit your résumé to one page. If you must use two pages, the content of the information must justify the extra page. The exception to the two-page rule is for a curriculum vitae. These can be quite lengthy to highlight the many papers the applicant has published and the number of committees on which the applicant has served. Unless you are preparing a c.v. for an academic position, limit yourself to one or two pages.

Don't provide too little information. Too many people neglect to discuss important job skills, important secondary academic strengths, important achievements or awards, and other kinds of potentially relevant or helpful information.

TIPS: Writing Your Print Résumé

- Use a larger point size for your name to make it stand out from the rest of the text (see the tips list, "Designing Your Print Résumé," page 374), and avoid using wording such as "Résumé of Jane Doe." Just use your name. Center your name or left-justify it.

- At the top of the résumé, below your name, provide your address, phone number, e-mail address, and other contact information. You should use a smaller point size for this information.

- Next, provide a job objective statement if you want your readers to know what your immediate employment goal is. Use a specific statement such as "Seeking a position as a technical writer or editor in the software industry" or "Seeking a position as a programmer analyst in the defense contract industry." Notice that these job objectives are specific about the type of position and the industry. When appropriate, you may also be specific about the department in which you would like to work. Many people who read résumés say that they are bothered when someone doesn't provide a job objective. For them a lack of a job objective suggests that the applicant doesn't know what he or she wants to do in a company. Others suggest that a job objective is optional. Remember that even though you will provide a job objective on your résumé, you can tailor that description to better match the particular job for which you are applying. Next time you send out your résumé, you can send it with a different objective, depending on the type of job for which you are applying.

- If your work experience isn't extensive (more than three or four career-related jobs), discuss your education before you discuss your work experience. Unless you attended a special technical high school, you usually omit high-school education from your résumé. If you have any special certificates or licenses, you could list them here. If you have an Associate of Arts degree or an Associate of Science degree, list it on your résumé and provide the name of the institution, the location, and the month and year you received the degree. Provide the same information for any four-year college degree. If you won't be graduating for a semester or two, simply use wording such as "Expected month/year." If you don't have a lot of work experience to emphasize, you may want to call attention to secondary academic strengths by listing, for example, a minor in addition to your major, specialized courses, an internship or co-op experience (or discuss this under work experience instead), and grade point average (if good).

- For your work experience you should provide the job title, the company name and location, and the start and end date for each position. (Use "To present" to indicate the end date of your current job.) Make sure there are no gaps in the dates you list. If there are, representing a time of unemployment for example, account for it honestly. A gap in dates sends up a red flag to your readers. They will wonder what you were doing and why you didn't want to mention it. Use action verbs (*managed, maintained, handled, supervised, directed, controlled*) to list in phrase form three or more specific job duties for each job, and emphasize the job duties that are related to the job for which you are currently applying. Also list any specific contributions or work-related achievements. It may also be helpful to indicate which jobs have been full-time and which have been part-time. This part of the résumé becomes easier to write as you acquire more experience.

- Of course, emphasize career-related jobs as much as possible or any job that specifically relates to the opening for which you are applying. If you are a college student and have had only part-time jobs that are unrelated to your career, it's still helpful to list these jobs and describe them at least briefly. For example, you may simply state "Held a variety of part-time jobs including deliveryman, cook, waiter, and salesperson to help pay for 75 percent of my education."

- If you are majoring in a writing field such as technical communication or journalism, you may want to have a separate main section titled "Writing Experience" or "Writing Projects." Here you should list in reverse chronological order the specific titles of your projects and publications. Place titles in italics (for manuals, brochures, and magazine titles) or in quotation marks (for article titles).

(continued)

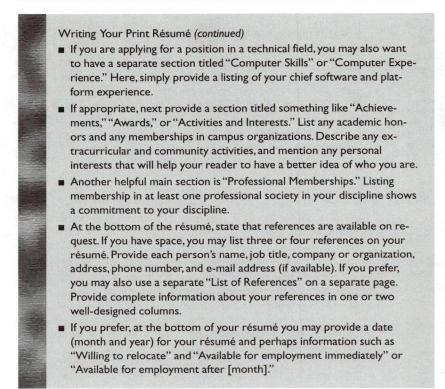

Writing Your Print Résumé *(continued)*

- If you are applying for a position in a technical field, you may also want to have a separate section titled "Computer Skills" or "Computer Experience." Here, simply provide a listing of your chief software and platform experience.

- If appropriate, next provide a section titled something like "Achievements," "Awards," or "Activities and Interests." List any academic honors and any memberships in campus organizations. Describe any extracurricular and community activities, and mention any personal interests that will help your reader to have a better idea of who you are.

- Another helpful main section is "Professional Memberships." Listing membership in at least one professional society in your discipline shows a commitment to your discipline.

- At the bottom of the résumé, state that references are available on request. If you have space, you may list three or four references on your résumé. Provide each person's name, job title, company or organization, address, phone number, and e-mail address (if available). If you prefer, you may also use a separate "List of References" on a separate page. Provide complete information about your references in one or two well-designed columns.

- If you prefer, at the bottom of your résumé you may provide a date (month and year) for your résumé and perhaps information such as "Willing to relocate" and "Available for employment immediately" or "Available for employment after [month]."

When you have finally listed all of the relevant information, then focus on the design of your résumé. The design choices that you make for this document are very important, and you should consider design elements as carefully as you considered content and organization elements.

TIPS: Designing Your Print Résumé

- Provide a minimum one-inch margin on all four sides. Too many résumés crowd the margins. You'll need to consider many factors—such as point sizes, spacing, punctuation, and styles—to help you place the essential information on one page within one-inch margins.

- Use easily readable fonts, and don't mix too many fonts and styles (italics, bold). Some fonts are difficult to read, and some are boring (Courier, for example). Devote some attention to the font you will use for your name;

consider a different font for the contact information and yet another font for major headings. Avoid mixing more than three fonts on the page.

- Don't use point sizes that are too large or too small. Any type that is smaller than 8 point will be too difficult to read, and point sizes larger than 24 point may be inappropriate.

- Place major headings flush left, or center them on the page. Use a consistent font and point size for these headings. Space appropriately before and after these headings to make it easy for your reader to skim the résumé.

- Use styles (bold, italics) to call attention to job titles, company names, or other important information. Use these styles consistently for the same kinds of information.

- Avoid clutter (too much text together). Placing too many lines of text together makes the résumé more difficult to read and less visually appealing.

- Use punctuation consistently throughout the résumé.

After you have printed a draft of your complete and well-designed résumé, ask several people to critique it for readability, information gaps, and design effectiveness, and then make any necessary changes. When you finish your final draft and are sure that the résumé is free of errors, print multiple copies on high-quality paper. Use conservative paper colors. Consider using matching stationery for your cover letters.

ONLINE RÉSUMÉS

Today, even if you have a well-designed traditional print résumé, you will find it useful to have several well-designed online résumés, too. Consider, for example, designing at least one online résumé for your home page on the Web (many companies access home pages to find qualified applicants) and another kind of online résumé for various electronic scanning technologies. One of the technologies that employment agencies use, résumé management systems, scans electronic résumé databases for certain kinds of résumés according to specific criteria. Basically, the résumés are read and categorized by a computer, not by a person. Because this practice is now common, you need to know how to make at least one online résumé as computer and scanner friendly as you can to increase the chances that you will be identified as a good candidate.

To make an online résumé for your home page effective and appealing, you can use a variety of design features. (See Part Three,

"Design and Illustration.") However, creating an online résumé for electronic scanning technologies presents additional challenges.

To create a scannable résumé, you don't abandon all of the good qualities of a good print résumé or a good online résumé designed for the Web. You just place more emphasis on some topics and take a different approach with others.

TIPS: Designing Scannable Résumés

- Realize that in scannable résumés, nouns dominate. Use nouns as much as possible so that your résumé will be more easily searchable by a computer that is looking for keywords.

- Use keywords effectively. Keywords are the key for this kind of résumé. Use keywords for education, experience, interpersonal skills, technical skills, and tool skills that are most likely to be searched for in your discipline. The more keywords you use, the more likely it is that your résumé will be selected. Try to use the keywords and phrases that are most often used in your profession. Stated another way, maximize the jargon of your profession. For example, in business, use *account management, manager, sales,* and *advertising.* In computer science, use specific computer program names such as Pascal, C++, or Visual Basic. In technical communication, use specific skills areas such as online help, Web-page design, graphic design, and technical editing, and tools skills such as RoboHELP, Adobe PageMaker, FrameMaker, and Microsoft Word. Look at help-wanted ads in your field to see the buzzwords that appeal to employers.

- Keep the design simple. Scanners will have difficulty with graphics and tables, for example.

- Help the scanner to read the résumé. For example, left-justify the entire résumé. Avoid tabs, hard returns, italics, underlining, or bold. Avoid horizontal or vertical lines. Avoid parentheses, brackets, or other unusual punctuation marks.

- Stick to common fonts such as Helvetica, Times, Palatino, and Courier.

- Restrict the size of your font to 10- to 14-point type.

- Use a single-column format.

- Keep abbreviations to a minimum.

- Place your name first and then contact information on a separate line.

- Use page length to your advantage. Limiting your résumé to a page is less of a concern in designing résumés to be scanned.

The traditional job market still requires you to have and use multiple versions of print résumés, most of which are read in thirty seconds or less by prospective employers. Adding the online résumé to your job-seeking effort increases your chances of being hired, especially in instances in which certain companies are doing at least their initial screening electronically.

TOOLS FOR CREATING RÉSUMÉS

Many computer programs are available to help you with both the content and the design of a résumé. Some of the major word-processing programs, such as Microsoft Word and Corel WordPerfect, have templates or wizards (helper applications) to help you get started or to guide you through the entire process of creating a résumé. Web Résumé Writer <http://www.jumbo.com/pages/upload/files.ASP?X_fileid =106240> offers similar features and includes additional features such as a contact database for potential employers and an e-mail program. Other programs are also available to help you design and print a professional-looking résumé.

Portfolios

Your portfolio is the tangible record of the projects and papers you have worked on. It is the permanent record of your accomplishments in school or on the job, and it can be your ticket to future employment. A piece of work needn't have been created entirely by you to be included in it, but do not incorporate someone else's work into your portfolio without giving credit to the other person.

An increasing trend among college students is to provide a version of their portfolio on the Web. Many students have personal Web pages, and in addition to providing information about themselves and their favorite links, they are providing online résumés and samples of their work. Samples may be provided as separate Web documents linked to the home page. They may include papers, group work, design work, or, for computer science students, for example, sample code that the student has written. (See "Design and Illustration," beginning on page 126 for a discussion of Web page design.)

TIPS: Creating a Portfolio

- Always ask for extra copies of documents you produce so that you can include them in your portfolio. Some writers actually put a statement in their contracts stipulating that they receive a certain number of copies of the documents they produce.

- Like your résumé, your portfolio should be updated periodically to keep it current.

- Organize the pieces in your portfolio logically, and use a table of contents to help readers locate pieces. You may choose to organize items to reflect different skills, such as writing, editing, illustrating, and designing documents. You may organize your portfolio to present different document types in different sections, such as brochures, manuals, and newsletters. If you have worked in several different fields, such as telecommunications, software development, and engineering, you might choose to organize your pieces to reflect them.

- Be sure to use only originals or good-quality copies of the documents you include. Poor-quality photocopies will seriously detract from a professional image.

- Consider including a section on awards and honors, particularly if you are a recent college graduate.

- Use graphic elements—color, rules, bullets, and other icons—to tie together the pages of your portfolio. Background paper in a muted color may help to set off your documents and make your entire portfolio appear more unified.

- Annotate the pieces that you include in your portfolio. White cards or labels can be affixed to the bottom corners of pages to highlight relevant information, such as your role in a group project or an award won by a document you produced. Do not annotate the obvious. You should not need to tell your reader that a brochure is a brochure.

- Consider showing more than one view of documents such as brochures so that the reader can see your design skills.

- Make sure documents are firmly placed in sleeves so that they do not slip around and look messy and disorganized.

- Never leave your portfolio with a prospective employer. Consider creating a small packet of documents and other relevant materials, such as your résumé and references, that you can leave behind after an interview.

- If given the chance, use your portfolio to talk your way through an interview and demonstrate what you have accomplished.

Employment Sources

Successful job hunting requires a great deal of time and energy. One way to save time is to know where and how to find jobs. The following sources are the major ones to explore for good job opportunities.

TRADE JOURNALS AND MAGAZINES

Become familiar with the major trade journals and magazines of your profession, and consult them frequently. It makes sense that employers will advertise job openings in a profession's key publications. In technical communication, for example, the journal of the Society for Technical Communication (STC) is *Technical Communication*. It is mailed quarterly to all 21,000 members of the society and contains a good variety of job ads. The society also publishes a monthly magazine called *Intercom,* which has job ads as well. STC also has a variety of special interest groups that publish newsletters and often have their own discussion group on the Internet.

NEWSPAPERS

Many college students make the mistake of relying chiefly on their local newspaper or the local newspaper of the city where they hope to be employed. Typically, fewer than 20 percent of availiable job openings will be advertised in the paper. However, this fact does not mean that you should skip looking at ads in the paper altogether. Newspapers can be a valuable source for finding a possible job opening.

EMPLOYMENT AGENCIES

Employment agencies are another choice, but be aware that there are different kinds of employment agencies. Many perfectly legitimate companies find and subcontract employees to other companies. For example, many technical communicators at Walt Disney World Company are subcontracted to the company by a job agency. The agency receives a percentage of salary—all paid by the employer—for the employee's services. In some professions in which highly skilled people are in demand, employers are more than happy to pay an employment agency to search for and place the right people. Many students in computer science, engineering, the sciences, and technical communication, for example, find jobs this way.

Other employment agencies are called career-counseling or career-marketing companies. You can easily find their ads in any major city newspaper. You have to pay these companies a fee to help find you a job. They charge you a fee to help you find a job (either an hourly rate, such as $50 to $100 an hour, or a flat charge for the service provided), and many charge an employer if they place you in a particular job. Many of these companies are legitimate and do make a sincere effort to help you find a job; many others are unethical and are merely out to get as much money from you as they can without providing a useful service.

Some job hunters have had bad experiences with career-counseling companies that demand excessive up-front fees and promise more than they can deliver. You should be wary if you are asked to sign a contract requiring payment in advance. Avoid signing any contract with a company that offers help in exchange for 5 percent to 10 percent of your annual salary or one that refuses to let you contact its other clients. You should be allowed to talk to other people who are using the company's services. Be wary of companies that promise access to unadvertised jobs. Many career-marketing companies brag about their ability to tap into a supposedly hidden job market. Also be careful if you are offered unlimited access to counselors. Any good career-marketing company will give you an accurate assessment of the availability of its counselors, but it won't be unlimited.

There are pitfalls in working through job agencies of any kind. Even some of the companies that subcontract work to other companies are unethical and will tell you, for example, that you are obligated to wait until they find a job for you. In other words, some of these agencies don't want you to find a job on your own or through another agency because then they would lose the commission.

There are enough other legitimate job-hunting resources that college students typically shouldn't go to the expense of signing a contract with a career-counseling company. Searching for a job on your own may or may not take longer, but your own search will usually be far less expensive. Of course, if a legitimate contracting company wants to help you find a job at no cost to you, then take advantage of the extra help.

NETWORKING

Networking can be an extremely valuable source for finding out about job opportunities in your field. (See Part Seven, "Interpersonal Skills.") The longer you are in a profession, the easier it is to network. Your list of contacts will be extensive and will continue to grow. However, college seniors looking for their first full-time job in their profession can find networking a little intimidating or difficult.

Though you have yet to begin your career, you can still network in a variety of ways. For a start, you can get to know several of your professors. They often know many people in industry, and if these professors know that you are looking, they can be very helpful in placing you in the right job or at least putting you in touch with people who might be hiring.

Involvement with campus activities, especially job fairs held on campus, are another valuable networking source. Many college campuses host a variety of job fairs, and many job fairs are held in major cities off campus but in a convenient location. Don't attend these job fairs empty-handed. At the very least, take some copies of your résumé, and take advantage of every opportunity to drop one off at a booth. Networking at a job fair requires preparation, but with some practice you can soon become an old hand at navigating the booths, the representatives, and all of the rest.

Don't forget about developing ties with your campus job placement office. These offices help students to acquire part-time employment while they attend college classes, but many also list full-time job opportunities. Knowing the director of this campus office can be an asset.

PROFESSIONAL MEETINGS

Attending professional meetings, both on and off campus, is another excellent way to network. Many professional organizations have student chapters, and meetings are often held on campus. Don't wait until you graduate to join the chapter of your professional organization. Become involved while you're still a student, and you will have a good basis for job searching later. The Society for Technical Communication, for example, has many student chapters; it also has professional chapters that allow students to join. Typically, thirty or so professional members and students attend the monthly meetings of an average-sized chapter. The professional members get to know the students, and the students become more involved in the profession before they graduate.

In addition to joining the key organization of your society, you can attend various local affiliated conferences, trade shows, and the like. For example, many students in technical communication should attend not only the annual regional conference or national conference of this organization, but also various meetings of other technical societies. Most major cities have a variety of professional societies of people who share common interests.

THE INTERNET

In addition to attending the local and national conferences of a professional society, many people are now contributing to the knowledge of their organizations through various mailing lists or discussion groups on the Internet. In technical communication, many job openings are posted on the mailing list TECHWR-L. For those who are searching for a job teaching technical communication, job ads can often be found on the mailing lists for the Association of Teachers of Technical Writing (ATTW-L) and the Council for Programs in Technical and Scientific Communication (CPTSC-L). There are also numerous discussion groups on computer science, engineering, and the sciences. (See "Joining Professional Organizations," page 298, for a listing of some professional organizations, many of which have at least one discussion group on the Internet.)

The Internet is proving to be an invaluable resource for finding out about jobs. Just about every major company now has a Web site, and many such Web sites contain links related to job openings. The Internet is proving valuable for the job search in other ways. More than a few students have been hired on the basis of their own personal Web pages and the portfolios of their work they have posted at these sites. Finally, various search engines and indexes are making it increasingly easy to track down job openings listed on Web sites or discussed in Usenet newsgroups.

JOB DATABASES

Another helpful source of job information is a job database, which is essentially a list of job openings that is available via telephone or the Internet. Some job databases are simply job announcements made available through voicemail. You call the number and listen to one announcement after another. Others are job openings on various computer databases that are available to you by mail or computer access. Still others are computer databases that are searchable on the Internet. Some common job databases available on the Internet are:

<http://www.careerpath.com> for major newspapers
<http://www.cweb.com> for nationwide jobs
<http://www.computerjobs.com> for The Computer Jobs Store
<http://www.careermag.com> for the Career Magazine
<http://www.Careermosaic.com> for Career Mosaic
<http://www.espan.com> for E-Span employment connection
<http://www.monster.com> for the Monster Board

The amount of information that is available at these sites can be overwhelming. You will need to hone your job-searching skills so that you don't waste a lot of time looking at positions for which you are not qualified.

RESOURCES

Atkinson, Toby D. *Merriam-Webster's Guide to International Business Communications*. New York: Macmillan, 1996.

Bolles, Richard Nelson. *The 1999 What Color Is Your Parachute: A Practical Manual for Job-Hunters and Career Changers*. Berkeley: Ten Speed, 1998.

De Vries, Mary E. *The Elements of Correspondence*. New York: Macmillan, 1996.

————. *Internationally Yours: Writing and Communicating Successfully in Today's Global Marketplace*. Boston: Houghton Mifflin, 1994.

Gonyea, James C., and Tom Jackson. *The On-Line Job Search Companion: A Complete Guide to Hundreds of Career Planning and Job Hunting Resources*. New York: McGraw-Hill, 1995.

Kramer, Marc. *Power Networking: Using the Contacts You Don't Even Know You Have to Succeed in the Job You Want*. Lincolnwood: NTC/Contemporary, 1997.

Merriam-Webster's Guide to Business Correspondence. 2nd ed. Springfield: Merriam-Webster, 1996.

Moore, David J. *Job Search for the Technical Professional*. New York: Wiley, 1991.

Murdick, William. *The Portable Business Writer*. Boston: Houghton Mifflin, 1999.

Poe, Ann. *McGraw-Hill Handbook of More Business Letters*. New York: McGraw-Hill, 1998.

Poe, Roy W. *McGraw-Hill Handbook of Business Letters*. 3rd ed. New York: McGraw-Hill, 1994.

PART **9**

Technical Documents

Illustrations

Tips

Technical Documents

*I*n this handbook the word *document* refers to any genre or form of technical communication. This section will treat some of the most common types of technical documents: instructions, manuals, newsletters, proposals, reports, and specifications. These documents are first defined or described in this section, and then tip lists are provided to help you create each type more effectively. Of course, you will be creating and reading many other types of documents. Documents that are discussed elsewhere in this handbook include letters, memos, e-mail messages, minutes, meeting agendas, résumés, literature reviews, and annotated bibliographies.

INSTRUCTIONS

Overview

Instructions inform readers how to assemble, make, create, or otherwise manipulate something. They are one of the most common types of document in technical prose. Instructions are different from process descriptions or policies and procedures. Process descriptions will tell you how something works. Policies and procedures will tell you about standard operating procedures that an organization follows. But if you want to perform the task yourself, you need instructions.

As a college student, you've no doubt repeatedly followed instructions to obtain money from an ATM, log onto a computer in a computer lab, or place a parking sticker on the rear bumper or rear window of your car. These and many other simple instructions are a regular part of campus life. As an entry-level employee, you no doubt have received instructions on everything from how to obtain your employee badge to how to use the network printer.

Instructions are also commonplace outside of college campuses and the workplace. Most taxpayers are familiar with the federal government's instructions for filling out a tax return. Homeowners typically have dozens of appliance manuals filed away for occasional reference concerning appliance maintenance and use; hobbyists are constantly using instructions to further their skills in their particular crafts or activities; and computer enthusiasts are continually consulting printed or online instructions for troubleshooting or for learning new computer skills.

The content and length of instructions vary, of course, depending on the subject, purpose, scope, audience, and context. In some cases a sentence or two will be all that is required. For example, a note placed in a classroom that states "Please shut off the lights when you leave" is an instruction. For complex topics, instructions may occupy several or more volumes.

To write instructions effectively, you'll have to go through the processes of planning, researching, writing, editing, designing and illustrating, reviewing and evaluating, and producing, all of which are discussed in other sections of this handbook.

Elements of Instructions

The most common elements of instructions are the introduction, a theory section, a list of equipment and tools or materials needed, a discussion of the steps that must be followed to complete the instructions, a section on troubleshooting, and, depending on the topic, warnings. Some instructions will need only a few of these elements; others will need all of them.

INTRODUCTION

The introduction to instructions isn't always required. The title of your instructions often will suffice to tell your readers what your document is about. However, sometimes you will want to provide a paragraph or more providing context or background information, identifying the audience for which the instructions are intended, stating the purpose of your instructions (if it isn't obvious), and identifying the scope or depth of detail. It's helpful to tell your reader what won't be covered in your instructions. Sometimes you may also want to set your readers at ease by reviewing the kinds of knowledge and background that are necessary to use the item you are describing, motivate your readers to carry out the tasks in the order provided, tell them how to use the instructions effectively, or comment on the organization of the instructions. If possible, provide an overview of what the instructions cover.

THEORY SECTION

Sometimes your readers require a discussion of how certain equipment works; in these instances you may need to provide a theory section. Often your readers will have a better understanding of how to assemble something or how to carry out other kinds of instructions if they have an understanding of the underlying processes that are involved. For example, if you are writing instructions for how to clean a VCR, it would be helpful to discuss briefly how and why the heads of the VCR come into contact with the tape and require regular cleaning. The theory section of your instructions provides your readers with necessary background or context so that the remainder of the instructions make more sense. If your instructions require someone to repair or to use equipment, you will need to explain how the equipment works and identify the essential parts. A diagram or photograph, with accompanying labels or call-outs, can visually depict the various component parts.

LIST OF MATERIALS, TOOLS, OR EQUIPMENT

A list of materials or tools and equipment needed is one of the most common elements of many instructions. This list may simply be a one- or two-column listing, or the list may be part of a diagram (see "Drawings, Diagrams, and Maps," page 156) illustrating all of the materials, tools, or equipment needed. You'll need to be as specific as possible so that your readers will have everything they need to carry out the instructions.

STEPS

The actual steps of the process are another common part of instructions. These steps are the directions for assembling, creating, or using something. For relatively simple instructions a concise series of steps often suffices. For more complex instructions you may have to perform a task analysis to determine the actual number of steps involved. In a task analysis you identify all of the steps your readers must perform to carry out your instructions. Once you have done this, several strategies are available that will help to make your instructions useful. These include clearly identifying each step with a number, bullet, icon, or other symbol; using illustrations when readers may have difficulty understanding what your written instructions require them to do; giving your readers enough information to do what is required of them; adequately covering alternative steps readers may perform when necessary; and warning your readers of potential dangers.

There are many ways to order steps. A fixed order of steps means that the steps must be performed in the order in which they are discussed. A numbered list indicating the precise sequence is the best choice for this situation. A variable order of steps means that the steps may be performed in different sequences. A bulleted list is often helpful for indicating this kind of order. Sometimes an alternative order of steps is useful. For example, if one kind of condition exists, the reader may follow one order; if a different condition exists, the reader should follow a different order. Finally, a series of substeps is often necessary for complex instructions. Sometimes you have to take one major step and break it down into smaller substeps, indicating these substeps either by indenting further or by creating a substep sequence of a, b, c, and so on.

TROUBLESHOOTING SECTION

Many instructions require a troubleshooting section. In this section you tell your readers how to check or retry various tasks if their first efforts are unsuccessful. A thorough troubleshooting section can be one of the most important parts of a set of instructions. Troubleshooting sections may be organized in many different ways. Sometimes they are simply titled "Troubleshooting," and sometimes they are titled "Frequently Asked Questions (FAQs)," although typically FAQs include all sorts of information, not just solutions to problems. Information in FAQs is often provided in a table, chart, or list of bulleted questions followed by answers. Troubleshooting information is usually not numbered, unless you are providing a series of steps for a specific task to help address a problem. (See Figure 9.1 for an example of a troubleshooting chart.)

WARNINGS

Provide warnings in your instructions to notify users of the possibility of undesired consequences. Warnings may be an essential part of your instructions, used to help ensure that instructions can be successfully carried out. Or they may alert users to ways to prevent damage to equipment or injury or death to themselves or others. Many kinds of

Troubleshooting for LView Pro

Problem: I can't open an image.

Suggestion: In the **Open** option from the **File** menu, make sure you are selecting the correct directory and that you are looking for the proper file extension. If your file is in a .jpg format, for instance, you must have **JPEG (.jpg)** selected to open a .jpg image.

Problem: The image I opened is grainy.

Suggestion: Purchase a 256-color or 16.7 million-color VGA card. Sixteen-color VGA cards are often incapable of superior reproduction quality, especially with scanned photographs.

Problem: My image is too dark.

Suggestion: Increase the **Contrast Enhance, Gamma Correction, Exp Enhance,** or **SineH Enhance.** These options are found in the **Retouch** menu.

Problem: My image is too bright.

Suggestion: Decrease the Value in the **HSV Adjust,** or increase the luminance (Y) in the **YCbCr Adjust.** These options are found in the **Retouch** menu.

Problem: One color is dominant in my image.

Suggestion: Reduce the dominant color component with the **Color Balance,** or alter color characteristics using **Palette Entry.** These options are found in the **Retouch** menu.

Problem: I get a message that I don't have enough memory to open an image.

Suggestion: Make sure that you don't have any other images open, and close all other applications you may be running simultaneously with **LView Pro 1.6.**

Problem: My image looked good when I first opened it, but it keeps getting worse and worse.

Suggestion: There is a certain amount of loss that occurs every time you save (and thus compress) an image. You can reduce the amount of degradation by limiting the number of times you open and close the same image. You may also want to save your original copy of important images on diskette. Do this from the **File** menu.

FIGURE 9.1
Troubleshooting Chart

instructions are written without any warnings. If no risk of ruining the results, harming the equipment, or injuring users or others is involved, then no warnings are necessary. However, if damage or injury is a possibility, use warnings in your instructions.

Warnings are typically classified into three major kinds, depending on their severity: caution, warning, and danger. *Caution* makes readers aware that they may damage or ruin a project if they do not follow the steps exactly as they are listed. *Warning* tells readers that they may damage equipment or tools if they do not follow the instructions exactly. *Danger* warns readers of potential injury or death to themselves or others. It's important to place safety warnings where they will do the most good. Make sure you place them in the instructions where the reader will see and read them *before* carrying out the potentially risky step or steps.

TIPS: Creating Instructions

- Use the imperative mood in the steps of the instructions. Many of the steps will consist of brief sentences that tell your readers what to do: *To check for viruses on a disk in the A: drive, first place the disk in the drive. Second, open your antivirus program. Third, find the A: drive icon on the dialogue screen.*

- Avoid a telegraphic prose style. The need for brevity in instructions sometimes leads writers to omit articles (*a, an, the*), conjunctions (*and, but, or*), and prepositions (*to, of*), creating a style that reads like a telegram: *Boot up computer. Turn on monitor. Click on "Start."* This style can easily become annoying and distracting.

- Use visual aids wherever possible. Good instructions include illustrations to help readers grasp a concept, a step, or a context much more quickly.

- Focus on a design that will help your readers to follow your instructions. For instructions to work well—and for the best usability—follow the principles for effective page design throughout your instructions. Use consistent and appropriate fonts, headings, margins, spacing, and other design elements. (See Part Three, "Design and Illustration.")

- Perform a task analysis to help you identify all necessary steps. A task analysis involves taking a major task and breaking it down into smaller steps until all parts are covered. The major task may already be a relatively simple one, in which case your task analysis would not involve many substeps. Complex instructions will require a much more thorough task analysis. You must identify all of the major tasks that are involved in the complex set of instructions; then you must break each major task down into smaller tasks. Each of those may need further refinement until all the actions have been broken down into their component steps.

- Anticipate in any earlier step what readers may need to know to perform later steps. Many of us have been victimized by instructions that either omit essential steps or put steps in the wrong order.

- Keep the steps brief and simple. Each should typically be one or two sentences long. Provide more steps if you find that one part of the instructions is too involved or long. Your task analysis should help you to keep your steps manageable.

- Keep the style informal whenever possible. The formality of your prose depends on your audience and your purpose. Many technical instructions are formal, but many are not. For instructions aimed at homeowners, novice hobbyists, and other kinds of first-time users, an informal style can help the readers to relax and feel less intimidated.

- Provide a reader-friendly tone. A friendly tone, like an informal style, can help your readers to relax and feel less intimidated. (See "Developing an Appropriate Tone," page 85, for a discussion of various strategies and a tip list on using humor.)

- Make certain your instructions are complete. Don't omit an essential tool, part, step, illustration, warning, or helpful troubleshooting explanation.

- Test your instructions. One of the most valuable things you can do is to test your instructions on the intended audience. If the intended audience isn't available, then test your instructions on other members of your class or other employees in your office. (See Part Four, "Editing," for a discussion of usability testing.)

MANUALS

Manuals of many kinds are used at every university, company, or organization. Many are written by professional technical communicators; others are written by subject matter experts, programmers, or engineers. Manuals typically cover a wide range of instructions, procedures, or reference information necessary for accomplishing work in an efficient manner or for obtaining needed information. Technical manuals help readers to understand, use, maintain, and repair products. Typical technical manuals include software user manuals, reference manuals, tutorials, training manuals, safety manuals, maintenance manuals, and procedural manuals.

TIPS: Writing Manuals

- Strive for consistency throughout your manual in both format and style. For example, do not refer sometimes to a "manual" and other times to a "guide" when you are talking about the same document. Choose one term and stick with it.

- Give your reader as many methods for locating information as possible. For example, in addition to a table of contents and an index, include sec-

(continued)

Writing Manuals *(continued)*
tion contents pages, tabs, different colors or icons to highlight each section, headers and footers, a glossary, a sectional page numbering system,
and numerous heads and subheads.

■ Always keep a clear picture of your audience in mind, and make editing
decisions on the basis of how well the information will work for your
particular audience. For example, you may choose to delete a complex
description—even one that is well done—because it is beyond the
scope of your audience's needs and interests.

■ When writing a style manual for a technical audience, keep all your examples scientific and technical.

■ Define terms within the context of a sentence or paragraph if possible.
Also consider including a glossary of key terms in your front or back
matter.

■ Whenever possible, provide specific examples to support the general information you introduce in your manual.

■ Whenever possible, place tables and charts and other graphic examples
on the same page as the text to which they relate or on the facing page.

■ Carefully integrate graphics into your text and edit them for consistency in their use, placement, numbering, and labeling.

■ Be sure that graphics are sized large enough to be read easily.

■ Select no more than two (or three at the most) typefaces, and use them
to differentiate body text from headings.

■ Use different type styles (for example, bold, italic, condensed, outline) judiciously, and clearly signal to the reader that each type represents a
particular kind of information. For example, italic may be used to present only captions or words that are defined in a glossary.

■ Choose type sizes, colors, graphic elements and accents, paper, and
printing method with ergonomics in mind. How will the reader actually
use the manual? What will readers be trying to do? follow step-by-step
instructions? look up reference information? learn new concepts?

■ When making production decisions, always consider how your readers
will be using your manual. For example, a binding that allows the manual
to be folded over would be best for a manual that readers may want to
use while performing tasks.

NEWSLETTERS

Newsletters are excellent promotional, informational, and motiva-
tional documents for both internal and external use. They offer numerous advantages for organizations or companies. They provide

valuable information about your business and your services. They are a good way to introduce new staff, services, and products to employees and current and potential clients. They give your company added exposure, potentially increasing your chances for new business. They let your employees know what's going on in other departments. They are a great way to present informal news about your employees.

External Newsletters

External newsletters, targeted to customers and potential customers, can be used to introduce new services and products, emphasize the benefits your customers receive by doing business with you, demonstrate your strengths with testimonials from and articles about customers, and provide useful information about the company. Your newsletter essentially represents your company, so it should present a professional image, be more informational than sales oriented, and motivate readers to read beyond the front page by packaging information in such a way that it appears useful and interesting.

Internal Newsletters

Internal newsletters, targeted to employees, are used to promote a team spirit and build morale within the company. The typical house newsletter should explain your company's goals and plans for the future; introduce new staff, services, and products; educate your employees about various issues; improve relations within your company; recognize employees for achievements; promote higher morale among employees; and highlight your company's involvement in the community. The in-house newsletter reminds employees that they are all working as a team toward the same goals, even though they may not know all the other employees personally. The sense of participation fostered by the newsletter will promote higher morale among employees and boost job satisfaction.

Electronic Newsletters

Increasingly, many companies and organizations are making electronic versions of their newsletters available. Some of these are called electronic magazines, e-zines for short. Many of the principles for producing effective print newsletters still apply, and the special design challenges of designing online information also apply. (See Part Three, "Design and Illustration," for a discussion of designing and illustrating for online information and the Web.) If you send your newsletter by e-mail, make it easy for people to cancel their subscription, and don't send it unless your readers really want it. For customers and prospects, send one free issue; then let them decide whether they want to subscribe.

A newsletter is an important document for your organization or company, and you should use the best tools available to help you achieve a professional-looking document. Today's full-featured word-processing programs make it easier than ever before to create your own newsletter. Many programs have wizards (software helpers) that guide you through the entire process of placing your text in text boxes, selecting line art, and choosing and displaying graphics of all kinds. See Part Five, "Production," for a more detailed discussion of various word-processing and desktop-publishing programs.

TIPS: Newsletter Content

- Determine the purpose of your newsletter. It will be your guide to how your newsletter is written, the information you want to include, and how you will distribute it. Some points to ponder: Should you send the newsletter to existing customers only? to customers and prospects? to anyone who wants to receive it? Should you send it free, free with a sponsor, or only with a paid subscription? Should you send a text-only version by e-mail? a text-only version by surface mail? Should you create a graphics version for a Web site? a graphics version for surface mail or statement stuffer?

- Select content carefully. Include useful information, not just fluff. Effective newsletters usually have informative news about your company, humorous or brief stories, brief facts about your topic, information or spotlights about various people receiving awards or other kinds of recognition, letters or other comments from readers, information about new products and sales, and so on.

- Verify that your information is up to date and accurate.

- Design the newsletter so that it is between two and twelve pages long. If it is going to be shorter or longer, you should reconsider whether a newsletter is your best choice to convey the information.

- Keep the articles and news brief. Remember that you're writing a newsletter, not a newspaper. If you make each article brief, people will be more likely to read it. Nonetheless, be careful not to sacrifice information just for the sake of brevity. If you have a lot of information to share, write a series of small articles instead of one huge one.

■ Use a conversational tone and a reader-friendly style.

■ Carefully edit; carefully proofread. Then proofread again. If possible, have someone else proofread it too.

TIPS: Newsletter Design

■ Begin planning your newsletter by designing the nameplate and selecting the underlying grid.

■ Provide a table of contents on the first page.

■ Make it easy to skim to find the most useful information.

■ Strive for either symmetrical or asymmetrical balance.

■ Choose your column format carefully to allow for flexibility. Two-column grids tend to be more static than multicolumn grids. Avoid a layout in which everything is perfectly symmetrical and there is no visual interest.

■ Allow sufficient margins and gutters (spaces between columns). Judicious use of white space can mean the difference between an attractive, readable publication and one that is discarded unread.

■ Avoid excess white space within headlines and subheads. Be sure it is clear to readers which columns your headlines and subheads are attached to. There should always be more space between the end of one column and the headline for the new column than between the headline and the column to which it is attached.

■ Be sure to break up long columns of text with adequate (but not excessive) use of graphic elements and accents. Nothing is more boring than straight text, so include a good mix of photographs, other graphics and artwork, and text. Follow the principles of effective design for placement of text, line art, graphics, headings, fonts, and so on.

■ Focus your readers' attention with creative graphic elements—art, photos, boxes, screen tints. These will help them better understand your message.

■ Avoid overuse of clip art as it will make your publication appear less than professional. Use rules, screened boxes, and bullets rather than clip art. If photographs are within your budget, use them where appropriate.

(continued)

Newsletter Design *(continued)*

- Use only one focal point per page—that is, one headline or photograph that is larger than the rest and to which the readers' eyes will be drawn first.

- Design your publication with two-page spreads in mind. That is, be aware of pairs of pages that will be viewed facing one another. They should work together just as the elements on individual pages must work together.

- Avoid getting too fancy with printing. Avoid using colored text other than black or blue. Do use a second or third color sparingly for screen tints, large drop caps at the beginning of an article, page numbers, and any other graphic that is repeated throughout the newsletter. Too much use of another color is distracting to readers, and each color adds significantly to the cost of producing the newsletter.

- Use type sizes to establish a hierarchy of information.

- Avoid using more than two or three typefaces throughout your newsletter.

- Keep the fonts simple enough and large enough to encourage readers to read.

- Print your newsletter on an easy-to-read white, off-white, light gray, beige, or similar paper. Avoid red, green, blue, yellow, orange, and the like. Glossy, uncoated, or matte finishes are acceptable.

PROPOSALS

A proposal is a persuasive offer or bid to do something, whether it's an attempt to convince your professor to accept your topic choice for a technical report or a project due later in your writing course or an effort to convince your supervisor at work to provide new computer equipment for your office. Proposals ask your audience to approve or fund your request. They are a common kind of technical document, and they come in all sizes and forms. They may be as brief as a few sentences in a memorandum format or a few pages in a letter format, or they may be as detailed as thousands of pages in many volumes. As a college student you may have to write a few proposals, and as an employee you probably will have to write many.

Kinds of Proposals

Proposals may be internal or external, and they may be solicited or unsolicited. Internal proposals are submitted within a company or organization to help the company or organization in some way. External proposals are submitted to outside or independent companies or organizations. A typical external proposal is written by a consultant who is trying to obtain a contract with a company or organization. Solicited proposals are proposals that are sent in response to a request for a proposal (RFP) from a person, company, or government agency. For example, if a specific agency in the federal government needs computers, it will provide the specifications for them to various suppliers. Usually, the supplier with the lowest bid obtains the contract. An RFP may list numerous required specifications, or it may allow suppliers to create their own designs according to certain guidelines. Unsolicited proposals are proposals that the reader—a person, company, or government agency—has not requested. Unsolicited proposals must convince their reader that a need or a problem exists. They may be submitted within a company by an employee to address a problem, or they may be submitted by a company to a client in an effort to make a business deal.

Elements of a Proposal

Proposals vary a great deal in the kinds of elements that are covered. For many proposals, what must be covered is specified. For other kinds of proposals you must cover what your readers need to know, which varies according to the particular situation. What you include depends on what kind of proposal you are writing and your subject, purpose, scope, audience, and the context. If you are a consultant who is trying to obtain a client's business, you will take a different approach in your proposal from that of a person who is suggesting a course of action to a supervisor. If you are a college student writing a proposal concerning a class assignment, you may be required to take an approach different from those mentioned above.

Some proposals simply have an introduction, a body section, and various attachments. Other proposals may have the following eight sections and more.

INTRODUCTION

In your introduction you should state that you are providing a proposal. Your introduction may be no more than an acknowledgment that the reader has invited the proposal and that you are responding to that invitation. Provide some context, for example, stating that you have had previous contact with the reader or that you are responding to an RFP from the reader or the reader's company. A good introduction for a proposal should not overwhelm the reader with details. At this point you should provide only a general overview of what you want to do and of what the proposal contains. You want to generate interest in your idea, motivate the reader to read further, and help the reader to understand what will be covered.

BACKGROUND ON THE PROBLEM OR NEED

The background section most often follows the introduction or is sometimes part of the introduction. This section offers a thorough review of the problem your proposal is addressing. Here you provide the appropriate context to show that there is a need for your proposed service, product, or solution. If, as in many cases, your audience is familiar with the background issues, you may omit this section or make it very brief. However, it's a good idea to include this section, if only to show your audience that you understand the problem. Discuss your understanding of the requirement in detail, and provide a thorough analysis of the requirement. If your readers need to be persuaded that there is a problem, this is the section where you must persuade them.

DESCRIPTION OF WHAT IS PROPOSED

In this section you cover the specifics of what you plan to do or what you have to offer. In the description of the proposed service, product, or solution you discuss the specific advantages of your service, product, or solution. If your proposal is unsolicited, this section helps to persuade your audience that the need you have argued for in the background section of your proposal has its satisfaction in your described solution. So in addition to selling the need for your solution, you must persuade your audience that your proposed course of action will be effective.

SPECIFIC PLAN FOR PROVIDING WHAT IS PROPOSED

In this section you provide your detailed plans for implementing your approach. You may cover many topics here, including, for example, how the project will be organized, who will manage it, and what the specific plans and procedures are. Specify your approach to satisfying the requirement. In this section you should persuade your audience that you are able to manage the project. For some kinds of proposals this section discusses in detail the method (how you will bring about the result you have described) and resources (facilities, equipment, and other resources) necessary for your approach. In a grant proposal, this part is sometimes called the methodology section.

SCHEDULE FOR CARRYING OUT THE PLAN

In the schedule or timetable you should discuss in detail how you plan to achieve each phase of the work proposed, and you would typically attach a schedule showing how various deadlines will be met. Sometimes a brief schedule is included as part of a memo or letter simply by discussing the major tasks and expected dates of completion. Sometimes a more thorough schedule is created, for example, in the form of a Gantt chart (see Part Three, "Design and Illustration"). The more complex the proposal, the more complex your schedule will be. For very detailed proposals you will need to indicate your proposed work on every phase of the project. A thoroughly detailed schedule is one helpful element for persuading your audience that you can deliver your service, production, or solution on time. (See Part One, "Planning and Research," for a discussion of estimating schedules.)

QUALIFICATIONS OF THOSE WHO WILL CARRY OUT THE PLAN

In the qualifications section you need to discuss your specific credentials if you are writing your own proposal for an assignment or project. If your proposal is part of a team effort, you need to discuss the staff and the specific credentials of all involved. If appropriate, you may also have to provide an overview on your organization or company. For this part you will often provide a prose discussion of the credentials and qualifications of the team members, and you will often include curricula vitae (in academe) or résumés (in the corporate world) as an appendix to the proposal. Where appropriate, identify those who are specifically responsible for management of the project if more than a few people are involved. In a very detailed proposal you will also discuss the management structure of the group.

BUDGET

Many proposals require no budget. If you are writing a proposal for your professor concerning an assignment, you may not have any budget considerations to cover. However, sometimes in college assignments and often in company proposals there are production, research, or equipment costs. For grant proposals requesting funding for research projects, this is an extremely important section. In the budget section you list and explain the costs related to the project. Your explanation must persuade your audience that what you propose is necessary. You will have to justify every item and verify each and every cost. If you are proposing an expenditure to your company, you may need to discuss how it will be funded. The budget is literally the bottom line, and decisions to go ahead with a project are often heavily influenced by the information in this section.

CONCLUSION

For some proposals, no formal conclusion is necessary. The last item may simply be your budget or various attachments, such as a schedule or résumés for those involved. However, many proposals could also benefit from having a conclusion that discusses, for example, testimonials about past successes on other, possibly related projects. A conclusion can also be helpful for once more appealing to your audience for approval of the project and the requested funding. Also, your conclusion may reaffirm the overall confident tone of your proposal. Whether your proposal ends with a formal conclusion or not, review it to ensure that it closes appropriately, giving the reader a sense that it is a complete entity.

TIPS: Writing Proposals

- Think imaginatively and creatively when it comes to proposals, and don't be afraid to submit a proposal if you have a good idea. Many people have good ideas but are reluctant to write a proposal that will enable them to implement their idea.

- Make sure your proposal adds value or is worthwhile. Many ideas for proposals are bad ideas and should not be submitted at all. Don't just submit a proposal for the sake of submitting one. Make sure that you and perhaps some peers you have consulted think your idea is a good one.

- Make sure your proposal idea is adequately focused. Many proposal ideas are too broad or too vague. It requires considerable effort to narrow a proposal idea adequately.

- If your proposal is a collaborative effort, make sure everyone has the opportunity to contribute equally. (See Part Seven, "Interpersonal Skills," for tips on working collaboratively.)

- Suit your proposal to the format. Again, some proposals are brief memos, some are letters, some are e-mail messages, and some are many pages or perhaps even several volumes long. Regardless of the format, your proposal should have an introduction, a body (with some or all of the standard sections: background, description, plan, schedule, qualifications, budget), and a conclusion.

- If the proposal requests funds, carefully scrutinize the costs and account for any variables (delays, increasing costs of materials, salaries or stipends, support staff, photocopying and other services, office space, equipment, and so on). You don't want to request too little funding to carry out the tasks you are proposing to accomplish.

- Carefully scrutinize your schedule. Make sure you give yourself and, if applicable, your co-applicants plenty of time to do the proposed work. Be especially careful to account for delays due to holidays, illness, vacation time, equipment failure, and so on.

- If appropriate, review the literature on your subject. Many proposals require that you have reviewed the literature extensively before the proposal will be convincing. Your proposal should reflect that you are thoroughly familiar with the literature.

- Always keep your audience in mind. Many proposals are submitted by peers to other peers, but more often, proposals are submitted to a combined audience. (See Part One, "Planning and Research," for a discussion of various kinds of audiences.) If even one reader may not be familiar with your subject, you'll have to address this possibility in your proposal.

- Make certain your proposal is persuasive before you submit it. Have you made your case for the need to solve the problem or for the request? Have you been convincing about how and when the work will be done? Have you effectively represented the qualifications of all of those involved? Have you persuasively justified your expenses? These are just some of the questions you need to ask.

REPORTS

Reports are documents about something that was done or something that needs to be done. Reports are some of the most common documents written for organizations or companies in most disciplines or professions. As a college student, you may be asked to discuss some problem in your field and to explore a solution. You may be asked to provide background information about some new technology, trend, practice, or problem. Or you may be asked to analyze a problem in your discipline and to offer recommendations for addressing the problem. If you are an employee, you may be asked to give a progress report on your work on a project, you may be asked to submit a trip report listing your activities and expenses for a recent company trip, or you may be asked to submit a feasibility report concerning whether one course of action is preferable to another. Whether you are in academe or in industry, if you are in the technical professions, you will write many kinds of reports.

There are many kinds of reports, each with a different purpose, content, organization, format, and, often, audience. There are various ways to describe the many kinds of reports you may be asked to write. Many reports are informal; these are often written as memos, forms, e-mail, or letters. They may simply be routine communications on one kind of everyday issue or another—for example, a memo approving the transfer of funds from one department to another or an e-mail message announcing the details of a recent company purchase. Sometimes a report is simply a form that must be filled out and filed.

If the topic requires in-depth discussion, a formal report is provided instead of a memo, letter, or form. Formal reports are more commonly done in book format and include many of the elements discussed in "Common Elements of a Formal Report" on page 407. Formal reports are either *informational* or *analytical,* although sometimes these aims overlap.

Informational Reports

Informational reports chiefly give readers an understanding of a problem, an approach, a method, an issue, a product, a company, a trend, a technology, or some other topic. Informational reports are contextual rather than evaluative. They provide essential background or other kinds of details but do not typically analyze or dissect a problem, issue, trend, or other aspect.

BACKGROUND REPORTS

Background reports (also called general reports) are the most common type of informational technical report. They provide information to individuals or companies who need it. They are written to provide information only—an update on background or events—by summarizing a current situation, reviewing its history, or discussing trends. These reports provide information to individuals or companies that need it. Since your chief concern is not with how your readers will use the information, background reports seldom offer any recommendations.

LABORATORY REPORTS

Laboratory reports describe the testing and research that are performed in a laboratory. They seldom evaluate or interpret the information or data. They essentially just document work that is in progress in the lab. Information in these reports is often helpful for the methodology section of a research report. (See "Analytical Reports," on page 405.) Often lab reports are submitted on a standard or in-house form.

TECHNICAL DESCRIPTIONS

Technical descriptions provide information about a product to readers who may use it, buy it, or assemble it. Technical descriptions may be brief or they may be quite detailed, depending on the complexity of the product or mechanism being described.

A description of a relatively simple mechanism typically includes a description of the overall appearance and component parts, visuals, a description of the function of each part, and appropriate details about each part. A description of a complex mechanism may cover such introductory items as a definition, function, and background of the item; purpose; an overall description; principles of operation; and a list of major parts. The body would contain a detailed description of each major component, including definition, shape, dimensions, and material; subparts (if any); function; and relationship to other parts. The conclusion of a complex description might contain a summary (for quite complex mechanisms); a discussion of the interrelation of parts; and perhaps a walk-through of one complete operating cycle.

One of many challenges is deciding on the best order or sequence for describing the mechanism: spatially, functionally, chronologically, or a combination method. The choice will depend on the audience's needs. Although it is tempting to describe an object or process according to the writer's own understanding, viewing the same object from the user's perspective will often result in a far different decision about organizing the document.

If it's necessary to provide a definition as part of a technical description (or for some other document), you should be aware that there are informal and formal definitions and various strategies for creating both of these kinds of definitions. An informal definition simply provides other words that have meanings similar to that of the term you are trying to define. For example, to define *search engine,* you might write "a tool used to locate information and resources on the Web." An informal definition often appears as a parenthetical definition, a phrase definition, or a sentence definition.

A formal definition offers a more structured approach to a term. In a formal definition you may provide a formal sentence definition in which you identify the term, the class, and how the term is different from other members in the class. For example, you may write, "An animated GIF is a multi-image GIF where images appear to create motion." The term is *GIF,* the class is *multi-image GIF,* and the distinguishing feature is *where images appear to create motion.* Alternatively, you may provide an extended definition. An extended definition typically provides several paragraphs that offer first a sentence definition and then a detailed discussion of the term. In the detailed discussion of the term you may provide graphics, examples, comparisons, analogies, negation (defining the term by discussing what it is not), and etymology (origins of the term) to define the term further.

PROCESS DESCRIPTIONS

Process descriptions focus on how things are done or how they work. They may focus on processes concerning products made by people— for example, how televisions, computers, radios, car engines, or cellular phones work. They may also cover, for example, natural processes such as how volcanoes erupt, how tornadoes are formed, or how photosynthesis works. Process descriptions are not limited just to reports. They are often an integral part of many other kinds of technical documents, including instructions.

POLICIES AND PROCEDURES

Policies and procedures, also called directives, provide information for your readers on how things are done in an organization. They are the standardized ways of doing things in many organizations—for example, taking inventory, organizing incoming mail, stamping a time card, applying for a sabbatical, or taking vacation time. Many organizations cover all kinds of policies and procedures thoroughly in policies and procedures manuals or employee handbooks. But policies and procedures do not always have to be very elaborate. Even a brief e-mail directive about employee holidays during the holiday season qualifies as a policy.

Standard operating procedures (SOPs) are widely used in some industries to help workers complete a job safely, to ensure that production operations are handled consistently, to prevent failures in manufacturing that cause harm to people, and to ensure that official procedures are properly followed to comply with company or government regulations. An SOP may also function as a training document for teaching users about a process, serve as a checklist for coworkers or auditors, provide a historical record, or help in the investigation of an accident. Essentially, SOPs help employees to produce high-quality products so that their companies can be more competitive in the workplace.

Analytical Reports

Analytical reports chiefly evaluate some issue, problem, condition, situation, trend, or other topic. Analytical reports try to understand underlying relationships, causes and effects, and potential solutions of a particular topic. They are more than just informational, since their aim is to provide more than simply an understanding. Analytical reports may or may not lead to recommendations. Reports offering recommendations are sometimes referred to as recommendation reports, and because these reports offer recommendations, they are sometimes distinguished from informational and analytical reports.

FORMAL ANALYTICAL REPORTS

Formal analytical reports analyze a situation, problem, trend, issue, event, or some other topic. They break down a topic to study the inner nature of the interrelationships of the subcomponents. Analytical reports try to determine, for example, whether one product is better than another for a particular purpose, why a particular event occurred, how a product may be improved, how a situation may be avoided, whether a solution will work for a specific purpose, what the effects will be of a particular decision or action, and whether a solution is practical. Of course, analytical reports may combine several of these approaches.

RECOMMENDATION REPORTS

Whereas many analytical reports do not offer a specific course of action to take, recommendation reports always suggest a particular course of action or several alternative courses of action. Typically, in a recommendation report you analyze a problem, determine the possible solutions to the problem, and then recommend the best solution or solutions.

PROGRESS REPORTS

Progress reports provide your readers with updates on work completed and work remaining on a particular assignment, task, product, or process. In school you may write a progress report on the status of your paper or class project. In industry you may provide a progress report on a product, schedule, or budget. The most common way to organize a progress report is according to time: work completed and work remaining. You discuss the project and any problems, summarize work completed during this particular reporting period, and then provide an overview of the work that remains. You may also organize the progress report according to the tasks performed, describing, in order, each task that is necessary to complete the project and what has been accomplished on each major task.

FEASIBILITY STUDIES

Feasibility studies are written to determine the best course of action concerning, for example, a product, purchase, service, or market. A feasibility report lets readers know whether it is practical or possible to pursue one or more courses of action. Sometimes feasibility studies recommend that no course of action be taken. Typical feasibility studies include market analyses concerning opening a new business in a particular area, purchasing one product rather than a competing product, committing funding to one area of research instead of another, and so on.

RESEARCH REPORTS

Research reports—also called empirical research reports—relay information about tests, surveys, or experiments of some kind. Research reports analyze information or data and follow a rigid structure of introduction, methods, results, and discussion. Often they rely on data provided in a laboratory report (see "Laboratory Reports," page 402) or information from a questionnaire or survey (see "Questionnaires and Surveys," page 57). Many scientific publications are basically research reports. They analyze a problem in the context of the existing research on the topic, propose a solution, subject the proposed solution to various tests, and then discuss whether or not the solution, tested through experimentation, is viable.

TRIP REPORTS

Trip reports, also called travel or field reports, provide a detailed summary of the important events and other details concerning a business visit to another destination. A trip report is typically written after the employee returns to the office. Usually in memorandum format, a trip report emphasizes what is important instead of providing a narration of all of the events that occurred. In a trip report you may want to discuss the most important things you learned or consider the questions that weren't answered. Your analysis of what you learned or what remains to be determined makes this type of report analytical rather than purely informational. If the report describes an inspection, basic maintenance, or the study of a site, it is more commonly referred to as a field report.

WHITE PAPERS

A white paper is a brief document (usually no more than three pages, though some white papers are book-length studies) that is essentially a position paper on an issue or problem. In some ways it is similar to an unsolicited proposal, but unlike a proposal, its purpose is not to win a contract or funding. Its main purpose is to inform or to help shape policy. For example, a white paper may be written to influence how a project-to-be will take shape. If successful, a white paper establishes a climate for an eventual solution to a problem or issue.

Most white papers have four major elements: a statement of the problem, a discussion of major issues, a discussion of an approach that resolves the issues, and a suggested timeline. In the statement of the problem you provide a brief discussion of your understanding of the problem or issue. If the white paper is for a client, you should discuss your understanding of the problem based on what you have found out from talking to the client. In the discussion of the major issues, identify the main themes that must be addressed. Identify various potential obstacles and explain how these obstacles will be resolved. In the section on resolving the issues you should discuss the approach that puts your proposed solutions to work. In effect, you are providing a discussion of scope here and, optionally, discussing various tasks that must be completed. Finally, in your discussion of the timeline you should offer an estimate of how long it will take to complete the project. Discuss which tasks must be done and how.

Common Elements of a Formal Report

Although the specific internal structure of each type of report may vary, all of these reports have an introduction, a body, and a conclusion.

The introduction of a report discusses the subject, purpose, scope, and audience. (See Part One, "Planning and Research," for a discussion of all four of these topics.) The body of the report varies depending on the kind of report being written. The conclusion is where you put together all the information that you set out in the previous sections.

Like the body of the report, the conclusion varies depending on the kind of report. Some conclusions simply summarize and interpret the key data that are covered in the body of the report. Others are more elaborate and provide a persuasive summing-up to convince the reader to concur with the writer's viewpoint, decisions, or solution. In any case, data that are not covered in the body of the report should not be considered in the conclusion.

Some kinds of reports contain recommendations in addition to conclusions. Recommendations are specific courses of action that your readers may take concerning the topic covered in your report.

Once you have written a draft of your report, you'll need to decide how you will format it. Reports may be in memo, letter, or formal format. Memo reports are for internal use within a company or organization. They are generally more informal in format and style than formal reports. Instead of providing the various parts of a formal report in a memo report, you simply address the necessary people in your To: line and cc: line. And you may use second person point of view. Letter reports are usually documents for audiences outside the organization or company; like memo reports, they are typically brief—four or five pages at most.

If the purpose, audience, discourse community, and context require it, you may find that a formal report is more suited to your audience. A formal report may contain some or all of the following sections.

Front matter—letter of transmittal, cover, title page, abstract, executive summary, table of contents, and list of illustrations—helps readers to find the information they need, helps readers to decide whether they want to read the report, and gives readers the essential purpose, scope, and conclusions (and recommendations, if any) of the report.

LETTER OF TRANSMITTAL

A letter of transmittal is a letter that explains the purpose and content of the report. It accompanies most formal reports or proposals and usually precedes the title page. Sometimes it is bound in with the report; sometimes it is simply placed on top of the report. A letter of transmittal enables you to acknowledge those who helped with the report, highlight parts of the report that are of special interest, discuss any drawbacks of the study or any problems, and offer any personal observations. (Also see "Transmittal Letters," on page 342.)

COVER

A cover may be either hard or soft; its chief purpose is to protect the contents of the report. Covers should be used only for long documents. On the cover you may provide the document title, the writer's name, the date of submission, and the name or logo of the company or organization. Often all of these items are centered about four or five inches from the top of the page. Alternatively, the cover may contain no information.

TITLE PAGE

A title page provides the title, author(s), intended recipients (including the organization for whom the report is intended), and date the report was submitted. Sometimes it contains a report number or code, applicable contract numbers if the report was done under contract, and proprietary and security notices. Make your title as descriptive as possible. When appropriate, indicate whether the report is an analysis, a feasibility study, or whatever. The title page does not contain a page number, but it is counted in the pagination as page i.

ABSTRACT

An abstract is a condensed version of a longer piece of writing that highlights the major points covered, concisely describes the content and scope of the writing, and reviews the writing's contents in abbreviated form. Abstracts are commonly descriptive or informative.

The practice of using keywords in an abstract is crucial because of today's electronic information retrieval systems. Titles and abstracts are often filed electronically, and keywords are put in electronic storage. When people search for information, they enter keywords related to the subject, and the computer lists the titles of articles, papers, and reports containing those keywords. Consequently, an abstract must contain keywords that accurately reflect what is essential in an article, paper, or report so that someone else can retrieve information from it.

Descriptive Abstracts. Descriptive abstracts tell readers what information the report, article, or paper contains. This kind of abstract includes the purpose, methods, and scope of the report, article, or paper, but it does not provide results, conclusions, or recommendations. Descriptive abstracts introduce the subject to readers, who must then read the report, article, or paper to find out the author's results, conclusions, or recommendations. These abstracts are very short, usually under 100 words.

Informative Abstracts.   Informative abstracts communicate specific information from the report, article, or paper. They include the purpose, methods, and scope of the report, article, or paper, and they provide the report, article, or paper's results, conclusions, and recommendations. Informative abstracts allow readers to decide whether they want to read the report, article, or paper. Informative abstracts are short—from a paragraph to a page or two, depending on the length of the original work being abstracted. Usually, informative abstracts are 10 percent or less of the length of the original piece.

TIPS: Writing an Abstract

- Reread the article, paper, or report with the goal of abstracting in mind.
- Look specifically for these main parts of the article, paper, or report: purpose, methods, scope, results, conclusions, and recommendation.
- Use the headings, outline heads, and table of contents as a guide to writing your abstract.
- If you're writing an abstract about another person's article, paper, or report, the introduction and the summary are good places to begin. These areas generally cover what the article emphasizes.
- After you've finished rereading the article, paper, or report, write a rough draft without looking back at what you're abstracting.
- Don't merely copy key sentences from the article, paper, or report; you'll put in too much or too little information.
- Don't rely on the way material was phrased in the article, paper, or report; summarize information in a new way.
- Provide one or more well-developed paragraphs, and make sure that these paragraphs are unified, coherent, concise, and able to stand alone. (See Part Two, "Writing," for a discussion of paragraphs.)
- Provide an opening, a middle, and a closing in the abstract. In the opening of the abstract of an article, paper, or report you discuss the purpose. In the middle you discuss the results and conclusions. In the closing you discuss the recommendations.
- Follow the order of the topics of the article, paper, or report.

- Provide logical connections or transitions between the information included. Your abstract should not add any new information but should simply summarize. Your abstract should be accessible to a wide audience through its handling of language, specifically technical language.
- Use passive verbs where appropriate to downplay the author and emphasize the information.

EXECUTIVE SUMMARY

An executive summary reviews the essential points of a report. Executive summaries are useful for people who have neither the time nor the inclination to read a lengthy document but who want to scan the primary points quickly and then decide whether they need to read the entire version. These documents are called executive summaries because they are often geared to the usual reading style of busy managers. Managers typically do not want or need an in-depth discussion of the many details of one project or another. A good executive summary provides a broad understanding of a project. Executive summaries are parallel to an informative abstract. Like informative abstracts, executive summaries stand alone and offer key conclusions and recommendations. Executive summaries provide background, major findings, and, if applicable, chief recommendations.

TIPS: Writing an Executive Summary

- Read the entire original before writing a word. Get the complete picture. If you were the author of the report, review the material to refresh your memory of the overall picture.
- Reread and underline significant points (usually in the topic sentence of each paragraph).
- Rewrite the significant points in your own words.
- Don't sacrifice meaning for brevity. A summary should be short enough to be economical and long enough to be clear and comprehensive. A short, confusing summary will take more of a busy executive's time than a somewhat longer but clear one.
- Capture the essential meaning of the original document. A good summary will always tell the reader what the original says—its significant

(continued)

Writing an Executive Summary *(continued)*

points; primary findings; important names, numbers, and measurements; and major conclusions and recommendations. The essential message is the minimum that the reader needs to understand the shortened version of the whole. The essential meaning does not include background information, lengthy examples, visuals, or long definitions.

■ Write at the lowest level of specialization. If the executive summary is part of a report, more people may read the summary than the entire report. Write at the lowest level of technicality, translating specialized terms and complex data into plain English, because your summary will not include the supporting information for technical statements. If you know your audience, keep these people in mind. When in doubt, oversimplify.

■ Structure your summary to fit your audience's requirements. Some summaries follow the organization of the report, dealing briefly with the information in each chapter (or section) in order. Others highlight the findings, conclusions, and recommendations by summarizing them first, before going on to discuss procedures or methodologies. If you are writing a summary at the request of your manager, you may want to begin with the part that seemed to be of most interest to the manager.

■ Avoid introducing new data into the summary. Represent the original faithfully. An executive summary is not a book report. Avoid opinions such as "This report was very interesting" or "The author seems to think that...." You don't need to try to put the work into a particular perspective.

■ Write your summary so that it can stand alone. It should be a self-contained message. Your readers should read the original only if they want to get a fleshed-out view of the subject they should not have to read it to make sense of what you say in your summary.

■ Edit your draft, cutting unnecessary words and phrases.

TABLE OF CONTENTS

A table of contents is a list of the headings along with the page numbers where the headings can be found in the report. It is an essential reader aid for longer formal reports. It helps readers to find what they want and to understand quickly the overall organization and approach of the report. Not all levels of headings that appear in the report will appear in the table of contents. Generally, providing up to third-level headings in the table of contents suffices. Start by listing the letter of transmittal, then the list of illustrations, then the abstract or summary, then the introduction, then various body sections, the conclusion, appendixes, and references. Only headings that appear in the report may be included in the table of contents.

LIST OF ILLUSTRATIONS

A list of illustrations is essentially a table of contents for the tables and figures in the document. It is a listing of all formal illustrations used in the report. You may provide the list of illustrations on the same page as the table of contents. If your report has more than five figures or tables, place the list of illustrations on a separate page. The list of illustrations is called a list of figures if the report contains only figures and a list of tables if the report contains only tables.

The introduction, methods section, results section, conclusion, and sometimes the recommendation section make up the body of a formal report. The body is a discussion of the data—facts, figures, expert testimony, and other relevant factual information. This information must be adequately detailed.

INTRODUCTION

The introduction of a report is a discussion of the subject, purpose, scope, and audience. (See Part One, "Planning and Research," for a detailed discussion of these terms.) You should concisely identify what you are writing about, what your primary aim is, what your approach is, and for whom you are writing the report. For reports aimed at decision makers, you might also provide your major conclusions and recommendations as subsections of your introduction.

METHODS

A methods section, if you choose to provide one, tells your readers what you did. This section discusses how your study or project or test was set up and why.

RESULTS

The results in a report are the key data that you found or created. This part tells your readers what your observations are.

CONCLUSION

The conclusion is a concise interpretation of the facts that are covered in the body of the report. A good conclusion covers only what the data in the body of the report will support. There should be no conclusions in the report that are not derived from or built on the data covered in the body.

RECOMMENDATIONS

Recommendations are the particular actions the reader should take on the basis of the conclusions of the report. Reports do not necessarily include recommendations. Most informational reports and many analytical reports offer no recommendations or specific courses of action for the readers to follow.

The back matter of a report can include a glossary, a list of symbols (if any), any appendixes, reference lists, and an index. These important sections provide explanatory material, raw data, information on outside sources, and navigation tools to help the reader evaluate the value of the material in the report.

GLOSSARY AND LIST OF SYMBOLS

A glossary is an alphabetical listing of key terms in the report, providing definitions in the form of phrases or complete sentences. Glossaries are essential if you are writing for both technical and nontechnical readers, since words and expressions can have different meanings in general use and technical use. You should include and define all terms that you think will be unfamiliar to a general reader, being careful to coordinate the definition with the context the reader will encounter in the report itself. Glossary definitions for the same term can be quite different for different reports, depending on how the term is used in each report. Once you have decided on a format for your definitions, make sure that each entry conforms to that format.

For some reports, a list of symbols may also be necessary. Such a list provides a complete catalog of the symbols and other abbreviations that are used in a report. Standard symbols for most quantities have long been availiable in the sciences, for example. These standard symbols have been adopted by various international organizations, and you should use them whenever possible. Creating or inventing your own symbols may confuse the reader.

APPENDIXES

Appendixes include additional material that is useful but not essential to understanding the body of the report. You may think of an appendix as a reference section for the report. In this section you give the reader the opportunity to review and evaluate information you have used in forming your conclusions and recommendations. Readers who want to examine your raw data will find them in this section. Many reports have appendixes, and many do not. Typical items to include in an appendix are complex formulas, additional illustrations, questionnaires, and lab results.

REFERENCES

References are a listing of sources you consulted, provided in the documentation style that is preferred by the discourse community for whom you are writing. (See Part One, "Planning and Research," for a discussion of various documentation styles.)

INDEX

An index is a guide to the contents of a work. It differs from a table of contents in its level of detail and arrangement. While a table of contents normally lists major sections in the work arranged by page number, an index will list specific topics covered in the work, arranged according to some logical scheme—normally alphabetical or chronological. Readers check indexes when they want to find information that they have seen in a work but cannot locate, when they want an overview of the coverage on a particular topic, or when they need help locating material that might not be listed in a table of contents.

TIPS: Creating an Index

- Be aware of any specific requirements that your employer or your instructor has imposed. If there is a style sheet, follow its guidelines.
- Decide whether you will be providing a page number index, a chronological index, or something else.
- Edit your index. Once you have finished creating your entries, they will have to be edited, just as you would edit any other work you create. Eliminate duplicate or redundant entries, and make sure your entries are useful, are arranged logically, and correctly point to the appropriate location in the text.
- Ask yourself whether each entry is necessary, phrased usefully, and located where the reader will look for it. If a concept can be referred to in more than one way, make sure you have cross-references from the alternative forms to one indexed term.
- Avoid circular entries, that is, entries that point to each other without actually sending the reader to a location in the text.

(continued)

Creating an Index *(continued)*

■ Verify that your locators are correct. Select entries at random and check that their page numbers or other locators are accurate. If possible, sort your entries according to page number to check the correctness of locators more easily.

■ If possible, have someone else check your index for usability.

SPECIFICATIONS

Specifications refer both to the design and function requirements for technologies of all kinds—for everything from space shuttles to household appliances—and to the requirements for services. All of the technology around you is built, either well or poorly, according to specifications of one kind or another. Obviously, if the specifications are not carefully considered, then the technology won't be well designed, or it will be designed with flaws—even minor ones—that could have major consequences. So writing precise, accurate, and clear specifications becomes not only a desired accomplishment but a necessary one, requiring a great deal of skill.

Specifications consist of a collection of items to be included in the final product, along with detailed descriptions of each item. The specifications normally form a complete statement either of what you are requiring from someone else or of what you will provide to someone else. For example, if you want to build a house, you discuss your requirements with an architect. The architect then provides a design that includes a list specifying what type of each element you require and how many (for example, fourteen double-glazed, 2 foot by 3 foot single-casement windows with 15 percent bronze tint). You would then invite several general contractors to bid on the job. The specifications list would form the central list of materials of your house, and any quote that you receive would have to incorporate the materials you require.

TIPS: Writing Specifications

■ Specifications must be achievable. Complex technical specifications usually comprise a large number of individual requirements, which often affect the accomplishment of each other. When taken one at a time, such requirements may be easily met, but when combined—three, four, five, or twenty or more together—they may form a set that is mutually un-achievable. Making sure that this does not happen is the responsibility of the person who drafts the specifications.

- The satisfaction of specifications must be verifiable. Always specify things in such a way that you can verify for certain whether or not the requirements have been met. Verification methods include inspection, measurement, testing, and demonstrations. Definite pass-or-fail criteria, with no element of judgment or opinion involved, must be assigned to every verification method. Otherwise, disputes are likely to arise over the acceptability of the product.

- Avoid including explanations of things you have learned in the course of your work. Readers of your specifications can rely on such statements as true and correct and can act on them in preparing bids without any further investigation to verify them. If you feel that you must make ex-planations, be sure to check them carefully to make sure that they are true in all cases, regardless of what approach the bidders may take toward satisfying your requirements. Remember, specifications specify. They do not explain.

- Provide performance specifications rather than design specifications. Put simply, the difference is between telling someone what to do and telling them how to do it. Performance specifications are stated only in terms of the result required; design specifications stipulate what materials and methods are to be used in producing goods or rendering services. Most specifications that you encounter in practice are an unwitting combina-tion of the two types. All it takes to categorize a specification as a de-sign specification is one design requirement. In general, writers of mixed specifications are held responsible when a contractor adheres to the design requirements and claims that doing so prevented satisfaction of the performance requirements.

- Cite standardization documents. When you cite a published specifica-tion or standard, be sure that you have read the document and under-stand it fully. Be sure that you include all the information in your own specifications that is called for by the text of the document you are cit-ing. Avoid creating unnecessary work for your readers by citing docu-ments in their entirety. Instead, tell your readers specifically which para-graphs of the cited document apply. If only a few sentences of the cited document really apply to your specifications, then eliminate the citation by spelling out the requirements as part of your own specifications.

- Be consistent. Specifications must be rigorously consistent throughout. This means that every mention of each topic covered must be in logical conformance with everything else that is said about all the related top-

(continued)

Writing Specifications *(continued)*

ics. Editing a document for this degree of consistency requires a sharp memory and a great deal of concentration, especially when the document is a long one.

- Use terminology consistently. Even if your specifications sound odd when read aloud, you should use exactly the same terms every time you refer to the same item. Also, be very careful of pronouns, since readers sometimes choose a different antecedent for them from the one you intended.

- Carefully differentiate between *shall* and *will*. *Shall,* when used in specifications, imposes a binding requirement on the offerors. On the other hand, the word *will* is usually used only when the specifier wishes to make some kind of promise to the offerors. Careless errors in the *shall*s and *will*s often lead to very unpleasant surprises when work is being done under contract.

- Use *or* correctly. Be sure whenever you use it that you do not really mean *and.* Also be sure that you have not granted someone permission to make a choice of doing either one task or another when you intended for them to do both. Inspect the text of your specifications for occurrences of the word *or*, and check each one carefully to make sure that its logic is correct.

- Avoid using *any. Any* is an ambiguous word. Often it can be simply deleted from specifications without altering the meaning of the text; in other cases the words *each* and *every* may be useful in rewriting the requirement for better clarity.

RESOURCES

Allen, Jo. *Writing in the Workplace*. Boston: Allyn & Bacon, 1998.

Anderson, Paul V. *Technical Communication: A Reader-Centered Approach*. 4th ed. New York: Harcourt, 1998.

Brusaw, Charles T., Gerald J. Alred, and Walter E. Oliu. *Handbook of Technical Writing*. 5th ed. New York: St. Martin, 1997.

———. *The Technical Writer's Companion*. 2nd ed. New York: St. Martin, 1999.

Holtz, Herman. *The Consultant's Guide to Proposal Writing: How to Satisfy Your Clients and Double Your Income*. New York: Wiley, 1986.

Houp, Kenneth W., Thomas E. Pearsall, and Elizabeth Tebeaux. *Reporting Technical Information*. 9th ed. New York: Allyn and Bacon, 1998.

Killingsworth, Jimmie and Jacqueline S. Palmer. *Information in Action*. 2nd ed. Boston: Allyn & Bacon, 1999.

Lannon, John M. *Technical Writing*. 6th ed. New York: HarperCollins, 1994.

Lay, Mary, et al. *Technical Communication*. Chicago: Irwin, 1995.

Markel, Mike. *Technical Communication: Situations and Strategies*. 5th ed. New York: St. Martin, 1998.

McMurrey, David A. *Online Technical Writing*. <http://www.io.com/~hcexres/tcm1603/acchtml/acctoc.html>.

Perelman, Leslie C., James Paradis, and Edward Barrett. *The Mayfield Handbook of Technical and Scientific Writing*. Mountain View: Mayfield, 1998.

Pfeiffer, William S. *Pocket Guide to Technical Writing*. Upper Saddle River: Prentice-Hall, 1998.

Sims, Brenda. *Technical Writing for Readers and Writers*. New York: Houghton Mifflin, 1998.

Woolever, Kristin. *Writing for the Technical Professions*. New York: Longman, 1999.

Index

Page numbers ending in *t* refer to information in tip lists; page numbers ending in *f* refer to information in figures.